by Dickon Ross, Cathleen Shamieh, and Gordon McComb

A John Wiley and Sons, Ltd, Publication

Electronics For Dummies®

Published by
John Wiley & Sons, Ltd
The Atrium
Southern Gate
Chichester
West Sussex
PO19 8SQ
England

E-mail (for orders and customer service enquires): cs-books@wiley.co.uk

Visit our Home Page on www.wiley.com

Published by John Wiley & Sons, Ltd, Chichester, West Sussex

For general information on our other products and services, please contact our Customer Care Department within the U.S. at 877-762-2974, outside the U.S. at 317-572-3993, or fax 317-572-4002.

For technical support, please visit www.wiley.com/techsupport.

Wiley also publishes its books in a variety of electronic formats. Some content that appears in print may not be available in electronic books.

British Library Cataloguing in Publication Data: A catalogue record for this book is available from the British Library

ISBN: 978-0-470-68178-7

Printed and bound in Great Britain by TJ International, Padstow, Cornwall

10 9

About the Authors

Dickon Ross, adapting author of the UK edition, has been a science and technology journalist for 20 years, working on titles ranging from *Electronics Times* to *Focus*. Dickon is now Editor-in-Chief of *Engineering & Technology* magazine and *Flipside* – the magazine he launched for teenagers – for the Institution of Engineering and Technology.

Cathleen Shamieh is a writer with an engineering background who specialises in creating communication materials focused on technology and its business benefits. She received an outstanding education in electrical engineering at Manhattan College and MIT, and enjoyed working as an engineer for several years in the medical electronics and telecommunications industries. Accepting a challenge from a respected colleague, she shifted her career into business consulting with a focus on technology implementation, eventually migrating into marketing and communications consulting for high-tech companies. Cathleen enjoys leveraging her technical and business background to create white papers and other materials for not-so-technical audiences.

Gordon McComb has penned 60 books and over a thousand magazine articles. More than a million copies of his books are in print, in over a dozen languages. For 13 years, Gordon wrote a weekly syndicated newspaper column on personal computers. When not writing about hobby electronics and other fun topics, he serves as a consultant on digital cinema to several notable Hollywood clients.

Dedication

To my parents, Beth and Jim Corbett, who taught me that I can do anything I put my mind to; to Sister Eustelle, who made a writer out of me; to my wonderful husband, Bill, who's always there to support me; and to my four fantastic sons, Kevin, Peter, Brendan, and Patrick, who make life a fun, loving adventure every single day.

C.S.

To my father, Wally McComb, who instilled in me a fascination with electronics; and to Forrest Mims, who taught me a thing or two about it.

G.M.

Authors' Acknowledgements

Dickon Ross would like to thank his son Edmund for his help in building and testing the circuits in this book.

Cathleen Shamieh extends her thanks to the excellent editors at Wiley, especially Katie Feltman and Christopher Morris, for their hard work, support, and gentle reminders, and to Kirk Kleinschmidt for his intense technical scrutiny of the material. She is also grateful to Linda Hammer and Ken Donoghue, who kindly recommended her work to Wiley. Finally, Cathleen thanks her family and friends, whose support, assistance, and understanding helped make her goal of becoming a *Dummies* author a reality.

Gordon McComb gives heartfelt thanks to Wiley and the hard-working editors at Wiley, especially Katie Feldman, Nancy Stevenson, Carol Sheehan, Laura Miller and Amanda Foxworth. Many thanks also to Ward Silver, for his excellent and thorough technical review, and Matt Wagner at Waterside Productions for always having a positive outlook. **Gordon** also wishes to thank his family, who once again put their lives on hold while he finished another book.

Publisher's Acknowledgements

We're proud of this book; please send us your comments through our Dummies online registration form located at www.dummies.com/register/.

Some of the people who helped bring this book to market include the following:

Commissioning, Editorial, and Media Development

Project Editor: Steve Edwards

Content Editor: Jo Theedom

Commissioning Editor: Nicole Hermitage

Assistant Editor: Jennifer Prytherch

Development Editor: Andy Finch

Copy Editor: Anne O'Rorke

Technical Editor: Roger Dettmer

Proofreader: Kelly Cattermole

Production Manager: Daniel Mersey

Cover Photos: © Tombaky/Fotolia

Cartoons: Ed McLachlan

Composition Services

Project Coordinator: Lynsey Stanford

Layout and Graphics: Carrie A. Cesavice, Joyce Haughey, Melissa K. Jester, Mark Pinto, Christine Williams

Proofreaders: Melissa Cossell, Rebecca Denoncour

Indexer: Ty Koontz

Special Help

Brand Reviewer: Rev Mengle

Contents at a Glance

Table of Contents

Part IV: The Part of Tens 329

Appendix: Internet Resources 345

Index 349

Introduction

Are you curious to know what makes your iPod tick? How about your mobile phone, laptop, stereo system, digital camera, plasma TV – or, well, just about every piece of electronics you use for work or play, in the office, at home or on the move?

Perhaps you've even thought that you could design and build your own little electronic circuit or gadget to do something you want it to do?

If you've ever wondered how transistors, capacitors and other building blocks of electronics work, or if you've been tempted to try building your own electronic devices, you've come to the right place!

Electronics For Dummies is your entry into the electrifying world of modern electronics. No dry, boring or incomprehensible tome, this; what you hold in your hands is *the* book that enables you to understand, create and troubleshoot your own electronic devices. We're getting excited already!

Why Buy This Book?

All too often, electronics seems like a mystery, because it involves controlling something you can't see – electric current – which you've been warned repeatedly not to touch. That's enough to scare most people away. But as you continue to experience the benefits of electronics on a daily basis, you may begin to wonder how so many incredible things can happen in such tight spaces.

This book is designed to explain electronics in ways you can relate to. It gives you a basic understanding of exactly what electronics is, provides down-to-earth explanations of how major electronic components work and gives you just what you need to build and test working electronic circuits and projects. Although this book doesn't pretend to answer all your questions about electronics, it does give you a good grounding in the essentials.

We hope that when you're done with this book, you'll realise that electronics isn't as complicated as you may have thought. And we want to arm you with the knowledge and confidence you need to go deeper into the exciting world of electronics.

Why Electronics?

Electronics is everywhere. You find electronics in your phones, audio and video systems, and kitchen appliances. Electronic systems control traffic lights, Internet commerce, medical devices – even many toys. You can't see most of them, but electronic systems also proliferate throughout your car. Try for just one minute to imagine your life without electronics; you may as well be living in the Dark Ages!

So what does all this mean to you as you peruse this book? After all, you don't expect to be able to design satellite communication systems after a sit-down session with this humble *For Dummies* book.

Remember though that even the most complicated electronics systems consist of no more than a handful of different electronic component types governed by the same set of rules that make simple circuits work. So if you want to understand complex electronic systems, you start with the basics – just like the designers of those systems did when they started out.

More importantly, understanding the basics of electronics can enable you to create some really useful, albeit somewhat simple, electronic devices. You can build circuits that flash lights at just the right time, sound a buzzer upon sensing an intruder or even move an object around the room. And when you know how to use integrated circuit (IC) chips, which are populated with easy-to-use fully functioning circuits, you can create some rather clever designs for just a few well-spent pounds.

Technology development being what it is – lightning fast, smaller and cheaper year after year – you can now hold the ingredients for very advanced electronic systems in the palm of your hand. With a little knowledge and a willingness to experiment, you can build a unique musical birthday card, fantastic flashing decorations or an alarm that senses someone trying to get into your bedroom or biscuit tin.

Also, you may have another hobby that can be enriched by electronics. If you're into model railways, you can build your own automated switching points. If your hobby is racing radio-controlled cars, electronics know-how may enable you to improve the performance of your car and win the next championship. Knowing more about electronics can really enhance your hobbies.

Last but not least, electronics is fun. Finding out about and messing with electronics is its own reward.

Foolish Assumptions

This book assumes that you're curious about electronics, but you really don't know much, if anything, about its inner workings. You chose this book, rather than a book consisting exclusively of recipes for electronic circuits, and therefore we assume that you want to discover more about how parts such as resistors, capacitors and transistors actually work.

So we take the time (and more than half the book) to explain the basics to you, distilling fairly technical information down into easy-to-understand concepts. You don't need to be well-versed in physics or mathematics to benefit from reading this book, although a little bit of school algebra is helpful (but we do our best to refresh that possibly painful memory).

We assume you may want to jump around this book a bit, diving deep into a topic or two that holds special interest for you, and possibly skimming through other topics. For this reason, we provide loads of chapter cross-references to point you to information that can fill in any gaps or refresh your memory on a topic. And although the first half of the book is devoted to how electronic circuits and individual parts work, we include cross-references to simple circuits and projects that appear later in the book. That way, as soon as you find out about a component, you can jump ahead, if you like, and build a circuit that uses that very component.

The table of contents at the front of the book provides an excellent resource that you can use to find quickly exactly what you're looking for. Finally, the good people at Wiley have thoughtfully provided a thorough index at the back of the book to help you find what you want fast.

Safety Is Number 1

Reading about electronics is pretty safe. About the worst that can happen is that your eyes get tired from too many late nights with this book. But actually building electronic projects is another matter. Lurking behind the fun of your electronics hobby are high voltages that can electrocute you, soldering irons that can burn you and little bits of wire that can fly into your eyes when you snip them off with sharp cutters. Aaaagh!

Safety comes first in electronics. It's so important, in fact, that we devote a major section of Chapter 9 to it – and continually refer you to this section. If you're brand new to electronics, please be sure to read this section thoroughly. Don't skip over it, even if you think you're the safest person on earth.

Even if you've dabbled in electronics before, we still say you should read this bit as you may be surprised by some of the information. When you follow proper precautions, electronics is a very safe and sane hobby. Be sure to keep it that way!

Although we try to give you great advice about safety throughout, we can't possibly give you every safety precaution in the world in one book. In addition to reading our advice, use your own common sense, read manufacturer's instructions for parts and tools that you work with and always stay alert.

How This Book Is Organised

Electronics For Dummies is organised so that you can quickly find, read and understand the information that you want. Also, if you have some experience with electronics, or want to deepen your knowledge of one particular topic, you can skip around and focus on the chapters that interest you.

The chapters in this book are divided into parts to help you zero in on the information that you're looking for quickly and easily.

Part I: Understanding the Fundamentals of Electronics

Turn to Part I if you want to get a thorough grounding in basic electronics theory. Chapter 1 gives you the big picture of exactly what electronics is and the amazing things it can do for you. You discover the fundamentals of electronic circuits and get introduced to voltage, current and sources of electrical energy in Chapter 2.

In Chapters 3–6, you dive deep into the heart of all the major electronic components, including resistors, capacitors, inductors, transformers, diodes and transistors. You find out how each component works, how it handles electric current and what role it plays in electronic circuits.

Chapter 7 introduces you to integrated circuits (ICs) and explains a bit about digital logic and how three popular ICs function. Chapter 8 covers sensors, speakers, buzzers, switches, wires and connectors.

Throughout Part I, we point you to introductory circuits you can build in Part III to see what each component does.

Part II: Getting Your Hands Dirty

Part II is all about tooling-up, constructing real circuits and probing around working (and non-working) circuits – without electrocuting yourself.

In Chapter 9, you find out how to set up an electronics workbench, what electronic components, tools and other supplies you need to build circuits, and how to protect yourself and your electronic components as you work on circuits. Chapter 10 explains how to interpret circuit diagrams (known as *schematics*) so that you know how to connect components together when you build a circuit.

You discover various methods of wiring up temporary and permanent circuits in Chapter 11, including how to solder. Finally, Chapter 12 explains how to use the most important testing tool in electronics – the multimeter – to explore and analyse your circuits. This chapter also introduces you very briefly to two other tools: the logic probe and oscilloscope.

Part III: Putting Theory into Practice

If you're anxious to wire up some circuits and get your electronic juices flowing, Part III is the place to be.

Chapter 13 shows you some elementary circuits that you can build to demonstrate the principles of electronics and observe specific electronic components functioning as advertised. Turn to this chapter if you want to reinforce your theoretical knowledge of electronics or gain experience building simple circuits.

When you're ready for more involved circuits, explore Chapter 14. Here, you find several projects that you can have fun building and exploring. You may even decide to put one or two of them to good use in your home or office.

Part IV: The Part of Tens

As you may expect, Part IV is where you can find further information laid out in top-ten list format.

Chapter 15 offers pointers to help you expand your electronics horizons. Here, you can find information on all-inclusive project kits and circuit simulation software, suggestions for additional testing tools and tips on how to get great deals on electronics supplies.

When you're ready to shop for all things electronic, turn to Chapter 16 for a list of the top electronics suppliers in the UK and abroad.

Icons Used in This Book

We can't place dozens of Post-it notes in each and every copy of *Electronics For Dummies*, so we use icons to draw your attention to critical information.

Tips alert you to information that can really save you time, headaches or money (or all three!). If you use our tips, your electronics experience is that much more enjoyable.

When you tinker with electronics, you're bound to encounter situations that call for extreme caution. Enter the Warning icon: a not-so-gentle reminder to take extra precautions to avoid personal injury or prevent damage to your tools, components, circuits – or your bank balance.

This icon reminds you of important ideas or facts that you really need to keep in mind. Occasionally, we use this icon to note where in the book an important concept is originally introduced, so that you can flick back to more detailed information for a refresher, if you need one.

Even though this entire book is about technical stuff, we flag up some mini topics to alert you to deeper techie info that may require a little more brain power to digest. Of course, if you choose to skip over this info, that's absolutely fine; you can still follow along with no problem. Think of this techie stuff as extra material – a diversion off the main path, if you will – like bonus questions in a quiz.

Part I

Understanding the Fundamentals of Electronics

'The porch light packaging said it was the absolute ultimate in security lighting.'

In this part . . .

Do you ever wonder what makes electronic devices tick? Are you ever curious to know how speakers speak, motors move and computers compute? Well, then, you've come to the right place!

In the chapters ahead, we explain exactly what electronics is, what it can (and does) do for you and how all sorts of electronic things work. But don't worry. We don't bore you with long essays involving physics and mathematics. We use analogies and down-to-earth examples to make understanding electronics easy – fun, even. And while you're enjoying yourself, you're discovering how electronic components work and combine forces to make amazing things happen.

Chapter 1

What Is Electronics and What Can It Do for You?

In This Chapter

- Seeing electric current for what it really is
- Recognising the power of electrons
- Using conductors to go with the flow (of electrons)
- Making the right connections with a circuit
- Controlling the destiny of electrons with electronic components
- Applying electrical energy to loads of things

If you're like most people, you probably have some idea about what electronics is. You've been up close and personal with lots of so-called consumer electronics devices, such as iPods, stereo equipment, personal computers, digital cameras and televisions, but to you, they may seem like mysteriously magical boxes with buttons that respond to your every desire.

You know that underneath each sleek exterior nestles an amazing assortment of tiny components connected together in just the right way to make something happen. And now you want to understand how.

In this chapter, you discover that electrons moving in harmony constitute electric current, which is shaped by electronics. You take a look at what you need to keep the juice flowing, and you also get an overview of some of the things you can do with electronics.

Just What Is Electronics?

When you turn on a light in your home, you're connecting a source of electrical energy (usually supplied by your power company) to a light bulb in a complete path, known as an *electrical circuit.* If you add a dimmer or a timer to the light bulb circuit, you can control the operation of the light bulb in a more interesting way than simply switching it on and off.

Electrical systems, like the circuits in your house, use a standard electric current to make things such as light bulbs work. *Electronic systems* take this a step further: they *control* the electrical current, changing its fluctuations, direction and timing in various ways in order to accomplish a variety of functions, from dimming a light bulb to communicating with satellites (take a look at Figure 1-1). This control is what distinguishes electronic systems from electrical systems.

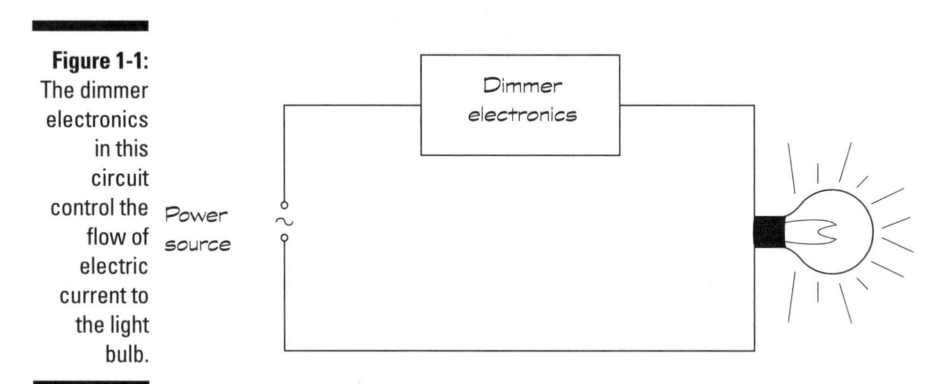

Figure 1-1: The dimmer electronics in this circuit control the flow of electric current to the light bulb.

To understand how electronics controls electricity, you need to first understand what electricity is and how it powers things like light bulbs.

Understanding Electric Current

Electric current is the flow of *electrical charges* carried by unbelievably small particles called *electrons*. So what on earth are electrical charges, where exactly do you find electrons and how do they move around? You find the answers by taking a peek inside the atom.

Getting a charge out of electrons

Atoms are the natural building blocks of everything. They're so tiny that you can find millions of them in a single speck of dust – so you can imagine how many exist in your average sumo wrestler! Electrons are found in every single atom in the universe, outside the atom's centre, or *nucleus*. All electrons have a negative electrical charge and are attracted to positively charged particles, known as *protons*, which exist inside the nucleus. *Electrical charge* is a kind of force within a particle, and the words 'positive' and 'negative' are somewhat arbitrary terms used to describe the two different forces that exhibit opposite effects. (We can call them 'north' and 'south' or 'Tom' and 'Jerry' instead, but those names are already taken.)

Under normal circumstances, an equal number of protons and electrons reside in each atom, and the atom is said to be *electrically neutral*. The attractive force between the protons and electrons, known as an *electromagnetic force*, acts like invisible glue, holding the atomic particles together, much as the gravitational force of the earth keeps the moon within sight. The electrons closest to the nucleus are held to the atom with a stronger force than the electrons farther from the nucleus, and some atoms hold on to their outer electrons with a vengeance whereas others are a bit more lax.

Moving electrons in conductors

Materials such as air and plastic, in which the electrons are all tightly bound to atoms, are *insulators* – they don't like to let their electrons move and so they don't easily carry an electric current. However, other materials, like the metal copper, are *conductors* because they have 'free' electrons wandering between the atoms, normally moving around at random. When you give these free electrons a push, they all tend to move in one direction and, hey presto, you have an electric current. This flow appears to be instantaneous because all those free electrons, including those at the ends, move at the same time.

A *coulomb* is defined as the charge carried by 6.24×10^{18} (that's 624 followed by 16 zeros) electrons. If a coulomb of charge moves past a point within a second, we say that the strength of the electric current is *one ampere,* or one amp (abbreviated to 1 A). That's a whole lot of electrons moving simultaneously, and much more than is typically found in electronic systems. You're more likely to see current measured in *milliamps* (mA). A milliamp is one one-thousandth of an amp.

Experiencing electricity

You can personally experience the flow of electrons by shuffling your feet across a carpet on a dry day and touching a doorknob; that zap you feel (and the spark you may see) is the result of electrical charges jumping from your fingertip to the doorknob, a form of electricity known as *static electricity.*

If you can get enough charges to move around and you can harness the energy they release, you can use that energy to power light bulbs and other things.

Lightning is another example of static electricity (but not one you want to experience personally), with electrical charges travelling from one cloud to another or from a cloud to the ground. When electrical charges move around, they release energy (hence the zaps and the sparks).

You may have seen in a film how a certain Doctor Frankenstein used this energy to dramatic effect, but explaining how to use thunderstorms to animate monsters assembled from human body parts is a little beyond the scope of this book!

Harnessing Electricity to Do Work

Benjamin Franklin was one of the first people to observe and experiment with electricity, and he came up with many of the terms and concepts (for instance, current) that we know and love today. Contrary to popular belief, Franklin didn't actually hold the key at the end of his kite during that storm in 1752. (If he had, he wouldn't have been around for the American Revolution.) He may have performed that experiment, but not by holding the key.

Franklin knew that electricity was both dangerous and powerful, and his work got people wondering whether a way existed to use the power of electricity for practical applications. Scientists such as Michael Faraday, Thomas Edison and others took Franklin's work a bit further and figured out ways to harness electricity and put it to good use.

Where Electrical Energy Comes From

In this section, we explore where electrical energy comes from and how you can apply that energy to make things work.

Tapping into electrical energy

An electric current flowing in a conductor moves energy from its source, such as a battery, to a place where it can do something useful. That place could be a light bulb, motor or loudspeaker, for example. These useful objects convert the electrical energy into another form of energy, such as light, heat or mechanical energy. In this way, you make the filament glow, the motor shaft rotate or the speaker diaphragm vibrate.

As you can't see – and don't necessarily want to touch – the masses of flowing electrons, try thinking about water to help make sense out of harnessing electricity. A single drop of water can't do much to help (or hurt) anyone, but get a whole group of water drops to work in unison, funnel them through a conduit, direct the flow of water towards an object (for example, a waterwheel) and you can put the resulting water energy to good use. Just as millions of drops of water moving in the same direction constitute a current, so too millions of electrons moving in the same direction make an electric current. In fact, Benjamin Franklin came up with the idea that electricity acts like a fluid and has similar properties, like current and pressure (but he probably would have cautioned you against drinking it).

Giving electrons a nudge

The force that gets the free electrons in a conductor moving is known as *voltage*, which is measured in units called *volts* (abbreviated to V). Think of voltage as electric pressure. Much like water pressure pushes water through pipes and valves, voltage pushes electrons through wires and other circuit components. The higher the pressure, the stronger the push, and so the higher the voltage, the stronger the electric current that is pushed through a circuit.

You may also hear the terms *potential difference, voltage potential, potential drop* or *voltage drop* used. Try not to let these different terms confuse you. We discuss this a bit more in Chapter 2.

Using conductors to make the circuit

Electric currents don't just flow anywhere. (If they did, you'd be getting shocked all the time.) Electrons only keep flowing if you provide a closed conductive path, or *circuit*, for them to move through and start that flow by applying a source of electrical energy such as a battery. Copper and other conductors are commonly formed into wire to provide a path for the flow of free electrons, so that you can direct electrical energy to a light bulb or other part that can use it. Just as with pipes and water, the wider the wire, the more freely the electrons flow.

If a break exists in the path (an *open circuit*), the electrons get stuck in a dead end. Picture water flowing through an open pipe. The water flows for a short time, but then stops when all the water exits the pipe. If you pump water through a closed pipe system, the water continues to flow as long as you keep forcing it to move.To keep the electric current flowing, you need to connect everything together into one big happy *electrical circuit*. As shown in Figure 1-2, every circuit needs at least three basic things to ensure that electrons get energised and deliver their energy to something that needs work done:

- **A source of electricity (or electrical energy):** The source provides the force that nudges the electrons in the chain reaction. You may also hear the terms *electrical source*, *power source*, *voltage source* and *energy source* used to describe a source of electricity. We discuss sources of electricity in Chapter 2.
- **A load:** The load is the thing that uses the energy in a circuit (for instance, a light bulb or a speaker). Think of the load as the destination for the electrical energy.
- **A path:** A conductive path provides a conduit for electric current to flow between the source and the load.

Working electrons create power

Work is a measure of the energy that a device like a light bulb or a motor uses over a certain amount of time when you apply a voltage to it. The more electrons you push, and the harder you push them, the more electrical energy is available and the more work can be done. The total energy used in doing work over some period of time is known as *power* and is measured in *watts.* Power is calculated by multiplying the force (voltage) by the strength of the electron flow (current):

$$\text{Power} = \text{voltage} \times \text{current}$$

Power calculations are really important in electronics, because they help you understand just how much energy electronic parts are willing (and able) to handle without complaining. If you energise too many electrons in the same electronic part, you generate a lot of heat energy and may fry that part. Many electronic parts come with maximum power ratings so that you can avoid getting into a heated situation. We remind you about this in Chapters 3 to 8 when we discuss specific components and their power ratings.

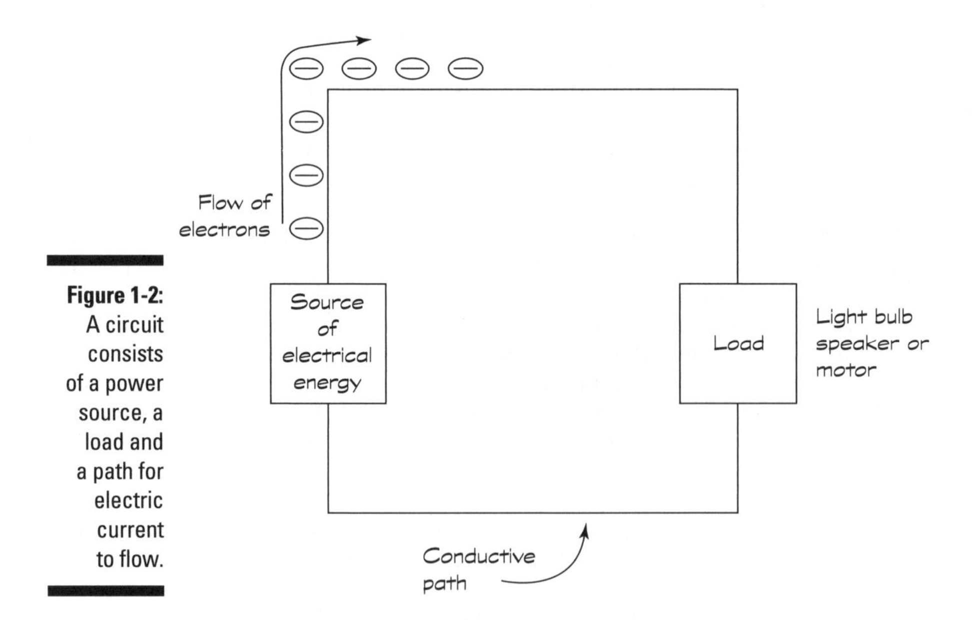

Figure 1-2: A circuit consists of a power source, a load and a path for electric current to flow.

An electric current starts with a push from the energy source and flows through the wire path to the load, where energy is released to make something happen, for instance, emitting light.

Oh, the Things Electrons Can Do!

Imagine applying an electric current to a pair of speakers without using anything to control or shape the current. What would you hear? It certainly wouldn't be music! By using the proper combination of electronics assembled in just the right way, you can control the way each speaker diaphragm vibrates, producing recognisable sounds, like speech or music (well, certain music anyway). And you can do so much more with electric current when you know how to control the flow of electrons.

Electronics is all about using specialised devices, known as *electronic components* (for example, resistors, capacitors, inductors and transistors, which we discuss in Chapters 3, 4, 5 and 6 respectively) to control current (also known as the flow of electrons) in such a way that it performs a specific function.

Simple electronic devices use a few components to control current flow. The dimmer switch that controls current flowing into a light bulb is one such example. But most electronic systems are a lot more complicated than that; they connect lots of individual components together in one or more circuits to achieve their ultimate goal. The great thing is that when you understand how a few individual electronic components work and find out how to apply some basic principles, you can begin to understand and build interesting electronic circuits.

This section provides just a sampling of the sorts of things you can do by controlling electrons with electronic circuits.

Creating good vibrations

Electronic components in your iPod, car stereo and other audio systems convert electrical energy into sound energy. In each case, the system's speakers are the load, or destination, for electrical energy, and the job of the electronic components within the system is to shape the current flowing to the speakers so that the diaphragm within each speaker moves in such a way as to reproduce the original sound.

Seeing is believing

In visual systems, electronic components control the timing and intensity of light emissions. Many remote control devices, such as the one wedged down the back of your sofa, emit invisible infrared light when you press a button, and the specific pattern of the emitted light acts as a sort of code to the device you're controlling, telling it what to do.

Cathode ray tube (CRT) TV sets (the sort we all used before flat-panel sets) are coated with phosphors that glow when struck by electron beams within the tube. The electronic circuits within the TV set control the direction and intensity of the electron beams, thus controlling the pattern painted across the TV screen, which is the image you see. Enlightening, isn't it?

Sensing and alarming

Electronics can also be used to make something happen in response to a specific level of light, heat, sound or motion. Electronic *sensors* generate or change an electrical current in response to a stimulus. Microphones, motion detectors, temperature sensors and light sensors can be used to trigger other electronic components to perform some action, such as activating an automatic door opener or sounding an alarm.

Controlling motion

A common use of electronics is to control the on/off activity and speed of motors. By attaching various objects, from wheels to aeroplane flaps, to motors, you can use electronics to control their motion. Such electronics can be found in robotic systems, aircraft, spacecraft, elevators and lots of other places.

Solving problems (aka computing)

Much as the ancients (those living thousands of years ago, not your great-grandparents) used the abacus to perform arithmetic operations, so you use electronic calculators and computers to perform computations. With the abacus, beads were used to represent numbers and calculations were performed by manipulating those beads. In computing systems, different electrical signals are used to represent numbers, letters and other information, and computations are performed by manipulating those patterns using electronic components. Of course, the worker-bee electrons inside have no idea they're crunching numbers!

Communicating with each other

Electronic circuits in your mobile phone work together to convert the sound of your voice into an electrical pattern, manipulate the pattern (to compress and encode it for transmission), convert it into a radio signal and send it out through the air to a communication tower. Other electronic circuits in your handset detect incoming messages from the tower, decode the messages and convert an electrical pattern within the message into the sound of your friend's voice (via a speaker).

Data communication systems, which you use every time you shop online, use electronics to convert your materialistic desires into shopping orders – and extract money from your bank account.

Chapter 2

Manipulating Electricity to Make Something Happen

In This Chapter

- Finding a source of electrical force
- Being positive about the direction of current
- Shedding light on a circuit in action
- Taking control of electron flow
- Sending current this way and that

Electronics is all about controlling the flow of electrons through conductors in a complete path (circuit) in order to shape the electrical energy delivered to a load, such as a light bulb, motor or speaker, in just the right way. By manipulating the flow of electrons, electronic components enable you to do some amazing things with electrical energy, such as vary the sound produced by speakers, change the direction and speed of motors and control the intensity and timing of lights, among many other things. In other words, electronics doesn't make electricity – it makes electricity better.

In this chapter, you discover how to get electrons flowing through a circuit and why conventional current can be thought of as electrons moving in reverse. You also explore the depths of a simple electronic circuit and look at different ways to connect electronic components so that you can begin to shape and direct current the way you see fit in your own circuits.

Supplying Electrical Energy

If you take a copper wire and arrange it in a circle by twisting the ends together, do you think that the free electrons flow? (Chapter 1 has all about electrons.) Well, the electrons may dance around a bit, because they're so easy to move, but unless a force is pulling them one way or another, you don't get current to flow.

Think about the motion of water that's just sitting in a closed pipe: the water may bounce up and down a bit, but it's not going to go whooshing through the pipe on its own. You need to introduce a force in order to deliver the energy needed to get a current flowing through the pipe.

Just as a car needs petrol to get it going, a circuit needs a source of electrical energy to get the electrons moving. Batteries and solar cells are common sources. The energy available at your wall sockets comes from many different power plants. But how do you conjure up electrical energy?

All sources of electrical energy take another form of energy (for instance, mechanical, chemical, heat or light) and convert it. Exactly how electrical energy is generated by your favourite source turns out to be important, because different sources produce different types of electric current: *direct current* (known as DC) and *alternating current* (known as AC).

- **Direct current** is a steady flow of electrons in one direction, with very little variation in the strength of the current. Cells (commonly known as batteries) produce DC; most electronic circuits use DC.
- **Alternating current** is a fluctuating flow of electrons that keeps changing direction back and forth. Power companies supply AC to your electrical sockets.

Getting direct current from a battery

A battery converts chemical energy into electrical energy through a process called an *electrochemical reaction*. When two different metals are immersed in a certain type of chemical, metal atoms react with chemical atoms to produce charged particles. Negative charges build up on one of the metal plates, while positive charges build up on the other metal plate. The difference in charge across the two metal terminals (a *terminal* is just a piece of metal to which you can hook up wires) creates the force needed to push electrons around a circuit. We give this electrical force the name *voltage*, and it's a measure of how strong a force the electrical energy source can supply.

To use a battery in a circuit, you connect one side of your load, for instance, a light bulb, to the negative terminal (known as the *anode*) and the other side of your load to the positive terminal (known as the *cathode*). In this way you create a path that allows the charges to move, and electrons flow from the anode, through the circuit, to the cathode (as we show in Figure 2-1). The passage of electrons through the wire filament of the light bulb releases electrical energy and the bulb lights up.

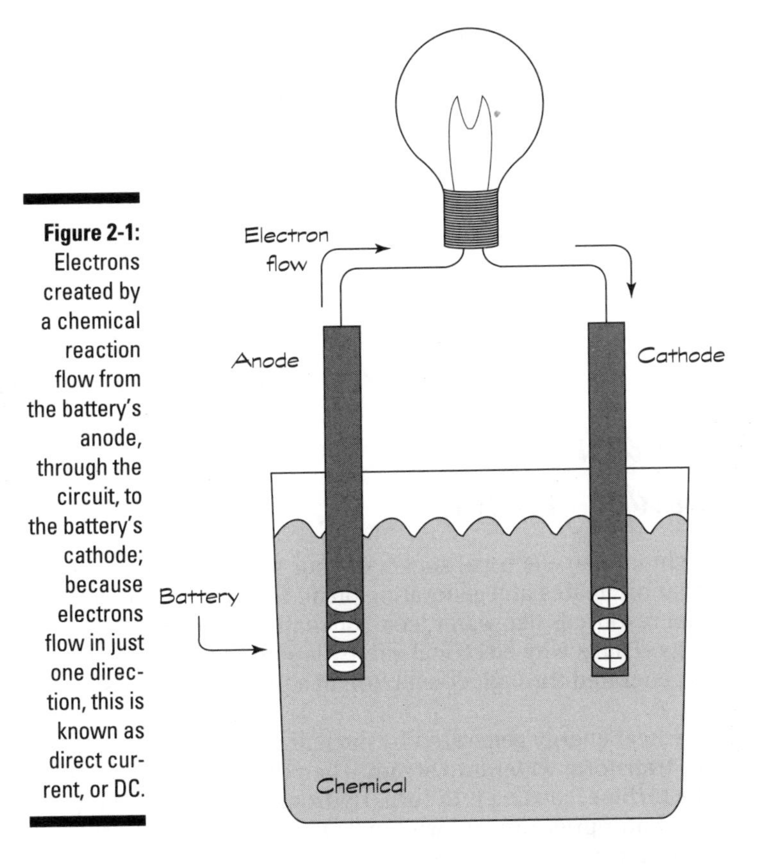

Figure 2-1: Electrons created by a chemical reaction flow from the battery's anode, through the circuit, to the battery's cathode; because electrons flow in just one direction, this is known as direct current, or DC.

Because the electrons move in only one direction (from the anode, through the circuit, to the cathode), the electric current generated by a battery is DC. A battery continues to generate current until all the chemicals inside it have been used up by the electrochemical process. The AAA-, AA-, C-, and D-size batteries that you can buy almost anywhere each generate about 1.5 volts – regardless of size. The difference in size amongst those batteries has to do with how much current can be drawn from them. The larger the battery, the more current can be drawn, and the longer it lasts. Larger batteries can handle heavier loads, which is just a way of saying that they can produce more power (remember, power = voltage × current), so that they can do more work.

Technically speaking, real batteries are only formed from connected sub-batteries called *cells*. When you connect several cells together, as you often do in many torches and children's toys, you create a battery. The battery in your car is made up of six cells, each generating 2.1 volts, connected together to produce 12.6 volts in total. We discuss various types of cells and how to connect them to create higher voltages in Chapter 8.

The symbol commonly used to represent a battery in a circuit diagram is shown below. The plus sign signifies the cathode and the minus sign signifies the anode. Usually the voltage is shown alongside the symbol.

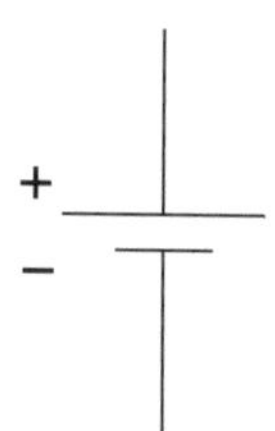

Using alternating current from the power station

When you plug a light into an electrical socket in your home, you're using electrical energy that originates at a generating plant. Generating plants convert energy from resources like water, coal, oil, natural gas or uranium into electrical energy. That's why electrical energy is said to be a *secondary* energy source: it's generated through conversion of a primary energy source.

Many plants use the heat energy generated by nuclear reactions or the burning of fossil fuels to transform water into steam. Then the steam exerts pressure on the fins of a turbine, causing it to turn. Hydroelectric power plants located at dams use water pressure and windmills use wind energy to rotate turbines. Power plant turbines are connected to electromechanical generators, which convert mechanical energy (the motion of the turbine) into electrical energy. A generator contains a coil of wire inside a huge magnet. As the turbine rotates, it turns the coil of wire, and – hey presto! – electrical current is *induced* in the wire. That's just a technical way of saying that something is causing electrons to flow, without any direct contact with the wire.

Electron flow can be induced by moving a wire near a magnet, or moving a magnet near a wire. This technique is called *electromagnetic induction*, and it works thanks to the close relationship between magnetism and electricity. We mention this relationship again in Chapter 5 when we discuss inductors.

As the coil rotates inside the magnet, the magnet first causes the electrons to flow in one direction, but when the coil has rotated 180 degrees, the magnet pulls the electrons in the other direction and they go right back again. This rotation creates *alternating current* (AC).

In UK power plants, the coil makes 50 complete rotations each second, so the electron flow changes direction 100 times per second. One complete rotation is called a *cycle*. The number of cycles per second in alternating current is known as *frequency* and is measured in units called *Hertz*, abbreviated to Hz. The UK and most of Europe generate AC at 50 Hertz, whereas the United States and many other countries use 60 Hertz as a standard.

AC is *stepped up*, or transformed to higher voltages, for transmission across long distances. When the current reaches its destination, it's *stepped down* to lower voltages (230 volts in the UK, for example) for distribution to homes and businesses. The symbol used in circuit diagrams for an AC voltage source is shown below:

Making (sine) waves

AC is constantly changing, so you can't describe its strength with a single number, as you can with DC. A common way to discuss its variations is to look at a *waveform*, or the pattern of the voltage over time. The AC waveform shows how the changing voltage causes electrons to flow in one direction (the positive region of the graph below) and then the other (the negative area). The *instantaneous current* is the strength of the current at a single point in time, and *peak current* is the magnitude (absolute value) of the current at its highest and lowest point. Because you can use the mathematical sine function to calculate the current for a given time, AC waveforms are often referred to as *sine waves.* (If you think that you can smell trigonometry, you're right, but don't worry – you don't need to wipe the dust off your school maths books! We just want you to be aware of the term 'sine' as it's used in electronics.)

The peak voltage (symbolised by V_p) is the magnitude of the highest voltage. You may hear the term *peak-to-peak voltage* (symbolised by V_{pp}), which is a measure of the difference between the highest and lowest voltage on the waveform, or twice the peak value. Another term thrown around is *rms voltage* (symbolised by V_{rms}), which is short for *root-mean-square voltage* and is used to work out how much power is in use in AC circuits. To get the rms voltage, you multiply the peak value by 0.7071.

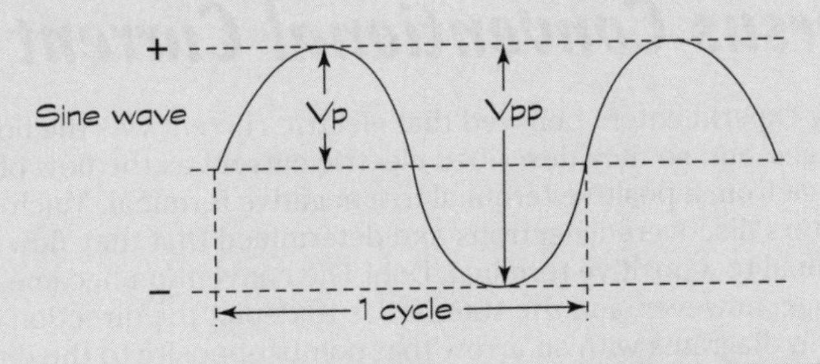

Many electronic devices, such as laptop computers for instance, require a steady DC supply, so if you're using AC to supply an electronic device or circuit, you need to convert AC to DC. *Regulated power supplies*, also known as AC-to-DC adapters, or AC adapters, convert AC to DC and are commonly included with electronic devices when purchased. Think of your mobile phone charger; this little device essentially converts AC power into DC power that the battery in your phone uses to charge itself back up.

Transforming light into electricity

Solar cells, also known as *photovoltaic cells*, produce a small voltage when you shine light on them. They're made from *semiconductors*, which are materials that are somewhere between conductors and insulators in terms of their willingness to conduct electrons. (We discuss semiconductors in detail in Chapter 6.) The amount of voltage produced by a solar cell is fairly constant no matter how much light you shine on it, but the strength of the current you can draw depends on the intensity of the light: the brighter the light, the higher the available current.

Solar cells have wires attached to two terminals for conducting electrons through circuits, so that you can power your calculator or garden lights. You may notice panels of solar cells powering road signs or parking meters or see pictures of satellites with large solar cell arrays on each side. Solar panels are becoming increasingly popular for supplying electrical power to homes and businesses as a way to reduce utility costs. If you scour the Internet you can find lots of information on how to make your own solar panels for just a couple of hundred pounds – and a willingness to try. You can read more about this subject in *Solar Power Your Home For Dummies* by Rik DeGunther (Wiley).

Understanding Directions: Real Electron Flow versus Conventional Current Flow

Early experimenters believed that electric current was the flow of positive charges, and so they described electric current as the flow of a positive charge from a positive terminal to a negative terminal. Much later, experimenters discovered electrons and determined that they flow from a negative terminal to a positive terminal. Doh! The convention became too well used to change, however, and the standard is to depict the direction of electric current in diagrams with an arrow that points opposite to the direction of the actual electron flow (see Figure 2-2).

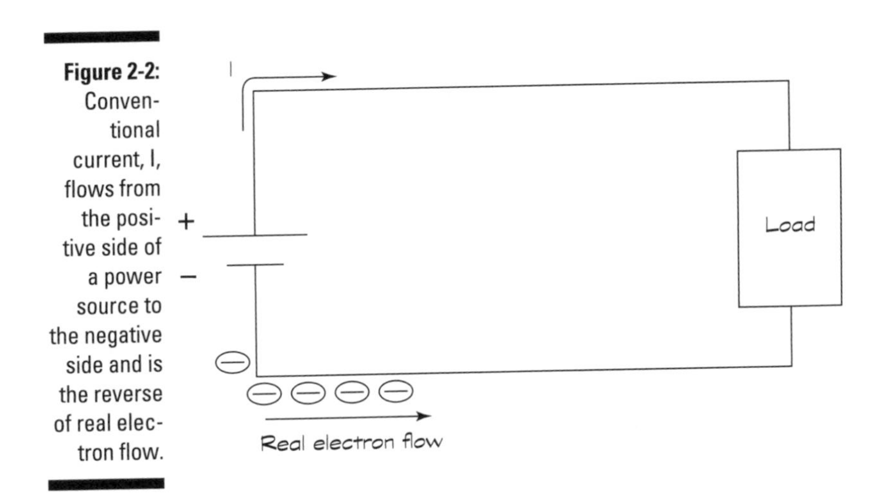

Figure 2-2: Conventional current, I, flows from the positive side of a power source to the negative side and is the reverse of real electron flow.

Conventional current is the flow of a positive charge from positive to negative voltage and is just the reverse of real electron flow. All descriptions of electronic circuits use conventional current, so if you see an arrow depicting current flow in a circuit diagram, you know that it's showing the direction of conventional current flow. The symbol, I, is used to represent conventional current, which is measured in amperes (or amps, abbreviated to A). You're more likely to encounter *milliamps* (mA) in circuits you build at home. A milliamp is one one-thousandth of an amp.

In AC circuits, current is constantly reversing direction. So how do you show current flow in a circuit diagram? Which way should the arrow point? The answer is that it doesn't matter. You arbitrarily pick a direction for current flow (known as the *reference direction*), and you label that current I. The value of I fluctuates up and down as the current alternates. If the value of I is negative, that just means that the (conventional) current is flowing in the opposite direction to the way the arrow is pointing.

Examining a Simple Light Bulb Circuit

The diagram in Figure 2-3 depicts a battery-operated circuit that lights a bulb, much like you find in a torch. What you see in the figure is a circuit diagram, or *schematic*, that shows all the components of the circuit and how they're connected. We discuss schematics in detail in Chapter 10.

The battery is supplying 1.5 volts DC to the circuit. That just means it supplies a steady 1.5 volts (until the battery begins to run out). The plus sign near the battery symbol indicates the positive terminal of the battery, from which current flows (conventional current, of course). The negative sign near the battery symbol indicates the negative terminal of the battery, to which the current flows when it makes its way around the circuit. The arrow in the circuit indicates the reference direction of current flow, and because it's pointing away from the positive terminal of the battery in a DC circuit, you can expect the value of the current to be positive all the time.

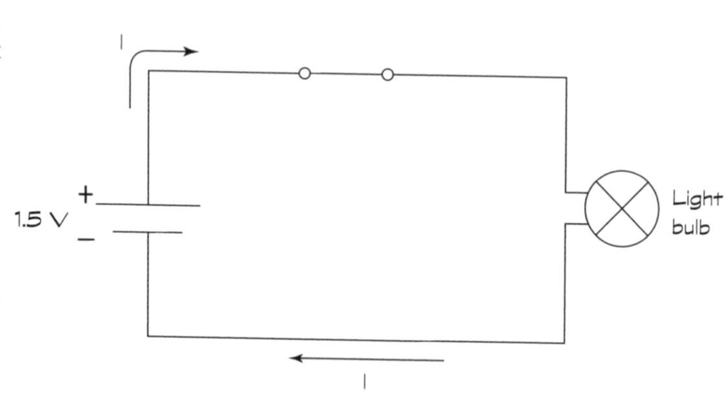

Figure 2-3: Current from a battery flows through the circuit, delivering electrical energy to the light bulb as voltage is applied to the bulb.

The lines in the circuit diagram show how the circuit components are connected, using wire or other connectors. (We discuss various kinds of wire and connectors in Chapter 8.) Switches and other circuit components are usually made with *leads*, consisting of protruding wires connected to the innards of the component that provide the means to connect the component to other circuit elements.

Next to the battery is a switch, which simply opens and closes the circuit, allowing current to flow out of the battery or stopping it dead in its tracks. If the switch is closed, current flows out of the battery through the light bulb, where electrical energy is dissipated as light and heat, and back into the negative terminal of the battery. If the switch is open, current doesn't flow at all through this *open circuit.*

The battery *supplies* electrical energy and the light bulb *uses* electrical energy. A give-and-take relationship exists between the two: voltage is the push the battery gives to get current moving, and the energy for that push is used up when current moves through the light bulb. As current flows through the bulb, voltage 'drops' across the bulb. Think of it as if the bulb is using up the force (voltage) that pushes the current through it.

When you drop voltage across a light bulb or other component, the voltage where the current enters the component is higher (or more positive) than the voltage where the current exits the component. Voltage is a relative measurement, comparing one point in a circuit to another, which is why you sometimes see voltage called *voltage drop*, *potential difference*, or *potential drop*.

The voltage at any point in a circuit is always relative to the voltage at some other point – usually the negative terminal of the battery.

An analogy that may help you understand: measuring voltage is much like measuring distance. If someone asks you, 'What's the distance?', you probably say, 'Distance from what?' Similarly, if you're asked, 'What's the voltage at the point in the circuit where the current enters the light bulb?' you should ask, 'With respect to what point in the circuit?' On the other hand, you may say, 'I'm five miles from home', stating your distance from a reference point (home). So if you say, 'The voltage where the current enters the bulb is 1.5 volts with respect to ground', that makes perfect sense. (See the sidebar 'Standing your ground' for more about the term 'ground'.)

If you start at the negative terminal of the battery in our simple light bulb circuit and travel all the way around the circuit measuring voltages, you can see that the voltage at the positive terminal of the battery is 1.5 volts, and all that 1.5 volts is dropped across the light bulb. (In reality, the switch eats up a tiny bit of voltage, because even the best conductors use up some energy, but the amount is negligible compared to the voltage drop across the light bulb.)

The important thing to notice here is that as you travel around a DC circuit, you gain voltage going from the negative terminal of the battery to the positive terminal (that's known as a *voltage rise*), and you lose or drop voltage as you continue in the same direction across circuit components. By the time you get back to the negative terminal of the battery, all the battery voltage has been dropped and you're back to 0 volts. With all circuits (AC or DC), if you start at *any* point in the circuit and add up the voltage rises and drops going around the circuit, you end up back where you started. The net sum of the voltage rises and drops around a circuit is zero.

Keep in mind that these voltage drops have real physical meaning. The energy that a battery supplies is used up by the light bulb. The battery keeps supplying energy and pushing out current, and the light bulb keeps using up that energy, until the battery dies (runs out of energy). The battery runs out when all the chemicals inside it have been consumed in the chemical reactions that produced the positive and negative charges. When that happens, you know that all the chemical energy supplied by the battery has been converted into electrical energy and used up by the circuit.

Standing your ground

In electronics, the word *ground* can have two different meanings. *Earth ground* means pretty much what it says: it's a direct connection to the ground – real ground. The top pin in a three-pin plug is connected to earth ground. Behind each wall socket is a wire that runs through your house or office and eventually connects to a metal post that makes good contact with the ground. This provides extra protection for circuits that use large amounts of current, by providing a path to sink dangerous current directly into the earth. Benjamin Franklin's lightning rod provided a direct path for dangerous lightning to hit the ground – instead of a house or person (see Chapter 1). The term *floating ground* refers to a circuit that's not connected to earth ground and may be dangerous. You're wise to stay away from such a circuit until it's safely earthed!

The other type of ground is called a *common ground*, or simply *common*. Unlike earth ground, common ground isn't a physical ground; it's a just a reference point within a circuit for voltage measurements. Certain types of circuits, particularly the circuits commonly used in computers, label the negative terminal of a DC power supply the common ground, and connect the positive terminal of another DC power supply to the same point. That way, the circuit is said to have both positive and negative power supplies. The two physical power supplies may be identical, but the way you connect them in a circuit and the point you choose for the zero voltage reference determines whether a supply voltage is positive or negative. It's all relative!

You can measure the voltage drop across the light bulb using a *voltmeter* (which we discuss in detail in Chapter 12). And if you multiply the *voltage across the bulb* (that's a common way of saying voltage drop) by the strength of the current running through it, you get the power dissipated in the bulb (power = voltage × current) in watts.

Controlling Electrical Current with Basic Components

If you start to build the simple light bulb circuit discussed in the preceding section and you don't have a 1.5-volt battery available, you may think that it's okay to use the 9-volt battery you found in the kitchen drawer. After all, 9 volts is more than 1.5 volts, so the battery should provide enough energy to light the bulb. If you do use the 9-volt battery, however, your little circuit draws a lot more current – and you may overload your light bulb. If too many electrons are allowed to flow through a filament, the electrical energy dissipated in the filament creates so much heat that the bulb bursts.

Empowering you to make the right choices

Light bulbs and other electronic components have maximum power ratings for good reason. Send too much current through them, and they overheat and burn or melt. Remember that power is the product of voltage and current, so when you understand how to figure out voltage drops and current through these components, you have the ability to estimate the power rating (how many watts the part can handle before blowing up in your face) you need for the components you select for your circuits.

What you can do is insert a little electronic device called a *resistor* in between the battery and the light bulb. Resistors restrict the flow of current through a circuit and are commonly used to protect other circuit elements, such as light bulbs, from receiving more electrons than they can handle. The resistor is just one electronic component that controls the flow of current in a circuit, but many more exist.

Ways to control current

Controlling electrical current is similar in many ways to controlling water current. How many different ways can you control the flow of water using various plumbing devices and other components? Among the things you can do are restrict the flow, cut off the flow completely, adjust the pressure, allow water to flow in one direction only and store water. (Although this water analogy may help, it isn't 100 per cent valid because you don't need a closed system for water to flow, as you do with electricity.)

Many, many electronic components help control the electrical energy in circuits (see Figure 2-4). Among the most popular components are:

- *Resistors*, which restrict current flow. (We discuss resistors in Chapter 3.)
- *Capacitors*, which store electrical energy. (We discuss capacitors in Chapter 4.)
- *Inductors* and *transformers*, which are devices that store electrical energy in magnetic fields. (You can find out more about inductors and transformers in Chapter 5.)
- *Diodes*, used to restrict current flow in one direction, much like valves. (We discuss diodes in Chapter 6.)
- *Transistors*, which are versatile components that you can use to switch circuits on and off, or to amplify current. We cover transistors in Chapter 6.)

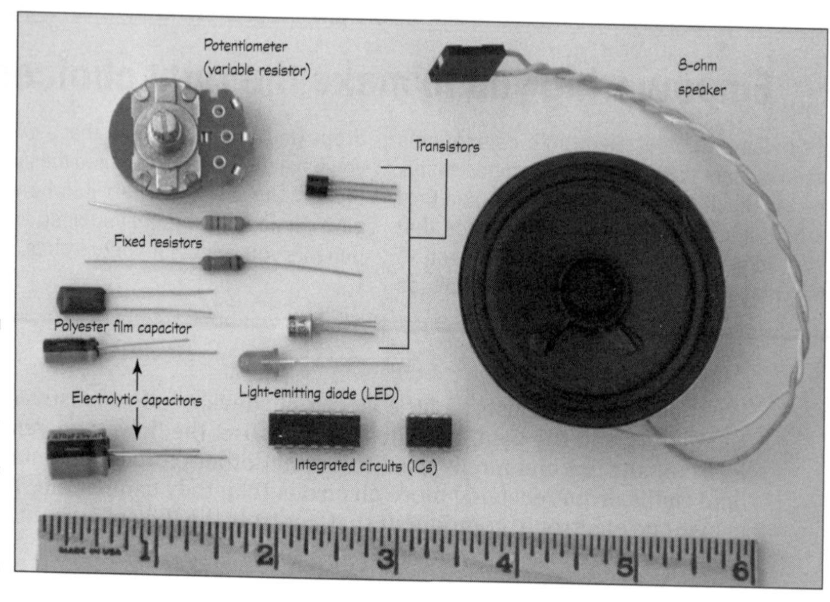

Figure 2-4: Electronic components come in a variety of shapes and packages.

Active versus passive components

You may see the terms active components and passive components used as category headings for types of electronic components. *Active components* are parts that provide gain or direction, such as transistors and diodes. (They may also be categorised as *semiconductors*, which is the type of material they're made from.) *Passive components* provide neither gain (amplification) nor direction, so resistors, capacitors, inductors and transformers are all passive components. A circuit that contains only passive components is called a *passive circuit*, and one that contains at least one active component is an *active circuit*.

Making Connections: Series and Parallel

Just as you can build structures of all shapes and sizes by connecting Lego bricks or Meccano parts together in various ways, so too you can build many different kinds of circuits by connecting electronic components in various ways. Exactly how you connect components together dictates how current flows through your circuit and how voltage is dropped throughout the circuit.

Series connections

In the simple light bulb circuit we examined earlier in this chapter (Figure 2-3), current flows from the positive battery terminal through the closed switch and then through the light bulb and back to the negative terminal of the battery. You call this arrangement a *series circuit*, which just means that the current runs through each component sequentially, or in series.

Two important things you need to remember about series circuits are:

- Each component has the same current.
- Voltage supplied by the source is divided (not necessarily evenly) among the components. If you add up the voltage drops across each component, you get the total supply voltage.

You may run into a problem with series circuits: if one component fails, it creates an open circuit, stopping the flow of current to every component in the circuit. So, if your flash new sign sports 200 light bulbs wired together in series to say 'GREAT PUB GRUB' and a stray cricket ball from the village green knocks one bulb for six, every one of the light bulbs goes out.

Parallel connections

You can fix the problem of all components in a series circuit blacking out when one component fails by wiring the components using parallel connections, such as in the circuit in Figure 2-5. With a parallel circuit, if several cricket balls take out a few bulbs in your sign, the rest of it stays lit. (You may be left with a sign advertising 'EAT PU RUB' but you can't have everything!)

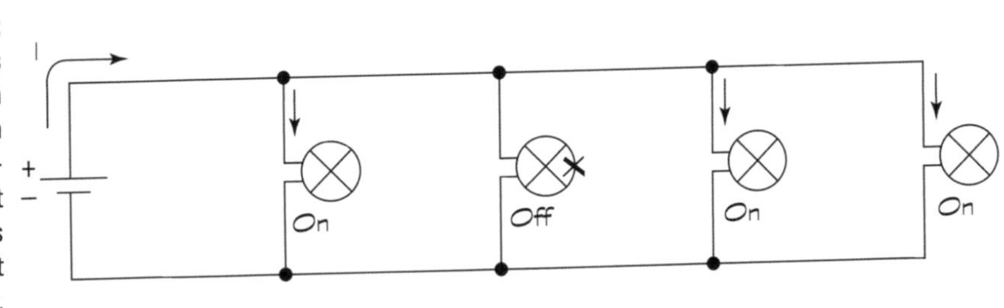

Figure 2-5: Light bulbs are often arranged in a parallel circuit so that if one burns out, the rest stay lit.

Here's how the parallel circuit in Figure 2-5 works: current flows from the positive battery terminal and then splits at each branch of the circuit, so that each light bulb gets a share of the supply current. The current flowing through one light bulb doesn't flow through the other light bulbs. So, if your restaurant sign has 200 light bulbs wired together in parallel and one burns out, light still shines from the other 199 bulbs.

In parallel circuits, the voltage across each parallel branch of the circuit is the same. And when you know how to calculate the current flowing in each branch of the circuit (which we discuss in Chapter 3), you can see that if you add up all the branch currents, you get the total current supplied by the battery.

Two important things you need to remember about parallel circuits are:

- Voltage across each branch is the same.
- Current supplied by the source is divided among the branches. The currents in each branch add up to the total supply current.

For the same circuit components, connecting them in parallel draws more current from your source than connecting them in series. If your circuit is powered by a battery, you need to be aware of just how long your battery can supply the necessary current to your circuit. As we discuss in Chapter 8, batteries have ratings of *amp-hours*. A battery with a rating of one amp-hour lasts for just one hour in a circuit drawing one amp of current. Therefore, when you decide what power source to use for a circuit, you must take into account both the current a circuit draws and how long you want to run the circuit.

Combination circuits

Most circuits are combinations of series and parallel connections. How you arrange components in a circuit depends on what you're trying to do.

Take a look at the series-parallel circuit in Figure 2-6. You see a resistor (the zigzag symbol in the diagram) in series with the battery, and then three parallel branches, each containing a switch in series with a light bulb. If you close all three switches, the supply current travels through the resistor and then splits three different ways, with some current passing through each of the three bulbs. If you open all three switches, no complete path exists for current to flow, so no current flows out of the battery at all. If you close only one switch, all the supply current flows through just one bulb, and the other bulbs are off. By alternating which switch is open at any time, you can control which bulb is lit.

You can imagine such a circuit controlling the operation of a three-stage traffic light (with a few more parts to control the timing and sequencing of the switching action).

To analyse combination circuits, you have to apply voltage and current rules one step at a time, using series rules for components in series and parallel rules for components in parallel. At this point, you don't quite have enough information to calculate all the currents and voltages in the light bulb circuits we show here. You need to know about one more rule, called *Ohm's Law*, and then you have everything you need to analyse simple circuits. We cover Ohm's Law and basic circuit analysis in Chapter 3.

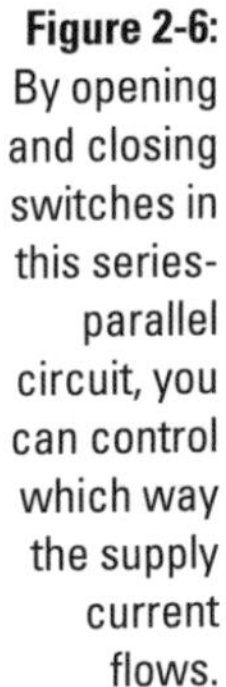

Figure 2-6: By opening and closing switches in this series-parallel circuit, you can control which way the supply current flows.

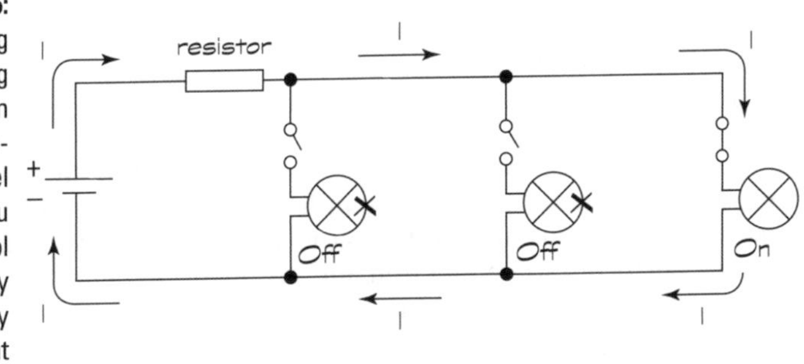

Creating Electronic Systems

Advanced electronic devices, such as personal computers, televisions and stereos, consist of a mind-boggling assortment of electronic components connected together in what appears to be a complex web of circuitry. All circuits – even ridiculously complicated ones – are designed to manipulate current in such a way as to perform one or more specific functions. Most complex electronic systems are made up of several stages, with each stage performing a specific function and the output of one stage feeding into the input of another stage.

In this section, we take a broad look at two such systems: your radio and your television set.

Making sound appear out of thin air

Electronic components in a radio control the current going to your speakers so that you hear the sounds your favourite radio station transmits. To get your speakers to reproduce the sounds originally created in the studio, several stages of electronics in your stereo system perform distinct functions:

- **Antenna:** Detects radio waves (invisible signals transmitted from many different radio stations) in the air and transforms them into an *electrical signal*, which is a variable flow of electric current.
- **Tuner:** Picks out just one radio signal from all the signals detected by the antenna, rejecting all the others.
- **Detector:** Separates the audio signal (a replica of the original sound) from the radio signal (which, in addition to the audio signal, includes a carrier signal that transports the audio signal through the air).
- **Amplifier:** Boosts the tiny audio signal so that you can hear it.

Each stage contains a bunch of electronic components combined in a way that manipulates or shapes the signal. After all that signal manipulation, the signal is sent to the speakers to be turned into sound. The specific pattern and intensity with which the electrical signal moves the diaphragm determines what exactly you're hearing and its volume.

Technical types often use *block diagrams* to describe the functionality of complex electronic systems, such as the radio receiver shown in Figure 2-7. Each block represents a circuit that takes the output of the previous block as its input signal, performs some function and produces an *output signal*, which is fed into another stage of the system.

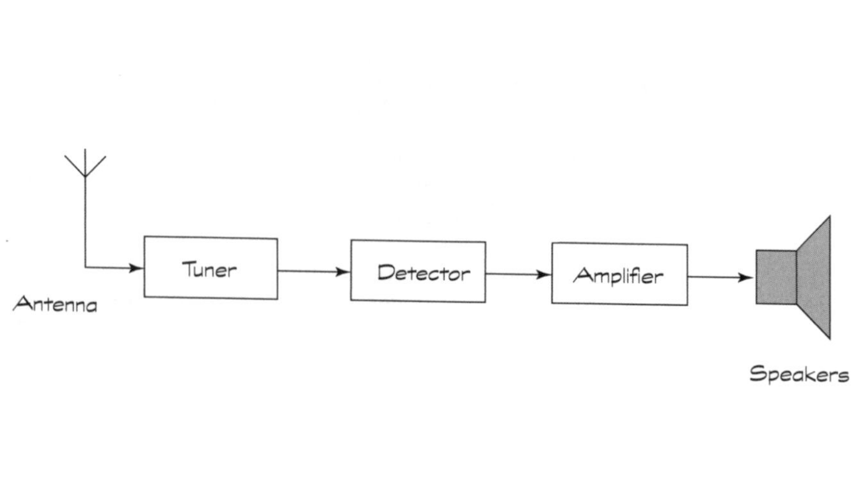

Figure 2-7: Block diagram representation of a radio receiver system. The electronics in the system perform several functions to shape the electric current that powers the speakers.

Deciphering electrical signals

An *electrical signal* is the pattern over time of an electrical current. Often, the way an electrical signal changes its shape conveys information about something physical, such as the intensity of light, heat or sound, or the position of an object, such as the diaphragm in a microphone or the shaft of a motor. Think of an electrical signal as a code, such as Morse code, sending and receiving secret messages that you can figure out – if you know the key.

An *analogue electrical signal*, or simply *analogue signal*, is so-named because it's an analogue, or one-to-one mapping, of the real physical quantity it represents. For instance, when a sound studio records a song, fluctuations in air pressure (that's what sound is) move the diaphragm of a microphone, which produces corresponding variations in electrical current. That fluctuating current is a representation of the original sound, or an analogue electrical signal.

Digital systems, such as computers, can't handle continuous analogue signals, so electrical signals must be converted into digital format before entering the depths of a digital system. *Digital format* is just another coding scheme, which uses only the binary values 1 and 0 to represent information. (Hey! That's just like Morse code's dot and dash!) A *digital signal* is created by sampling the value of an analogue signal at regular intervals in time and converting each value into a string of bits, or **bi**nary digi**ts**.

Painting pictures with electrons

Your television set, whether an old-fashioned cathode ray tube (CRT) device or a newer plasma or liquid crystal display (LCD) version, uses electronics to control which picture elements (pixels) on the screen should get fired up at any given time in order to paint a picture. The electrical signal that enters your set carries information about the picture to be displayed. (It may get that signal from your broadcast TV provider or the output of another electronic device, such as a DVD player). Electronic components within the set decode that electrical signal and apply the information carried by the signal to control the colour and intensity of each pixel.

Different types of TVs activate display pixels in different ways. For example, the electronics in a colour CRT TV steers three separate electron beams, positioning them to strike coloured phosphors on the inside surface of the screen. The electronics also controls whether each beam is on or off as it sweeps past each pixel, so that the phosphor it's aimed at gets bombarded by electrons or is left alone. When a beam of electrons strikes a phosphor, it glows. By co-ordinating the movement and on/off state of the so-called 'red', 'blue' and 'green' electron beams, the electronics in your TV creates colourful images on the screen.

Chapter 3

Meeting Up with Resistance

In This Chapter

- Using resistance to your advantage
- Creating just the right amount of resistance with fixed and variable resistors
- Understanding how current, voltage and resistance are governed by Ohm's Law
- Practising Ohm's Law by analysing circuits
- Using power to guide you in choosing circuit components

If you toss a marble into a sandpit, the marble doesn't go very far. But if you toss a marble onto the surface of a large frozen lake, the marble enjoys a nice little ride before eventually coming to a stop. A mechanical force called friction stops the marble on both surfaces – however, the sand provides more friction than the ice.

Resistance in electronics is a lot like friction in mechanical systems: it puts the brakes on electrons (those teeny-tiny moving particles that make up electrical current, or electricity) as they move through materials.

In this chapter, you find out exactly what resistance is, where you can find resistance (everywhere) and how you can use it to your advantage by selecting *resistors* (controlled amounts of resistance) for your electronic circuits. You explore the intimate relationship between voltage (the electrical force that pushes electrons) and current in components that have resistance, which is summed up quite nicely in a very simple equation with an authoritative name: Ohm's Law. We put Ohm's Law to work in analysing the goings-on in some basic circuits. Finally, you look at the role of Ohm's Law and related power calculations in the design of electronic circuits.

Resisting the Flow of Current

Resistance is a measure of an object's opposition to the flow of electrons. This opposition may sound like a bad thing, but in fact it's very useful. Resistance is what makes it possible to generate heat and light, restrict the flow of electrical current when necessary and ensure that the correct voltage is supplied to a device. For instance, as electrons travel through the filament of a light bulb, they meet up with so much resistance that they slow down a lot. Working to overcome resistance produces heat, as anyone who has burnt their hands sliding down a climbing rope knows. The filament has a very high resistance and produces so much heat that it lights up.

Everything – even the best conductor – exhibits a certain amount of resistance to the flow of electrons. The higher the resistance, the more restrictive the flow of current. So what determines how much resistance an object has? Resistance depends on several factors:

- **Material:** Some materials allow free electrons to roam freely, which makes them conductors, whereas others hold on tight to all their electrons, making them insulators. Conductors have lower *resistivity* than insulators because they provide less resistance to the flow of current in the same volume of material.
- **Cross-sectional area:** The larger the diameter, the more easily electrons move and the lower the resistance to their movement. Think of water flowing through a pipe: the wider the pipe, the easier the water flows. Resistance varies inversely with cross-sectional area. A copper wire with a large diameter has a lower resistance than a copper wire with a small diameter.
- **Length:** The longer the material, the more resistance it has because more opportunities exist for electrons to bump into other particles along the way. Resistance varies directly with length.
- **Temperature:** For most materials, the higher the temperature, the higher the resistance. Higher temperatures mean that the particles inside have more energy, so they move around a lot more, making it harder for the electrons to get through. Imagine trying to make your way through a crowded train station – doing so is always much harder if everyone there is rushing about. One notable exception is a device called a *thermistor*: increase the temperature of a thermistor, and it lowers its resistance in a very predictable way. (You can imagine how useful that is in temperature-sensing circuits.)

You use the symbol R to represent resistance in an electronic circuit. Sometimes you see a subscript next to a resistance, for instance, R_{bulb}. That just means that R_{bulb} represents the resistance of the light bulb (or whatever the subscript refers to). Resistance is measured in units called *ohms*, abbreviated with the Greek letter omega (Ω): the higher the ohm value, the higher the resistance.

An ohm is so small a resistance that you're likely to see resistance measured in larger quantities, such as kilohms (kilo + ohm), which is thousands of ohms and is abbreviated *k*Ω, or megohm (mega + ohm), which is millions of ohms and is abbreviated MΩ. So, 1 kΩ = 1000 Ω and 1 MΩ = 1,000,000 Ω.

Resistors: Passive Yet Powerful

Resistors are passive electronic components that are specially designed to have controlled amounts of resistance (for instance, 470 Ω or 1 kΩ). Although a resistor doesn't amplify or shape the electric current in any fancy way, it can be a powerful little device because it enables you to put the brakes on current flow in a very controlled way. By carefully choosing and arranging resistors in different parts of your circuit, you can control just how much – or how little – current each part of your circuit gets.

Discovering the usefulness of resistors

Resistors are among the most popular electronic components in town because they're simple and yet versatile. One of the most common uses of a resistor is to limit the amount of current in part of a circuit, but resistors can also be used to control the amount of voltage provided to part of a circuit.

Limiting current

The circuit in Figure 3-1 shows a 9-volt battery supplying current to a little device called a light-emitting diode (LED) through a resistor (shown as a zigzag). LEDs (like many other electronic parts) love current like children love chocolate: they try to gobble up as much as you give them. But LEDs run into a problem – they burn themselves out if they draw too much current. The resistor in the circuit serves the very useful function of limiting the amount of current sent to the LED (like a good parent restricts the intake of sweets).

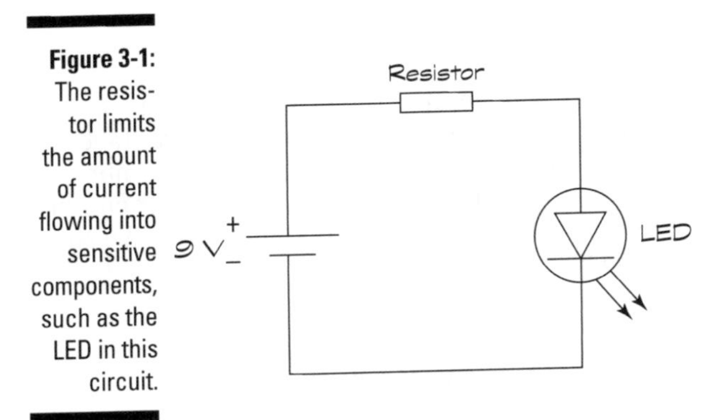

Figure 3-1: The resistor limits the amount of current flowing into sensitive components, such as the LED in this circuit.

Too much current can destroy many sensitive electronic components, such as transistors (which we discuss in Chapter 6) and integrated circuits (which we talk about in Chapter 7). By putting a resistor at the input to a sensitive part, you limit the current that reaches the part. This simple technique can save you a lot of time and money that you otherwise lose fixing accidental blow-ups of your circuits.

Reducing and controlling voltage

You can also use resistors to reduce the voltage supplied to different parts of a circuit. Say, for instance, you have a 9-volt power supply, but you need to provide 5 volts to power a particular integrated circuit you're using. You can set up a circuit, such as the one shown in Figure 3-2, to divide the voltage in a way that provides 5 volts at the output. Then – voilà – you can use the output voltage, V_{out}, of this *voltage divider* as the supply voltage for your integrated circuit. (We discuss exactly how this works in the later section 'Working out voltage across a component'.)

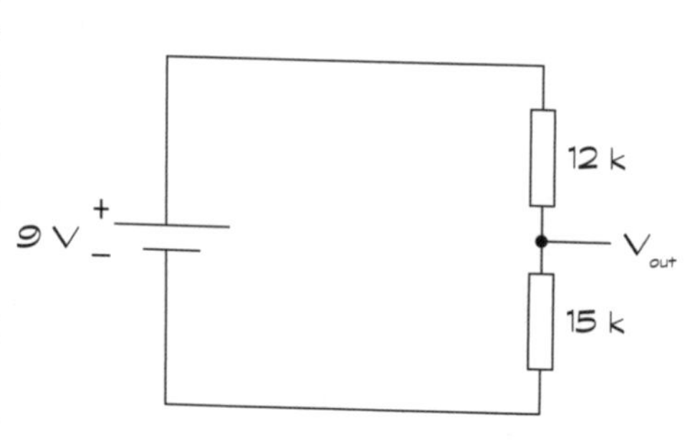

Figure 3-2: Using two resistors to create a voltage divider, a common technique for producing different voltages for different parts of a circuit.

You can also put a resistor to work with another popular component – a capacitor, which we discuss in Chapter 4 – to create up and down voltage swings in a predictable way. You'll find that the resistor-capacitor combo helps you create a kind of hourglass timer, which comes in handy for circuits that have time dependencies (for instance, a three-way traffic light). We show how this dynamic duo operates in Chapter 4.

Choosing a type: fixed or variable

Resistors come in two basic flavours: fixed and variable. You can see them in Figure 3-3. Here's the low-down on each type and why you would choose one or the other:

- A ***fixed resistor*** supplies a constant factory-determined resistance (no surprise there – that's what 'fixed' means). But the actual resistance of any given resistor may vary (up or down) from its nominal value by some percentage, known as the *resistor tolerance*. Therefore, when you choose a 1000Ω resistor with a 5 per cent tolerance, the actual resistance it provides can be anywhere from 950Ω to 1050Ω (because 5 per cent of 1000 is 50). Think of this as a resistance of 1000Ω, give or take 5 per cent. Two categories of fixed resistors exist:
  - ***High-precision resistors*** come within just 1 per cent of their nominal value. You use these in circuits where you need a resistor of a very specific value, such as in a precision timing circuit.
  - ***Standard-precision resistors*** can vary from 2 per cent to (gulp) 20 per cent of their nominal values. Markings on the resistor package tell you just how far off the actual resistance may be (for instance, ±2%, ±5%, ±10% or ±20%). Using standard-precision resistors in your circuits is okay 99 per cent of the time, because you're usually using resistors to limit current and don't need to ensure a precise current. Resistors with 5 or 10 per cent tolerance are commonly used in electronic circuits.
- A ***variable resistor,*** either a *potentiometer* (*pot* for short) or a *rheostat*, allows you to continually adjust the resistance from virtually zero ohms to a factory-determined maximum value. You use a potentiometer when you want to vary the amount of current you're supplying to part of your circuit. You can find potentiometers in light dimmer switches, audio system volume controls and joystick controls (for games and on aircraft).

What's in a name?

The word *potentiometer* is often used to categorise all variable resistors, but a difference exists between rheostats and potentiometers. Rheostats are two-terminal devices, with one lead connected to the wiper and the other lead connected to one end of the resistance track. Technically, a potentiometer is a three-terminal device, with leads connecting to the wiper and to both ends of the resistance track. You can use a potentiometer as a rheostat by connecting up only two leads, or you can connect all three leads in your circuit – and get a fixed and variable resistor for the price of one!

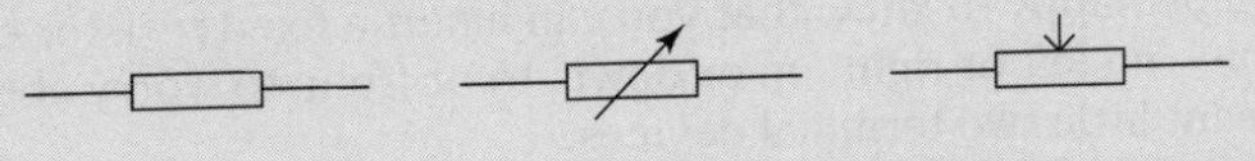

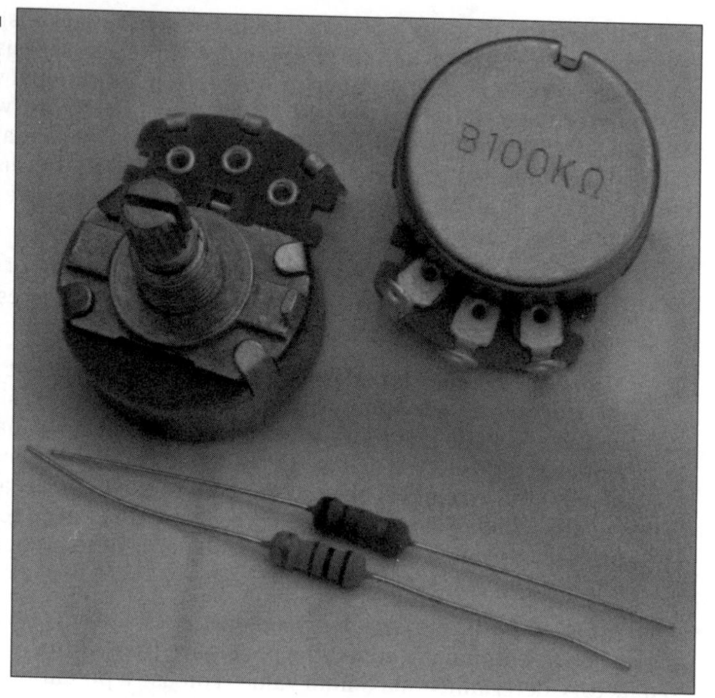

Figure 3-3: Colour-coded bands are used to denote the value of a fixed resistor, whereas variable resistors, known as potentiometers, are usually stamped with the highest value of resistance you can dial up.

In circuit diagrams, also called *schematics*, which we discuss in Chapter 10, you use a zigzag symbol to represent a fixed resistor. Resistors don't have polarity indicators (+ or –) on them, because current is happy to flow either way through a resistor. You add an arrow through the zigzag to create the schematic symbol for a rheostat (two-terminal variable resistor), and add an arrow pointing into the zigzag to create the symbol for a potentiometer (three-terminal variable resistor). We explain the difference between rheostats and potentiometers in the sidebar 'What's in a name?'.

Decoding fixed resistors

Most fixed resistors come in a cylindrical package with two leads sticking out so that you can connect them to other circuit elements. (See the later sidebar 'Recognising resistors on printed circuit boards' for exceptions to this.) You'll be happy to know that you can insert a fixed resistor either way in a circuit – no left or right, up or down, to or from, to worry about with these pleasant little two-terminal devices.

The attractive rainbow colours adorning most resistors serve a purpose beyond catching your eye (see Figure 3-3). Colour coding identifies the *nominal value* and *tolerance* of most resistors, whereas others are drab and boring, and have their values stamped on them. The colour code starts near the edge of one side of the resistor and consists of several stripes, or *bands*, of colour. Each colour represents a number and the position of the band indicates how you use that number. Standard-precision resistors use four colour bands: the first three bands indicate the nominal value of the resistor and the fourth indicates the tolerance. High-precision resistors use five colour bands: the first four bands indicate the value and the fifth indicates the tolerance (typically ±1%).

Using a special decoder ring (or the colour code shown in the first two columns of Table 3-1), you can decipher the nominal value of a standard-precision resistor as follows:

- The **first band** gives you the first digit.
- The **second band** gives you the second digit.
- The **third band** gives you the multiplier as a number of zeros, except when it's gold or silver.
  - If the third band is **gold**, you multiply by 0.1 (divide by 10).
  - If the third band is **silver**, you multiply by 0.01 (divide by 100).

  Result: You get the nominal value of the resistance by putting the first two digits together (side by side) and applying the multiplier.

The fourth band (tolerance) uses a different colour code, as shown in the third column of Table 3-1. If no fourth band exists, you can assume the tolerance is ±20%.

Table 3-1 Resistor Colour Coding

Colour	*Number*	*Tolerance*
Black	0	±20%
Brown	1	±1%
Red	2	±2%
Orange	3	±3%
Yellow	4	±4%
Green	5	n/a
Blue	6	n/a
Violet	7	n/a
Grey	8	n/a
White	9	n/a
Gold	0.1	±5%
Silver	0.01	±10%

Take a look at the following examples:

- Red-red-yellow-gold: A resistor with red (2), red (2), yellow (4 zeros) and gold (±5%) bands represents a nominal resistance of 220,000Ω, or 220 kΩ, which can vary up or down by as much as 5 per cent of that value. So it can have a resistance of anywhere between 209 kΩ and 231 kΩ.
- Orange-white-gold-silver: A resistor with orange (3), white (9), gold (0.1) and silver (±10%) bands represents a value of 39× 0.1, or 3.9Ω, which can vary by up to 10 per cent of that value. So the actual resistance can be anywhere from 3.5Ω to 4.3Ω.

For high-precision resistors, the first three bands of colour give you the first three digits, the fourth band gives you the multiplier and the fifth band represents the tolerance.

Most circuit designs tell you the safe resistor tolerance to use, whether for each individual resistor or for all the resistors in the circuit. Look for a notation in the parts list or as a footnote at the bottom of the circuit diagram. If the schematic doesn't state a tolerance, you can assume that using standard-precision resistors (±5% or ±10%) is okay.

Dialling with potentiometers

Potentiometers consist of a resistance track with connections at both ends and a *wiper* that moves along the track as you adjust the resistance across a range from 0 (zero) ohms to some maximum value (see Figure 3-4). Most often, potentiometers are marked with their maximum value – 10 K, 50 K, 100 K, 1 M and so forth – and they don't always include the little ohm symbol (Ω). For example, with a 50 K potentiometer, you can dial in any resistance from 0 to 50,000 ohms.

Bear in mind that the range on the potentiometer is approximate only. If the potentiometer lacks markings, you need to use a multimeter to figure out the component's value. (You can read about how to test resistances using a multimeter in Chapter 12.)

Potentiometers are available in various packages:

- *Dial pots* contain rotary resistance tracks and are controlled by turning a spindle or knob. Commonly used in electronics projects, dial pots are designed to be mounted through a hole cut in a case that houses a circuit, with the knob accessible from the outside of the case. They're popular for adjusting volume in sound circuits.
- *Slide pots* contain a linear resistance track and are controlled by moving a slide along the track. You see them on stereo equipment and some dimmer switches.

- *Trim pots* (also known as *preset pots*) are smaller, designed to be mounted on a circuit board and provide a screw for adjusting resistance. They're typically used to fine-tune a circuit design, for instance, to set the sensitivity of a light-sensitive circuit rather than to allow for variations, like volume adjustments, during the operation of a circuit.

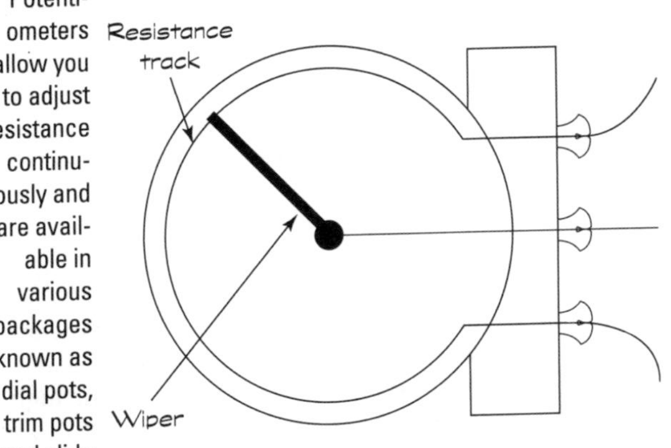

Figure 3-4: Potentiometers allow you to adjust resistance continuously and are available in various packages known as dial pots, trim pots and slide pots.

If you use a potentiometer (or rheostat) in a circuit, bear in mind that if the wiper is dialled down all the way, you've got zero resistance, and you're not limiting current at all with this device. Common practice is to insert a fixed resistor in series with a potentiometer as a safety net to limit current. You just choose a value for the fixed resistor so that it works together with your variable resistor to produce the range of resistances you need. (We discuss how to figure out the total resistance of multiple resistors in series in the later section 'Combining Resistors'.)

You can always use a multimeter to measure the resistance of an unknown resistor or the variable resistance provided by a potentiometer. We describe how to do this in Chapter 12.

Rating resistors according to power

Question: what do you get when you let too many electrons pass through a resistor at the same time? Yes, a horrible charred mess. Whenever electrons flow through something with resistance, they generate heat, and the more electrons, the higher the heat. Electronic components, such as resistors, can stand only so much heat (just how much depends on the size and type of component) before they have a meltdown. And because heat is a form of energy, and power is a measure of how much energy is being used each second, you can use the *power rating* of an electronic component to tell you how many watts (*watts* are units of power) a component can safely handle.

All resistors come with power ratings. Standard run-of-the-mill resistors are able to handle ⅛ watt or ¼ watt, but you can easily find ½-watt and 1-watt resistors. Of course, you don't see the power rating indicated on the resistor itself (that would make it too easy), so you have to figure it out by the size of the resistor (the bigger the resistor, the more power it can handle) or get it from the manufacturer or your parts supplier.

So how do you use the power rating to choose a particular resistor for your circuit? You estimate the peak power that your resistor is expected to handle, and pick a power rating that meets or exceeds it. Power is calculated as follows:

$$P = V \times I$$

In this equation (known as Joule's Law – see the later section 'The Power of Joule's Law' for more details), V represents the voltage (in volts) measured across the resistor and I represents the current (in amps) flowing through the resistor. For example, suppose the voltage is 5 volts and you want to pass 25 milliamps of current through the resistor. To calculate the power, you multiply 5 by 0.025 (remember, *milli*amps are thousandths of amps) and you get 0.125 watt, or ⅛ watt. So you know that although a ⅛-watt resistor may be okay, you can be sure that a ¼-watt resistor can take the heat just fine in your circuit.

Recognising resistors on printed circuit boards

As you find out more about electronics, you may get curious enough to take a look inside some of the electronics in your house. (Warning: Be careful! Follow the guidelines given in Chapter 9.) You may open up the remote control for your TV and see some components wired up between a touchpad and an LED, for instance. On some *printed circuit boards* (PCBs), which serve as the platform for building mass-produced circuits commonly found in computers and other electronic systems, you may have trouble recognising the individual circuit components. This is because manufacturers use fancy techniques to populate PCBs with components in order to eke out efficiencies and save space on the boards. One such technique, *surface mount technology* (SMT), allows components to be mounted directly to the surface of a board. *Surface mount devices*, such as SMT resistors, look a bit different from the components you would use to build a circuit in your garage, because they don't require long leads in order to connect them within a circuit. Such components use their own coding system to label the value of the part.

For most hobby electronics projects, ¼-watt or ⅛-watt resistors are just fine. You need high-wattage resistors for *high-load* applications, where loads, such as a motor or lamp control, require higher-than-hobby-level currents to operate.

High-wattage resistors take many forms, but you can bet that they are bigger and bulkier than your average resistor. Resistors with power ratings of over 5 watts are encased in epoxy or other waterproof and flameproof coating and have a rectangular rather than cylindrical shape. A high-wattage resistor may even include its own metal *heat sink*, with fins that conduct heat away from the resistor.

Combining Resistors

When you start shopping for resistors, you find that you can't always get exactly what you want. That's because it's impractical for manufacturers to make resistors with every possible value of resistance. So they make resistors with a limited set of resistance values, and you discover how to live with it. For instance, you can search far and wide for a 25 kΩ resistor, but you may never find it; however, 22 kΩ resistors are as common as muck.

Fortunately, you can combine resistors in various ways to create an *equivalent resistance* value that comes pretty darn close to whatever resistance you need. And because standard-precision resistors are accurate to 5 or 10 per cent of their nominal value anyway, combining resistors works out just fine.

Certain rules apply on how to combine resistances, which we cover in this section. You use these rules not only to help you choose off-the-shelf resistors for your own circuits, but also as a key part of your effort to analyse other people's electronic circuits. For instance, if you know that a light bulb has a certain amount of resistance and you place a resistor in series with the bulb to limit the current, you need to know what the total resistance of the two components is in order to calculate the current passing through them.

Using resistors in series

When you combine two resistors (or resistances) in series, you restrict the current somewhat with the first resistor, and then you restrict it even more with the next resistor. So the effect of the series combination is an *increase* in the overall resistance (see Figure 3-5). To calculate the combined (equivalent) resistance of two resistors in series, you simply add up the values of the two individual resistances. You can extend this rule to any number of resistances in series:

$$R_{series} = R1 + R2 + R3 + R4 + \ldots$$

R1, R2, R3 and so forth represent the values of the resistors and R_{series} represents the total equivalent resistance. The same current flows through all resistors connected in series.

You need to be careful with your units when adding up resistance values. For example, suppose you connect the following resistors in series: 1.2 kΩ, 680Ω and 470Ω. Before you add the resistances, you need to convert the values to the same units, for instance, ohms. In this case, the total resistance, R_{total}, is calculated as follows:

$$R_{total} = 1200\Omega + 680\Omega + 470\Omega = 2350\Omega \text{ or } 2.35 \text{ k}\Omega$$

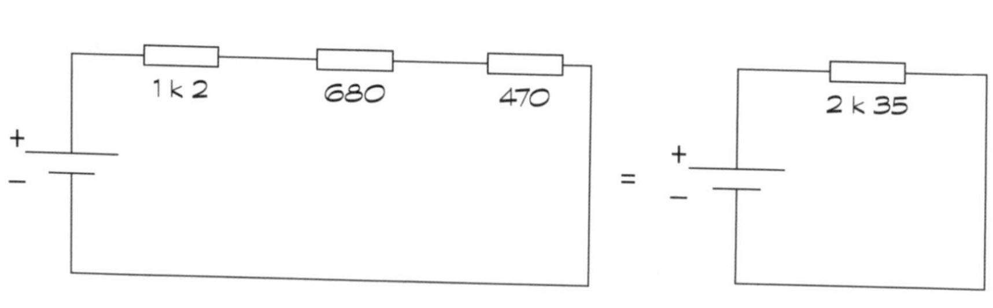

Figure 3-5: The combined resistance of two or more resistors in series is the sum of the individual resistances.

The combined resistance is *always* greater than any of the individual resistances.

This fact comes in handy when you're designing circuits. For example, if you want to limit current going into a light bulb, but you don't know the resistance of the bulb, you can place a resistor in series with the bulb and be secure in the knowledge that the total resistance to current flow is *at least* as much as the value of the resistor you added.

For circuits that use variable resistors, such as a light dimmer circuit, putting a fixed resistor in series with the variable resistor guarantees that the current is limited even if the pot is dialled down to zero ohms. (We discuss how to calculate just what the current is for a given voltage/resistance combo in the later section 'Using Ohm's Law to Analyse Circuits'.)

Employing resistors in parallel

When you combine two resistors in parallel, you provide two different paths for current to flow, so even though each resistor is restricting current flow through one circuit path, another path still exists that can draw additional current. From the perspective of the source voltage, the effect of arranging resistors in parallel is a *decrease* in the overall resistance. To calculate the equivalent resistance, $R_{parallel}$, of two resistors in parallel, you use the following formula:

$$R_{parallel} = \frac{R1 \times R2}{R1 + R2}$$

where R1 and R2 are the values of the individual resistors.

In the example in Figure 3-6, two 2 kΩ resistors are placed in parallel. The equivalent resistance is:

$$\begin{aligned} R_{parallel} &= \frac{2000 \times 2000}{2000 + 2000} \\ &= \frac{4,000,000}{4000} \\ &= 1000 \\ R_{parallel} &= 1\ k\Omega \end{aligned}$$

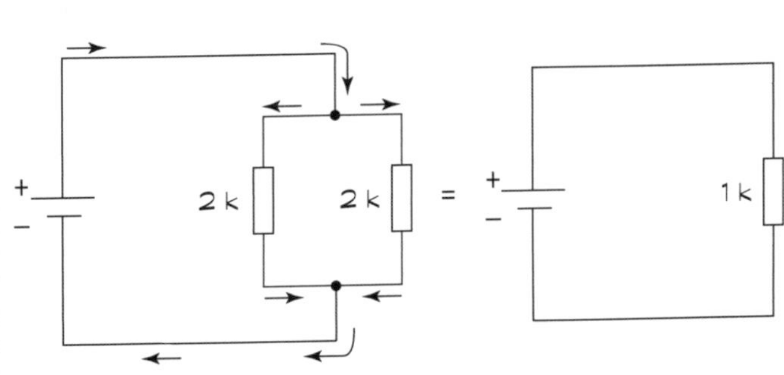

Figure 3-6: The combined resistance of two or more resistors in parallel is always lower than any of the individual resistances.

In this example, connecting two resistors of the same resistance in parallel results in an equivalent resistance of *half* the value of either one. Each resistor draws half the supply current. If two resistors of unequal value are placed in parallel, *more* current flows through the path with the *lower* resistance than the path with the higher resistance.

If you combine more than one resistor in parallel, the maths gets a little more complicated:

$$R_{parallel} = \frac{1}{\frac{1}{R1} + \frac{1}{R2} + \frac{1}{R3} \ldots \text{(and more as needed)}}$$

For multiple resistances in parallel, the amount of current flowing through any given branch is *inversely proportional* to the resistance within that branch. The higher the resistance, the less current goes that way; the lower the resistance, the more current goes that way. Just like water, electrical current favours the path of least resistance.

As shorthand in equations, you may see the symbol || used to represent the formula for resistors in parallel. For example:

$$R_{parallel} = R1 \,||\, R2 = \frac{R1 \times R2}{R1 + R2}$$

or

$$R_{parallel} = R1 \,||\, R2 \,||\, R3 = \frac{1}{\frac{1}{R1} + \frac{1}{R2} + \frac{1}{R3}}$$

Combining series and parallel resistors

Many circuits combine series resistors and parallel resistors in various ways so as to restrict current in some parts of the circuit while splitting current in other parts of the circuit. In certain cases, you can calculate equivalent resistance by combining the equations for resistors in series and resistors in parallel. For instance, in Figure 3-7 resistor R2 (2 kΩ) is in parallel with resistor R3 (2 kΩ), and that parallel combination is in series with resistor R1 (1 kΩ). You can calculate the total resistance (in kΩ) as follows:

$$\begin{aligned} R_{total} &= R1 + \left(R2 \,||\, R3\right) \\ &= R1 + \frac{R2 \times R2}{R2 + R3} \\ &= 1 + \frac{\left(2 \times 2\right)}{2 + 2} \\ &= 1 + 1 \\ &= 2 \\ R_{total} &= 2\text{ k}\Omega \end{aligned}$$

In this circuit, the current supplied by the battery is limited by the *total* resistance of the circuit, which is 2 kΩ. Supply current flows from the positive battery terminal through resistor R1 and then splits, with half flowing through resistor R2 and half flowing through resistor R3, and then combines again to flow into the negative battery terminal.

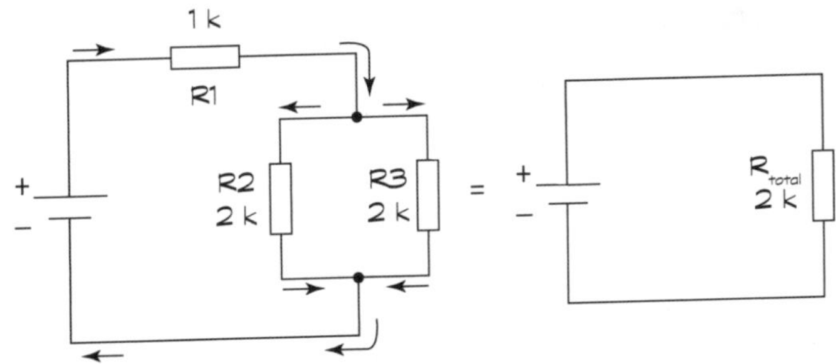

Figure 3-7: Many circuits include a combination of parallel and series resistances.

Circuits often have more complex combinations of resistances than simple series or parallel relationships, and figuring out equivalent resistances isn't always easy. You need to use matrix mathematics to analyse them, and because this book isn't a maths book, we're not going to get caught up in a matrix.

Obeying Ohm's Law

One of the most important concepts to understand in electronics is the relationship involving voltage, current and resistance in a circuit, summed up in a simple equation known as Ohm's Law. When you understand this thoroughly, you're well on your way to analysing circuits that other people have designed, as well as successfully designing your own circuits. Before diving into Ohm's Law, we take a quick look at the ebbs and flows of current.

Driving current through a resistance

If you place a voltage source across an electronic component that has measurable resistance, such as a light bulb or a resistor, the force of the voltage pushes electrons through the component. The movement of loads of electrons is what constitutes electrical current. By applying a greater voltage, you're exerting a stronger force on the electrons, so you're going to create a stronger flow of electrons, or a larger current, through the resistance: the stronger the force (voltage V), the stronger the flow of electrons (current I).

This effect is analogous to water flowing through a pipe of a certain diameter. If you exert a certain water pressure on the water in the pipe, the current flows at a certain rate. If you increase the water pressure, the current flows faster through that same pipe, and if you decrease the water pressure, the current flows slower through the pipe.

Understanding a relationship that's constantly proportional!

The relationship between voltage (V) and current (I) in a component with resistance (R) was discovered in the early 1800s by Georg Ohm, who gave his name to the unit of resistance. He figured out that for components with a fixed resistance, voltage and current vary in the same way: double the voltage, and the current is doubled; halve the voltage, and the current is halved. He summed this relationship up quite nicely in the simple mathematical equation that bears his name: Ohm's Law.

Ohm's Law states that voltage equals current multiplied by resistance, or

$$V = I \times R$$

What this equation really means is that the voltage (V) measured across a component with a fixed resistance is equal to the product of the current (I) flowing through the component and the value of the resistance (R).

For example, in the simple circuit in Figure 3-8, a 9-volt battery applied across a 1 kΩ resistor produces a current of 9 milliamps (which is 0.009 A) through the circuit:

$$9 \text{ volts} = 1000 \text{ ohms} \times 0.009 \text{ amps}$$

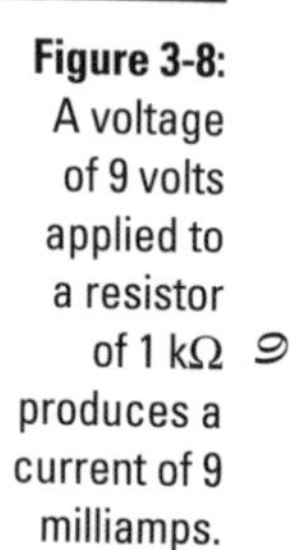
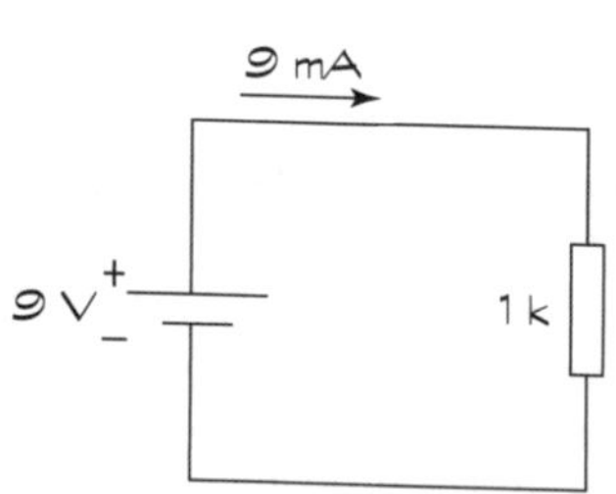

Figure 3-8: A voltage of 9 volts applied to a resistor of 1 kΩ produces a current of 9 milliamps.

More than just the law

Georg Ohm's name is associated with resistance values as well as the law. The definition of an ohm, or unit of resistance, comes from Georg Ohm's work. The *ohm* is defined as the resistance between two points on a conductor when one volt applied across those points produces one amp of current through the conductor. We just thought you may like to know that.

This little law is so important in electronics that you'd be wise to repeat it over and over again, like a mantra, until you've been transformed into an Ohm's Law follower. To help you remember, think of Ohm's Law as a **V**ery **I**mportant **R**ule.

When using Ohm's Law, watch your units carefully. Make sure that you convert any kilos and millis before you get out your calculator. If you think of Ohm's Law as volts = amps × ohms, you'll be okay. And if you're brave, you can also use volts = milliamps × kilohms, which works just as well (because the millis cancel out the kilos). But if you're not careful and you mix units, you may be in for a shock! For instance, a lamp with a resistance of 100Ω passes a current of 50 milliamps. If you forget to convert milliamps to amps, you multiply 100 by 50 to get 5000 volts as the voltage across the lamp! Ouch! The correct way to perform the calculation is to convert 50 milliamps to 0.05 amps, and then multiply by 100, to get 5 volts. (Much better!)

Working with one law, three equations

Remember your school algebra? Remember how you can rearrange the terms of an equation containing variables (such as the familiar x and y) to solve for one variable, as long as you know the values of the other variables? Well, the same rules apply to Ohm's Law. You can rearrange its terms to create two more equations, for a total of three equations from that one law!

$$V = I \times R \quad I = \frac{V}{R} \quad R = \frac{V}{I}$$

These three equations all say the same thing, but in different ways. You can use them to calculate one quantity when you know the other two. Which one you use at any given time depends on what you're trying to do. For example:

- **To calculate an unknown voltage**, multiply the current times the resistance (V = I × R). For instance, if you have a 2 milliamps current running through a 2 kΩ resistor, the voltage across the resistor is 2 milliamps × 2 kilohms (or 0.002 amps × 2000 ohms) = 4 volts.
- **To calculate an unknown current**, take the voltage and divide it by the resistance (I = V/R). For example, if 9 volts is applied across a 1 kΩ resistor, the current is 9 volts/1000 ohms = 0.009 amps or 9 milliamps.
- **To calculate an unknown resistance**, take the voltage and divide it by the current (R = V/I). For instance, if you have 3.5 volts across an unknown resistor with 10 milliamps of current running through it, the resistance is 3.5 volts/0.01 amps = 350 ohms.

Using Ohm's Law to Analyse Circuits

When you've mastered the theory of Ohm's Law, you're ready to put it into practice. Ohm's Law is like a master key, unlocking the secrets to electronic circuits. You can use it to understand circuit behaviour and to track down problems within a circuit (for instance, why the light isn't shining, the buzzer isn't buzzing or the resistor isn't resisting because it melted). You can also use it to design circuits and pick the right parts for use in your circuits. We get to that in the later section 'What is Ohm's Law Really Good For?'. In this section, we discuss how to apply Ohm's Law to analyse circuits.

Calculating current through a component

In the simple circuit in Figure 3-9, a 6-volt battery is applied across a 1 kΩ resistor. You calculate the current through the resistor as follows:

$$I = \frac{6 \text{ volts}}{1000 \text{ ohms}} = 0.006 \text{ amps or } 6 \text{ milliamps}$$

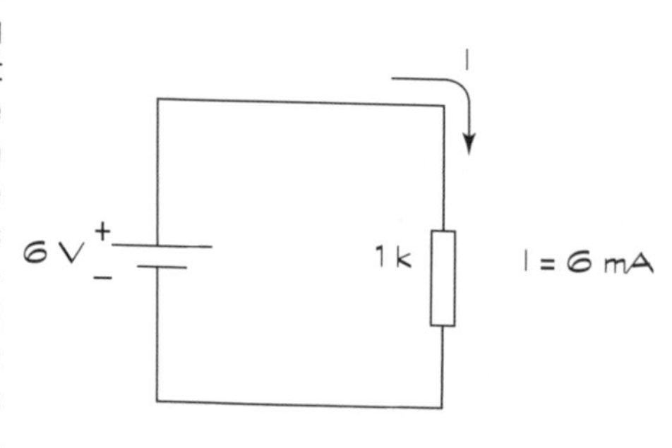

Figure 3-9: Calculating the current through the resistor in this simple circuit is a straight-forward application of Ohm's Law.

If you add a 220Ω resistor in series with the 1 kΩ, as shown in Figure 3-10, you're restricting the current even more. To calculate the current flowing through the circuit, you need to determine the total resistance that the 6-volt battery is facing in the circuit. Because the resistors are in series, the resistances add up, for a total equivalent resistance of 1.22 kΩ. The new current is:

$$I = \frac{6 \text{ volts}}{1220 \text{ ohms}} \approx 0.0049 \text{ amps or } 4.9 \text{ milliamps}$$

By adding the extra resistor, you've reduced the current in your circuit from 6 milliamps to 4.9 milliamps.

That double squiggle symbol (≈) in the equation just above means 'is approximately equal to', and we use that because we rounded off the current to the nearest tenth of a milliamp. Rounding off the tiny parts of values in electronics is usually okay – unless you're working on the electronics that control a subatomic particle smasher or other high-precision industrial device.

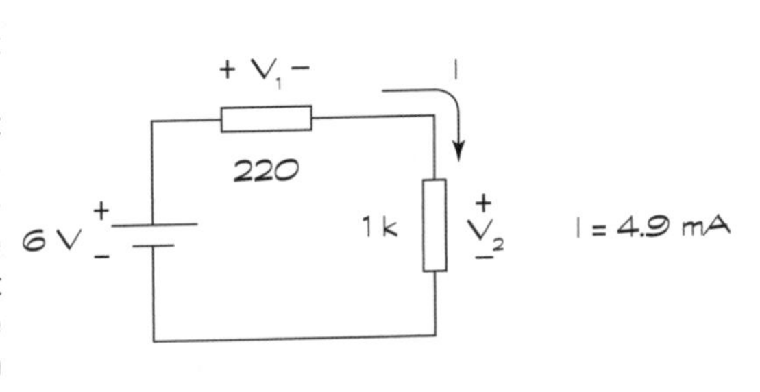

Figure 3-10: Calculating the current through this circuit requires determining the equivalent resistance and then applying Ohm's Law.

Working out voltage across a component

In the circuit in Figure 3-9, the voltage across the resistor is simply the voltage supplied by the battery: 6 volts. That's because the resistor is the only circuit element other than the battery. By adding a second resistor in series, as in Figure 3-10, the voltage picture changes. Now, some of the battery voltage drops across the 1 kΩ resistor and the rest of the battery voltage is dropped across the 220Ω resistor.

To calculate how much voltage is dropped across each resistor, you use Ohm's Law for each individual resistor. You know the value of each resistor, and *you know the current flowing through each resistor*. Remember that current is the battery voltage (6 volts) divided by the total resistance (1.22 kΩ), or approximately 4.9 milliamps. Now you can apply Ohm's Law to each resistor to calculate its voltage drop:

$$V_1 = 0.0049 \text{ amps} \times 220 \text{ ohms} = 1.078 \text{ volts} \approx 1.1 \text{ volts}$$

$$V_2 = 0.0049 \text{ amps} \times 1000 \text{ ohms} = 4.9 \text{ volts}$$

Using Ohm's Law

Ohm's Law is very useful in analysing voltage and current for resistors and other components that behave like resistors, such as light bulbs. But you have to be careful about applying Ohm's Law to other electronic components, such as capacitors (which we discuss in Chapter 4) and inductors (covered in Chapter 5), which don't have a constant resistance under all circumstances. For such components, the opposition to current, known as *impedance*, can and will vary depending on what's going on in the circuit. So you can't just use a multimeter to measure the resistance of a capacitor, for instance, and try to apply Ohm's Law willy-nilly. We discuss this subject a bit more in Chapter 4.

Notice that if you add up the voltage drops across the two resistors, you get 6 volts, which is the total voltage supplied by the battery. That's not a coincidence: the battery supplies voltage to the two resistors in the circuit, and the supply voltage is divided between the resistors proportionally based on the values of the resistors. This type of circuit is known as a *voltage divider.*

Many electronic systems use voltage dividers to bring down a supply voltage to a lower level and then feed that reduced voltage into the input of another part of the overall system.

Determining an unknown resistance

Say you have a large flashlight that you're running off a 12-volt battery, and you measure a current of 1.3 amps through the circuit (we discuss how to measure current in Chapter 12). You can calculate the resistance of the bulb by taking the voltage across the bulb (12 volts) and dividing it by the current through the bulb (1.3 amps):

$$R_{\text{bulb}} = \frac{12 \text{ volts}}{1.3 \text{ amps}} = 9 \text{ ohms}$$

What is Ohm's Law Really Good For?

Ohm's Law also comes in handy when you're analysing all kinds of circuits, whether simple or complex. You can use it in designing and altering electronic circuits, to make sure that you get the right current and voltage to the right places in your circuit. In fact, you're going to use Ohm's Law so much, it'll become second nature to you.

Analysing complex circuits

Ohm's Law really comes in handy when analysing more complex circuits. You often need to incorporate your knowledge of equivalent resistances in order to apply Ohm's Law and work out exactly where current is flowing and how voltages are being dropped throughout your circuit.

For example, imagine that you add to the circuit in Figure 3-10 by placing a 2.2 kΩ resistor in parallel with the 1 kΩ resistor, as shown in Figure 3-11. You can calculate the current running through each resistor step by step, as follows:

1. **Calculate the equivalent resistance of the circuit.**

   This figure is found by applying the rules for resistors in parallel and resistors in series:

$$
\begin{aligned}
R_{\text{equivalent}} &= 220 + \frac{1000 \times 2200}{1000 + 2200} \\
&\approx 220 + 688 \\
&\approx 908 \text{ ohms}
\end{aligned}
$$

2. **Calculate the total current supplied by the battery.**

   To do this, you apply Ohm's Law using the battery voltage and the equivalent resistance:

$$
\begin{aligned}
I_{\text{total}} &= \frac{6 \text{ volts}}{908 \text{ ohms}} \\
&\approx 0.0066 \text{ amps or } 6.6 \text{ milliamps}
\end{aligned}
$$

3. **Calculate the voltage dropped across the parallel resistors.**

   You can do this in one of two ways:

   - Apply Ohm's Law to the parallel resistors. You calculate the equivalent resistance of the two resistors in parallel, and then multiply that amount by the supply current. The equivalent resistance is 688 Ω, as shown in the first step above. So the voltage is:

$$
\begin{aligned}
V_2 &= 0.0066 \text{ amps} \times 688 \text{ ohms} \\
&\approx 4.55 \text{ volts}
\end{aligned}
$$

   - Apply Ohm's Law to the 220Ω resistor and subtract its voltage from the supply voltage. The voltage across the 220Ω resistor is:

$$
\begin{aligned}
V_1 &= 0.0066 \text{ amps} \times 220 \text{ ohms} \\
&\approx 1.45 \text{ volts}
\end{aligned}
$$

The voltage across the parallel resistors is:

$$V_2 = V_{supply} - V_1$$
$$= 6 \text{ volts} - 1.45 \text{ volts} = 4.55 \text{ volts}$$

4. **Calculate the current through each parallel resistor.**

   To do this, you apply Ohm's Law to each resistor, using the voltage you just calculated (V_2):

$$I_1 = \frac{4.55 \text{ volts}}{1000 \text{ ohms}} \approx 0.0046 \text{ amps or } 4.6 \text{ milliamps}$$
$$I_2 = \frac{4.55 \text{ volts}}{2200 \text{ ohms}} \approx 0.002 \text{ amps or } 2 \text{ milliamps}$$

   Notice that the two branch currents, I_1 and I_2, add up to the total supply current, I_{total}: 4.6 milliamps + 2 milliamps = 6.6 milliamps. That's a good thing (and a good way to check that you've done your calculations correctly).

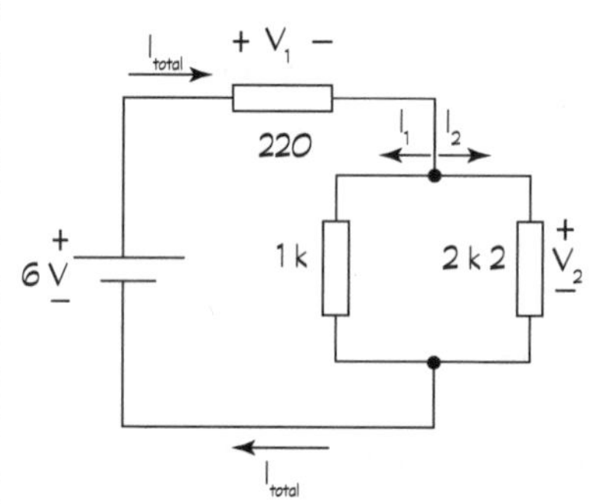

Figure 3-11: Complex circuits can be analysed by applying Ohm's Law and calculating equivalent resistances.

Designing and altering circuits

You can use Ohm's Law to determine what components to use in a circuit design. For instance, you may have a series circuit consisting of a 9-volt power supply, a resistor and an LED, as shown in Figure 3-1 earlier in this chapter. The voltage drop across an LED remains constant for a certain range of current passing through it (as we explain in Chapter 6), but if you try to pass too much current through the LED, it burns out. Say, for example, your LED voltage is 2.0 volts and the maximum current it can handle is 25 milliamps. What resistance should you put in series with the LED to limit the current so that it never exceeds 25 milliamps?

To work this figure out, you first have to calculate the voltage drop across the resistor when the LED is on. You already know that the supply voltage is

9 volts and the LED eats up 2 volts. The only other component in the circuit is the resistor, so you know that it eats up the remaining supply voltage, or 7 volts.

If you want to limit the current to be no more than 25 milliamps, you need to choose a resistor that's *at least* 7 volts/0.025 amps = 280 ohms. Because you can't find a 280Ω resistor, suppose you choose a 300Ω resistor. The current is 7 volts/300 ohms = 0.023 amps or 23 milliamps. The LED may burn a little less brightly, but that's okay.

Ohm's Law also comes in handy when tweaking an existing circuit. Say your spouse is trying to sleep, but you want to read, so you get out your big flashlight. The bulb in your flashlight has a resistance of 9Ω and is powered by a 6-volt battery, so you know that the current in the flashlight circuit is 6 volts/9 ohms = 0.65 amps.

Your spouse, however, thinks the light is too bright, and so to reduce the brightness (and save your marriage), you want to restrict the current flowing through the bulb a bit. You think that bringing it down to 0.45 amps will do the trick, and you know that inserting a resistor in series between the battery and the bulb is going to restrict the current. But what value of resistance do you need? You can use Ohm's Law to figure out the resistance value as follows:

1. **Use the desired new current to calculate the desired voltage drop across the bulb:**

   $V_{bulb} = 0.45 \text{ amps} \times 9 \text{ ohms} \approx 4.1 \text{ volts}$

2. **Calculate the portion of the supply voltage you want to apply across the new resistor. This voltage is the supply voltage less the voltage across the bulb:**

   $V_{resistor} = 6 \text{ volts} - 4.1 \text{ volts} = 1.9 \text{ volts}$

3. **Calculate the resistor value needed to create that voltage drop given the desired new current:**

   $$R = \frac{1.9 \text{ volts}}{0.45 \text{ amps}} \approx 4.2 \text{ ohms}$$

4. **Choose a resistor value that's close to the calculated value and make sure that it can handle the power dissipation:**

   $P_{resistor} = 1.9 \text{ volts} \times 0.45 \text{ amps} \approx 0.9 \text{ watts}$

   Result: As you can't find a 4.2-ohm resistor, you can use a 4.7-ohm 1-watt resistor to reduce the brightness of the light. This allows your spouse to sleep soundly, even if the snoring stops you reading!

The Power of Joule's Law

Another scientist hard at work in the early 1800s was the energetic James Prescott Joule. Joule is responsible for coming up with the equation that gives you power (as we discuss in the earlier section 'Rating resistors according to power'), known as *Joule's Law*:

$$P = V \times I$$

This equation states that the power (in watts) equals the voltage (in volts) across a component, times the current (in amps) passing through that component. The really nice thing about this equation is that it applies to every electronic component, whether it's a resistor, a light bulb, a capacitor or something else.

Using Joule's Law to choose components

In the earlier section 'Rating resistors according to power', we show you how to use Joule's Law to ensure that a resistor is big enough to resist a meltdown in a circuit. But you need to know that it also comes in handy when selecting other electronic parts. Lamps, diodes (which we discuss in Chapter 6) and other components also come with maximum power ratings. If you expect them to perform at power levels higher than their ratings, you're going to be disappointed. When you select the part, you need to consider the maximum possible power the part needs to handle by determining the maximum current that you're going to pass through the part and the voltage across the part, and then multiplying those quantities together. Then you choose a part with a power rating that exceeds that estimated maximum power.

Going perfectly together: Joule and Ohm

You can get creative and combine Joule's Law and Ohm's Law to derive more useful equations to help you calculate power for resistive components and circuits. For instance, if you substitute $I \times R$ for V in Joule's Law, you get:

$$P = (I \times R) \times I = I^2R$$

which gives you a way to calculate power when you know the current and resistance, but not the voltage. Similarly, you can substitute V/R for I in Joule's Law to get:

$$P = V \times \frac{V}{R} = \frac{V^2}{R}$$

Using that formula, you can calculate power if you know the voltage and the resistance, but not the current.

Joule's Law and Ohm's Law are so often used in combination that Georg Ohm sometimes gets the credit for both of those laws!

Trying Your Hand at Circuits with Resistors

If you want to get your hands dirty and experiment with some real resistor circuits, you can take a look at some of the circuits in the beginning of Chapter 13. These circuits enable you to experience Ohm's Law in action, play with a potentiometer to vary resistance and divide up voltages. But before you jump that far ahead, we recommend that you read through Part II, which explains how to set up your electronics workshop, stay safe, read schematics, construct circuits and measure everything in sight.

Chapter 4

Getting a Charge Out of Capacitors

In This Chapter

- Understanding the importance of capacitors
- Saying 'no' to DC and 'yes' to AC
- Using Ohm's Law (carefully!) for capacitive circuits
- Creating the dynamic duo: capacitors and resistors
- Exploiting capacitors to affect signals

Resistors may be the most commonly used electronic component, but capacitors come a close second. Skilled at storing electrical energy, capacitors are important in so many electronic circuits that the world would be a duller place without them.

Capacitors allow you to change the shape of electrical signals carried by current – a task that resistors alone can't perform. Although not as straightforward to understand as resistors, capacitors are essential in many of the electronic and industrial systems you enjoy today, such as radios, computer memory and car airbag systems, so investing your time and brain power in understanding how they work is worthwhile.

In this chapter, we look at what capacitors are made of, how they store electrical energy and how circuits use that energy. We watch a capacitor charge up and later release its energy, and see how it reacts to signals of different frequencies. We also show you how to use Ohm's Law to analyse capacitive circuits and how capacitors work closely with resistors to perform useful functions, as well as describing some uses of capacitors that prove their importance.

Reservoirs for Electrical Energy

When you're thirsty, you can get a drink of water that flows from the pipes when you turn on the tap or pour it from a storage container such as a bottle. You can think about electrical energy in a similar way: you get it from the flow around the circuit, driven by a battery, or from storage devices known as capacitors.

A *capacitor* is a passive electronic device that stores electrical energy transferred from a voltage source. Capacitors are useful because they can hold on to this energy until other parts of the circuit need it. A capacitor is made from two metal plates separated by an insulator, which is known as a *dielectric*. Two circuit symbols used for capacitors are shown here:

Charging and discharging

If you supply a DC voltage to a circuit containing a capacitor in series with a light bulb, as shown in Figure 4-1, current flow can't be sustained because a complete conductive path across the plates doesn't exist. However, electrons do move around this little circuit – temporarily – in an interesting way.

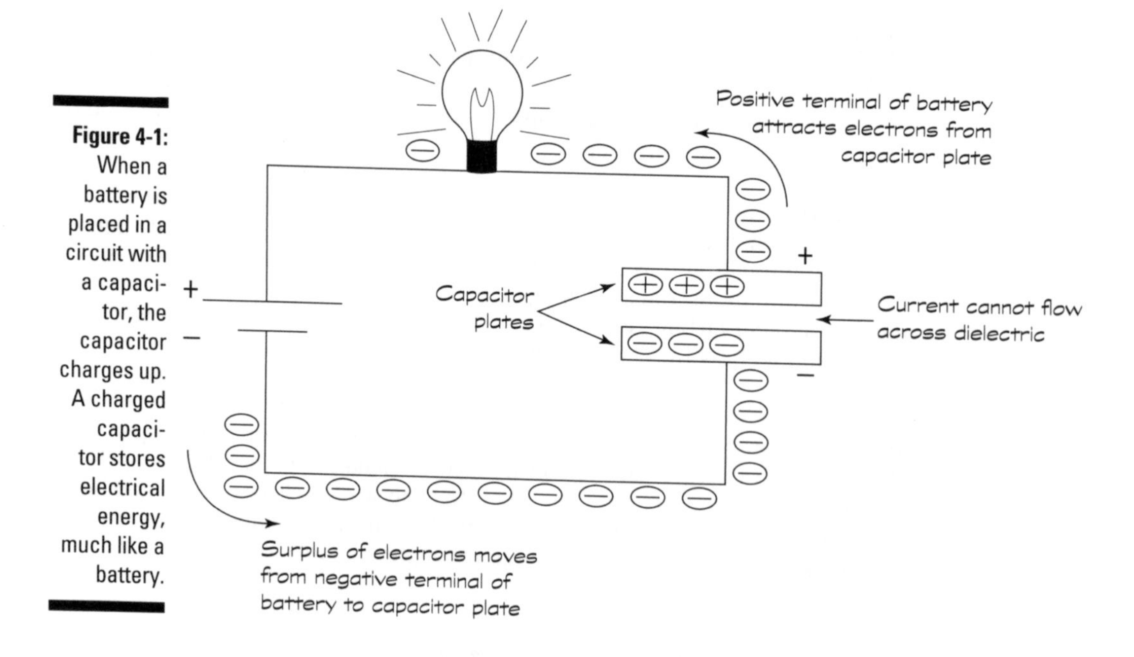

Figure 4-1: When a battery is placed in a circuit with a capacitor, the capacitor charges up. A charged capacitor stores electrical energy, much like a battery.

The negative terminal of a battery contains a surplus of electrons. Therefore, in the circuit in Figure 4-1, the surplus electrons begin to move away from the battery towards one side of the capacitor. The free electrons move towards one of the plates, where they're stopped in their tracks, with no conductive path to follow across the capacitor. This process results in an excess of electrons on that plate.

At the same time, the positive terminal of the battery attracts electrons from the other capacitor plate, so they begin to move. As they pass through the light bulb, they light it (but only for a split second, which we explain in the next paragraph). This activity produces a net positive charge (due to a deficiency of electrons) on that plate. A net negative charge on one plate and a net positive charge on the other plate results in a voltage difference across the two plates. This voltage difference represents the electrical energy stored in the capacitor.

The battery keeps pushing electrons onto one plate (and pulling electrons off the other plate) until the plate becomes saturated and can no longer accept any new charges (much like what happens when you reach the spending limit on your credit card). Electrons stop moving around through the circuit, and the light bulb goes out.

When the capacitor is saturated, the voltage drop across the plates then equals the battery voltage, and the capacitor is said to be *fully charged.* (In fact, the capacitor *plates* are charged; the capacitor as a whole has the same number of charges it started with.) If you remove the battery from the circuit, current doesn't flow and the charge remains on the capacitor plates. The capacitor looks like a voltage source, because it holds the charge, storing electrical energy.

If you replace the battery with a simple wire, you provide a path through the bulb for the surplus electrons on one plate to follow to the other (electron-deficient) plate. The capacitor plates *discharge* through the light bulb, lighting it up again briefly – even without a battery in the circuit – until the charge on both plates neutralises. The light bulb consumes the electrical energy that had been stored in the capacitor. When the capacitor is discharged (again, really the *plates* discharge), no more current flows.

A capacitor can store electrical energy for hours on end. You're wise to make sure that a capacitor is discharged before handling it, lest it discharges through you.

To discharge a capacitor, carefully place a bulb across its terminals, using insulated alligator clips (which we discuss in Chapter 9) to make the connection. If the bulb lights up, you know that the capacitor was charged, and the light should dim and go out in a few seconds as the capacitor discharges. If you don't have a bulb handy, place a 10-KΩ 1-watt resistor across the terminals and wait at least 30 seconds.

Opposing voltage change

Charges take time to build up on the capacitor plates when a DC voltage is applied and time to leave the plates when the DC voltage is replaced with an ordinary connecting wire; for this reason, capacitors are said to *oppose voltage change*. This phrase just means that if you suddenly change the voltage applied to a capacitor, it can't react straight away and the voltage across the capacitor changes more slowly than the voltage you applied.

Think of the process this way. Imagine that you've stopped at a red traffic light. When the light turns green, you set off, building up speed until you reach the speed limit. You take some time to get to that speed, just as a capacitor takes time to get to a certain voltage level.

This behaviour is quite different from what happens to resistors (which we discuss in Chapter 3). If you switch on a battery across a resistor, the voltage across the resistor changes almost instantly.

The capacitor voltage takes time to catch up to the source voltage, which is a useful attribute; in fact, many circuits use capacitors because they take time to charge up. For this reason, capacitors can change the shape of electrical signals.

Allowing alternating current

Capacitors can pass direct current (DC) only very briefly, because the dielectric provides a barrier to sustained electron flow. But they can pass alternating current (AC). How can that be?

An AC voltage source varies up and down, rising from 0 (zero) volts to its peak voltage, falling back through 0 volts and down to its negative peak voltage, and then rising back up through 0 volts to its peak voltage and so on.

Now, suppose that you apply an AC voltage source across a capacitor. Imagine being an atom on one of the capacitor plates and looking at the source terminal nearest you. You sometimes feel a force pulling your electrons away from you, and other times you feel a force pushing more electrons towards you. In each case, the strength of the force varies over time. You and the other atoms on the capacitor plate alternate between giving up electrons and getting electrons as the source voltage swings up and down.

What's really happening is that as the source voltage rises from 0 volts to its peak voltage, the capacitor charges up, just as it does when you apply a DC voltage. When the supply voltage reaches its peak, the capacitor may or may not be fully charged (depending on a bunch of factors, such as the size of the capacitor plates). Then the source voltage starts to decrease from its peak down to 0 volts. As it does, at some point, the source voltage becomes lower than the capacitor voltage. When this happens, the capacitor starts to discharge through the AC source. Then the source voltage reverses polarity and the capacitor discharges all the way.

As the source voltage keeps heading down towards its negative peak voltage, charges start to build up *in reverse* on the capacitor plates: the plate that previously held more negative charges now holds positive charges, and the plate that previously held more positive charges now holds negative charges. As the source voltage rises from its negative peak, the capacitor again discharges through the AC source, but in the opposite direction to its original discharge, and the cycle repeats itself.

The AC source constantly changes direction, and therefore the capacitor goes through a continuous cycle of charging, discharging and recharging. As a result, electrical charges move back and forth through the circuit, and even though the dielectric isn't conducting current, the effect is the same as if current is flowing through the capacitor. So, these amazing capacitors are said to pass AC even though they block DC.

Shining a light on capacitors

If you add a light bulb to your capacitor circuit powered by an AC voltage source, the bulb lights up and *stays lit* as long as the AC source is connected. Current alternates direction through the bulb, but the bulb doesn't care which way current flows through it. Although no current ever actually passes *through* the capacitor, the charging/discharging action of the capacitor creates the effect of current flowing back and forth through the circuit.

Characterising Capacitors

You can build capacitors in many ways, using different materials for the plates and dielectric and varying the size of the plates. The particular make-up of a capacitor determines its characteristics and influences its behaviour in a circuit.

Calculating the charge a capacitor can store

Capacitance is the ability of a body to store charge. The same term – capacitance – is used to describe just how much charge a capacitor can store on either one of its plates.

The capacitance of a capacitor depends on three things: the surface area of the metal plates, the thickness of the dielectric between the plates and the type of dielectric used (which we discuss in the later section 'Selecting dielectrics'). You don't need to know how to calculate capacitance (and, yes, a scary-looking formula does exist), because any capacitor worth its salt comes with a documented capacitance value. But knowing how to work out the capacitance helps you to understand that a capacitor's capacity to hold charge depends on how it's constructed.

Capacitance is measured in units called *farads*. One farad (abbreviated to F) is the capacitance needed to hold one *coulomb* of charge (see Chapter 1 for how many trillions of electrons that is) with a potential difference of 1V. Don't worry about the details of the definition; just know that one farad represents a very, very, very large amount of capacitance. You're more likely to run across capacitors with much smaller capacitances hovering in the microfarad (μF) or picofarad (pF) range. A microfarad is a millionth of a farad, or 0.000001 farad, and a picofarad is a millionth of a millionth of a farad, or 0.000000000001 farad.

Here are some examples:

- A 10-μF capacitor is 10 millionths of a farad.
- A 1-μF capacitor is 1 millionth of a farad.
- A 100-pF capacitor is 100 millionths of a millionth of a farad, or you can say that it's 100 millionths of a microfarad. Phew!

Reading capacitor values

Some capacitors have their values printed directly on them, in farads or portions of a farad, particularly larger capacitors that have enough space available to print the capacitance and working voltage.

Most smaller capacitors, such as 0.1- or 0.01-μF mica disc capacitors, use a three-digit marking system to indicate capacitance. Most people find the numbering system easy to use. But watch out! The system is based on picofarads, not microfarads. A number using this marking system, such as 103, means 10, followed by three zeros, as in 10,000, for a grand total of 10,000 picofarads. Some capacitors have a two-digit number printed on them that represents its value in picofarads. For instance, a value of 22 means 22 picofarads. No third digit means no zeroes to put on the end.

For values over 1,000 picofarads, your parts supplier very likely lists the capacitor in microfarads, even when the markings on it indicate picofarads. To convert the picofarad value you read on the capacitor into microfarads, just move the decimal point six places to the *left*. So, a capacitor marked with a 103, as in the example above, has a value of 10,000 pF or 0.01 μF.

Incredible shrinking capacitors

Making farad-range capacitors has become feasible only recently. Using older construction techniques, a one-farad capacitor would be bigger than a bread bin.

By using other technologies and materials, such as microscopically small carbon granules, manufacturers can now build capacitors of one farad and above that fit into the palm of your hand. Computer memories, clock radios and other electric devices that need to retain a small charge for extended periods of time when they have no access to power often use capacitors as substitute batteries.

Suppose that you're building a circuit that calls for a 0.1-μF disc capacitor. You can convert microfarads into picofarads to figure out what marking to look for on the capacitor package. Just move the decimal point six places to the *right*, and you get 100,000 pF. Because the three-digit marking consists of the first two digits of your pF value (10) followed by the additional number of zeros (4), you need a mica disc capacitor labelled '104'.

You can use Table 4-1 as a reference guide to common capacitor markings that use this numbering system.

Table 4-1 Capacitor Value Reference

Marking	*Value*
nn (a number from 01 to 99)	***nn pF***
101	100 pF
102	0.001 μF
103	0.01 μF
104	0.1 μF
221	220 pF
222	0.0022 μF
223	0.022 μF
224	0.22 μF
331	330 pF
332	0.0033 μF
333	0.033 μF

continued

Table 4-1 *(continued)*

Marking	*Value*
334	0.33 μF
471	470 pF
472	0.0047 μF
473	0.047 μF
474	0.47 μF

Another less-common numbering system uses both numbers and letters, as follows:

4R1

The placement of the letter R tells you the position of the decimal point: 4R1 is really 4.1. This numbering system doesn't indicate the units of measure, however, which can be in microfarads or picofarads.

You can test capacitance with a capacitor meter or a multimeter with a capacitance input. Most meters require that you plug the capacitor directly into the test instrument, because the capacitance can increase with longer leads, making the reading less accurate.

On many capacitors a single letter code indicates the tolerance. You may find that letter placed by itself on the body of the capacitor or placed after the three digit mark, such as:

103Z

The letter Z denotes a tolerance of +80 per cent to 20 per cent. This tolerance means that the capacitor, rated at 0.01 μF, may have an actual value as much as 80 per cent higher or 20 per cent lower than the stated value. Table 4-2 lists the meanings of common code letters used to indicate capacitor tolerance.

Capacitors rarely do what they say on the tin. The actual capacitance of the capacitor can vary quite a bit from the nominal capacitance. Manufacturing variations cause this problem; capacitor makers aren't trying to cheat you. Fortunately, this issue rarely creates a problem in home-made circuits. Still, you need to know about these variations so that if a circuit calls for a higher precision capacitor, you know what to buy. Like resistors, capacitors are rated by their tolerance, and this tolerance comes as a percentage.

Table 4-2 Capacitor Tolerance Markings

Code	*Tolerance*
B	+/– 0.1 pF
C	+/– 0.25 pF
D	+/– 0.5 pF
F	+/– 1 per cent
G	+/– 2 per cent
J	+/– 5 per cent
K	+/– 10 per cent
M	+/– 20 per cent
Z	+ 80 per cent, –20 per cent

Keeping an eye on the working voltage

The *working voltage*, sometimes abbreviated as WV, is the highest voltage that the manufacturer recommends placing across a capacitor safely. If you exceed the working voltage, you can damage the dielectric, which can result in current arcing between the plates, like a lightning strike during a storm. That means that you short out your capacitor and allow all sorts of unwanted current to flow – and maybe even damage other components.

Capacitors designed for DC circuits are typically rated for a WV of no more than 16–35 volts. That's plenty for DC circuits, which are usually powered by sources from 3.3 volts to 12 volts.

When you build circuits that use higher voltages, you need to concern yourself with the working voltage of a capacitor. If you do, be sure to select a capacitor that has a WV of at least 10–15 per cent more than the supply voltage in your circuit, just to be on the safe side.

Selecting dielectrics

Suppose that a child asks you to make her a banana split, but you don't have any bananas. So you improvise and use cucumbers instead. Yuck! It's just not the same. Similarly, designers of electronic circuitry specify capacitors for projects by the dielectric material in them. Some materials are better in certain applications: just like bananas in a banana split or cucumber in a sandwich, they provide a better match.

The most common dielectric materials are aluminium electrolytic, tantalum electrolytic, ceramic, mica, polypropylene, polyester (or Mylar), paper and polystyrene. If a circuit diagram calls for a capacitor of a certain type, you need to be sure to get one that matches.

Sizing up capacitor packaging

Capacitors come in a variety of shapes and sizes, as shown in Figure 4-2. Aluminium electrolytic and paper capacitors commonly come in a cylindrical shape. Tantalum electrolytic, ceramic, mica and polystyrene capacitors have a more bulbous shape because they typically get dipped into an epoxy or plastic bath to form their outside skin. However, not all capacitors of a particular type (such as mica or Mylar) get manufactured in the same way, so you can't always tell the component by its cover.

Your favourite parts supplier may label capacitors according to the way their leads are arranged: axially or radially. Axial leads extend out from each end of a cylindrically shaped capacitor, along its axis; radial leads extend from one end of a capacitor and are parallel to each other (until you bend them for use in a circuit).

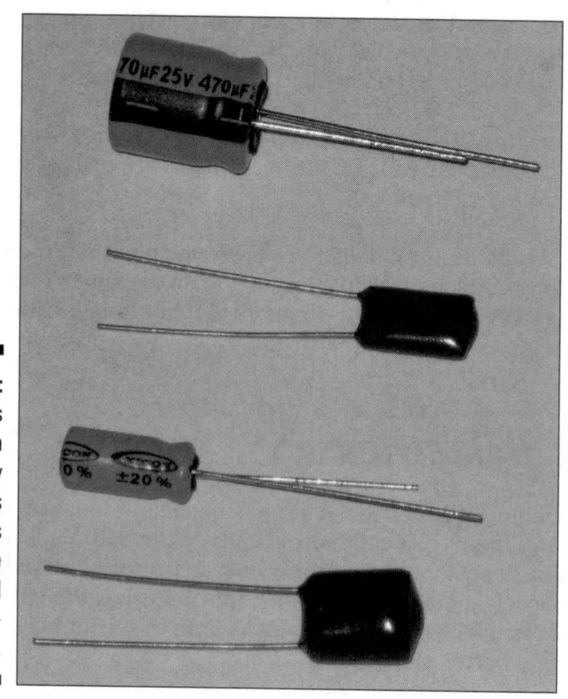

Figure 4-2: Capacitors come in a variety of shapes and sizes and may be polarised or non-polarised.

If you go searching for capacitors inside your personal computer (PC), you may not recognise them when you see them. That's because the capacitors in your PC don't have any leads at all! So-called *surface mount packages* for capacitors are extremely small and are designed to be soldered directly to printed circuit boards (PCBs), such as the ones in your PC.

Since the 1980s, high-volume manufacturing processes have been using surface mount technology (SMT) to connect capacitors and other components directly to the surface of PCBs, saving space and improving circuit performance.

Connecting with polarity

Some larger-value electrolytic capacitors (1μF and up) are polarised, meaning that the way in which you insert the capacitor into your circuit matters. Polarised capacitors are designed for use in DC circuits, and the positive terminal must be kept at a higher voltage than the negative terminal.

Many polarised capacitors sport a minus (–) sign or a large arrow pointing towards the negative terminal. For example, as the top capacitor in Figure 4-2 shows, the minus sign and arrow point to the negative lead of the aluminium electrolytic capacitor. For radial capacitors, the negative lead is often shorter than the positive lead.

The fact that capacitor packages generally point out the negative terminal doesn't mean circuit diagrams follow the same convention. Usually, if a circuit contains a polarised capacitor, you see a plus sign (+) on one side of the capacitor symbol, as shown below, showing you how to orient the capacitor in the circuit:

If a capacitor is polarised, you *really*, *really* must be sure to install it in the circuit the right way round. If you reverse the leads, say, by connecting the + side to the ground rail in your circuit, you can cause the dielectric inside the capacitor to break down, which can short-circuit the capacitor. You may damage other components in your circuit (by sending too much current their way), and your capacitor may even explode.

Varying capacitance

Variable capacitors allow you to adjust capacitance to suit your circuit's needs. The symbols for a variable capacitor are shown below:

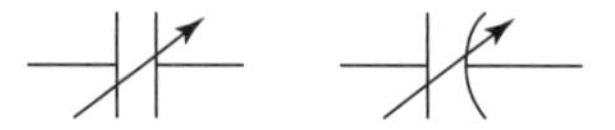

The most common type of variable capacitor is the air dielectric type, such as the one that you find in the tuning control of an AM radio.

Mechanically controlled variable capacitors work by changing the distance between the capacitor plates or the amount of overlap between the surfaces of the two plates. Specially designed diodes (a semiconductor device we discuss in Chapter 6) can function as electronically controlled variable capacitors, known as *varactors* or *varicaps*. By changing the DC voltage applied to these devices, you change the capacitance of the device.

You probably interact with variable capacitors more often than you think, because they're behind many touch-sensitive devices, such as the keys on computer keyboards and the buttons on your favourite remote control.

One type of microphone uses a variable capacitor to convert sound into electrical signals, with the diaphragm of the mike acting as a movable capacitor plate. Sound fluctuations make the diaphragm vibrate, which varies the capacitance, producing voltage fluctuations.

Combining Capacitors

If multiple capacitors are in a circuit you want to analyse, you can combine them to get the equivalent capacitance. But the rules for combining capacitors are different from the rules for combining resistors (which we discuss in Chapter 3).

Capacitors in parallel

Figure 4-3 shows two capacitors in parallel, with the common connection points labelled A and B. Notice that point A connects to one plate of capacitor C1 and one plate of capacitor C2. Electrically speaking, point A connects to a metal plate that's the size of the two plates combined. Likewise for point B, which connects to both the other plate of capacitor C1 and the other plate of capacitor C2. The larger the surface area of a capacitor plate, the larger the capacitance. Therefore, *capacitors in parallel add up.*

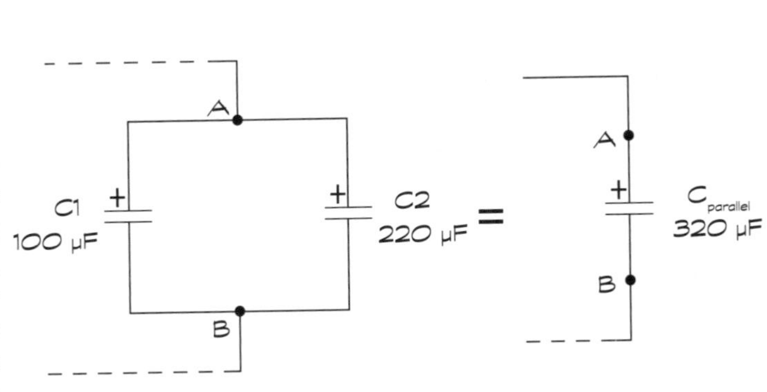

Figure 4-3: Capacitors in parallel add up. Each metal plate of one capacitor is tied electrically to one of the metal plates of the parallel capacitor. Each pair of plates behaves as a single larger plate, effectively resulting in a higher capacitance.

The equivalent capacitance of a set of capacitors in parallel is:

$$C_{parallel} = C1 + C2 + C3 + C4 + \ldots$$

C1, C2, C3 and so forth represent the values of the capacitors and $C_{parallel}$ represents the total equivalent capacitance.

For the capacitors in Figure 4-3, the total capacitance is:

$$\begin{aligned} C_{parallel} &= 100\mu F + 220\mu F \\ &= 320\mu F \end{aligned}$$

If you place the capacitors from Figure 4-3 in a working circuit, the voltage across each capacitor is the same, and current flowing in to point A splits up to travel through each capacitor and then joins together again at point B.

Capacitors in series

If placing capacitors in parallel increases the total plate area, then connecting capacitors in series increases the total dielectric thickness. So capacitors in series reduce the effective capacitance in the same way that resistors in parallel reduce the overall resistance:

$$C_{series} = \frac{C1 \times C2}{C1 + C2}$$

C1 and C2 are the values of the individual capacitors, and C_{series} is the equivalent capacitance. The total capacitance (in µF) of a 100-µF capacitor in series with a 220-µF capacitor, as shown in Figure 4-4, is:

$$\begin{aligned} C_{series} &= \frac{100 \times 220}{100 + 220} \\ &= \frac{22{,}000}{320} \\ &= 68.75 \\ C_{series} &= 68.75\mu F \end{aligned}$$

You can temporarily ignore the 'µ' part of 'µF' while you're performing the calculation above – as long as all the capacitance values are in µF and you remember that the resulting total capacitance is also in µF.

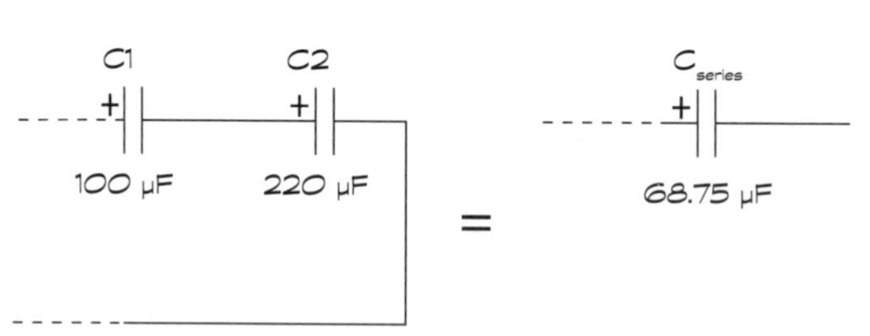

Figure 4-4: Capacitors in series work against each other, reducing the overall capacitance.

The equivalent capacitance of a set of capacitors in series is:

$$C_{series} = \frac{1}{\frac{1}{C1} + \frac{1}{C2} + \frac{1}{C3} \ldots \text{(and more as needed)}}$$

As for any components in series, the current running through each capacitor in series is the same, but the voltage dropped across each capacitor may be different.

Understanding Capacitive Reactance

In Chapter 3, we define resistance as a measure of an object's opposition to the flow of electrons, and we say that resistors have controlled amounts of resistance that remain the same no matter what the voltage or current situation. If you could measure a capacitor's opposition to the flow of electrons, you'd find that it varies, depending on the situation.

In the earlier section 'Charging and discharging', we say that capacitors block direct current except for a short time while they're charging and discharging, and allow AC. When a DC voltage is suddenly applied to a capacitor, such as in the light bulb circuit in Figure 4-1, at first, while the capacitor is charging, current flows through the circuit, lighting the bulb. As the capacitor nears its capacity to hold charges, less current flows until eventually the capacitor is fully charged, and current no longer flows. At first, when a sudden voltage change is applied, the capacitor provides very little opposition to electron flow, but it eventually behaves like an open circuit, opposing all electron flow.

Application of an AC voltage source to a capacitive circuit allows current to flow through the circuit. The faster the source voltage fluctuates, the less the capacitor opposes electron flow – in the same way as the sudden application of the battery to the discharged capacitor in Figure 4-1. The slower the source voltage fluctuates, the more the capacitor opposes electron flow.

The apparent AC resistance to current, known as the *capacitive reactance*, is measured in ohms (yes, ohms!). You calculate capacitive reactance, symbolised by X_c, using the following formula:

$$X_c = \frac{1}{2 \times \Pi \times f \times C}$$

In this formula, f represents the frequency in hertz (Hz) of the applied AC voltage, C is the capacitance in farads (*not* μF or pF) and Π is that constant you first met in geometry class that starts out as 3.14 and keeps going without repeating itself. You can approximate $2 \times \Pi$ as 6.28 and simplify the formula as follows:

$$X_c \approx \frac{1}{6.28 \times f \times C}$$

You can see from this formula that capacitive reactance *decreases* as the frequency of the applied voltage *increases*. (A higher value of frequency makes the denominator bigger, which makes the overall fraction smaller.) For instance, you can calculate the capacitive reactance of a 0.1-μF capacitor when an AC voltage source fluctuating at 20 KHz is applied as follows:

$$X_c \approx \frac{1}{6.28 \times 20,000 \times 0.0000001}$$
$$\approx 80 \text{ ohms}$$

If you slow down the frequency of the source voltage to 1 Hz, the capacitive reactance changes:

$$X_c \approx \frac{1}{6.28 \times 1 \times 0.0000001}$$
$$\approx 1.6 \text{ Mohms}$$

Note that this situation is *very* different from the done-and-dusted resistance of your average resistor. For capacitors, the faster the voltage fluctuates (higher frequency), the lower the reactance and the more freely current flows. The slower the voltage fluctuates (lower frequency), the higher the reactance and the less easily current flows. If the frequency is zero, which means no voltage fluctuations (or a constant DC voltage), the denominator is zero, and the reactance is infinite. That's the open circuit (infinite resistance) situation in which the capacitor blocks DC signals.

Using Ohm's Law for capacitive reactance

As capacitive reactance is measured in ohms, you may be wondering if you can use Ohm's Law for capacitors. The answer is yes – sort of. Ohm's Law works for capacitive reactance – but just *one frequency at a time*. If you change the frequency of the AC voltage, even if you don't change the magnitude of the voltage ups and downs, you have to recalculate Ohm's Law given the new frequency.

Suppose that you apply an AC voltage source with a peak value of 5 volts to your 0.1-μF capacitor. The current through the circuit alternates at the same frequency as the voltage, and yet Ohm's Law tells us that the peak value of the current depends on the peak value of the voltage and the capacitive reactance.

Say, for example, that the frequency is 1 Hz. From the calculations in the previous section, you know that the capacitive reactance of a 0.1-µF capacitor at 1 Hz is 1.6 MΩ. Now you can apply Ohm's Law to calculate the peak AC current 'through' the capacitor for a 1-Hz signal as follows:

$$\begin{aligned} I_{peak} &= \frac{V_{peak}}{X_c} \\ &= \frac{5 \text{ volts}}{1{,}600{,}000 \text{ ohms}} \\ &\approx 0.0000031 \text{ amps or } 3.1\mu A \end{aligned}$$

Now imagine that you change the frequency to 20 KHz, but keep the same peak voltage of 5 volts. Now the capacitive reactance is 80 Ω (as calculated in the previous section). Using Ohm's Law, the peak current 'through' the capacitor when you apply a 20-KHz voltage with a peak value of 5 volts is:

$$\begin{aligned} I_{peak} &= \frac{V_{peak}}{X_c} \\ &= \frac{5 \text{ volts}}{80 \text{ ohms}} \\ &\approx 0.0625 \text{ amps or } 62.5 \text{ milliamps} \end{aligned}$$

So for the same capacitor in the circuit, as you increase the frequency of the source voltage, you decrease the capacitive reactance, resulting in an increase in current flowing through the circuit. Likewise, if you decrease the frequency of the source voltage, you increase the capacitive reactance, resulting in a decreased current.

Understanding that behaviour depends on frequency

Unlike resistors, capacitors in an AC circuit act differently depending on the frequency of the voltage applied to them. This frequency-dependent behaviour means that pretty much everything about the circuit varies depending on the applied frequency.

As the following section shows, you can exploit this frequency-dependent behaviour to create circuits that perform useful functions, such as filters that favour high-frequency signals over low-frequency signals, and vice versa.

So What Have Capacitors Ever Done for Us?

Capacitors are in most electronic circuits you encounter every day. Their abilities to store electrical energy, block DC current and vary their opposition to current depending on applied frequency are commonly exploited by circuit designers to set the stage for extremely useful functionality. Capacitors (like the Romans) have made so many things possible that listing them all is difficult. In the following sections, we just make a start.

Storing electrical energy

Many devices use capacitors to store energy temporarily for later use. Uninterruptible power supplies (UPSs) and alarm clocks keep a charged capacitor on hand just in case of a power failure. The energy stored in the capacitor releases when the charging circuit disconnects (which it does when the power goes out!).

Similarly, cameras use capacitors to temporarily store the energy used to generate the flash, and many electronic devices use capacitors to supply energy while the batteries are being changed. Car audio systems commonly use a capacitor to supply energy when the amplifier needs more than the car's electrical system can give. Without that capacitor, every time you hear a heavy bass note, your headlights would dim!

Blocking DC current

When connected in series with a signal source, such as a microphone, capacitors block DC current but pass AC current. This process is known as *capacitive coupling* or *AC coupling*, and when used this way, the capacitor is known as a *coupling capacitor*. Multi-stage audio systems commonly use this functionality between stages so that only the AC signal from the first circuit passes to the second circuit, and any DC offset is removed prior to the amplification stage.

Smoothing out voltage

Power supplies that convert AC to DC often take advantage of the fact that capacitors don't react quickly to sudden changes in voltage. These devices use large electrolytic capacitors to smooth out varying DC supplies. These *smoothing capacitors* keep the output voltage at a relatively constant level by discharging through the load when the DC supply falls below a certain level. This situation is a classic example of using a capacitor to store electrical energy until you need it: when the DC supply can't maintain the voltage, the capacitor gives up some of its stored energy to take up the slack.

Creating timers

Because a capacitor takes time to charge and discharge, they're often used in timing circuits to create 'ticks' and 'tocks' when the voltage rises above or falls below a certain level. The value of the capacitor, in addition to other components in the circuit, controls the timing of the ticking. We discuss this aspect in more detail in the later section 'Teaming Up with Resistors'.

Tuning in (or out) frequencies

Capacitors are often used to help select or reject certain electrical signals, depending upon their frequency. For instance, a tuning circuit in a radio receiver system relies on capacitors and other components to allow the signal from just one radio station to pass through to the amplifier stage, while blocking signals from all other radio stations. Each radio station is assigned a specific broadcast frequency, and therefore the value of capacitors in tuning circuits is selected so that the target frequency is tuned in.

We describe simple filters in the later section 'Selecting Frequencies with Simple RC Filters'.

Teaming Up with Resistors

Capacitors are often found working hand in hand with resistors in electronic circuits, combining their talent for storing electrical energy with a resistor's ability to control the flow of electrons. Put these two together and you can control how fast electrons fill (or charge) a capacitor and how fast those electrons empty out (or discharge) from a capacitor. This dynamic duo is very popular, and circuits containing both resistors and capacitors are known as *RC circuits*.

Timing is everything

Take a look at the RC circuit in Figure 4-5. The battery charges the capacitor through the resistor when the switch is closed. At first, the voltage across the capacitor, V_c, is zero (assuming the capacitor was discharged to begin with).

When you close the switch, current starts to flow and charges start to build up on the capacitor plates. Ohm's Law tells you that the charging current, I, is determined by the voltage across the resistor, V_r, and the value of the resistor, R ($I = V_r/R$). And because the voltage drops around a circuit must equal the supply voltage, you know that the resistor voltage is the difference between the supply voltage, V_{supply}, and the capacitor voltage, V_c ($V_r = V_{supply} - V_c$). Using those two facts, you can analyse what's going on in this circuit over time, as follows:

- **Initially:** Because the capacitor voltage is initially zero, the resistor voltage initially equals the supply voltage.
- **Charging:** As the capacitor charges, it works together with the resistor to slow the current.
- **Fully charged:** When the capacitor is fully charged, current stops flowing, no voltage drops across the resistor and the entire supply voltage drops across the capacitor.

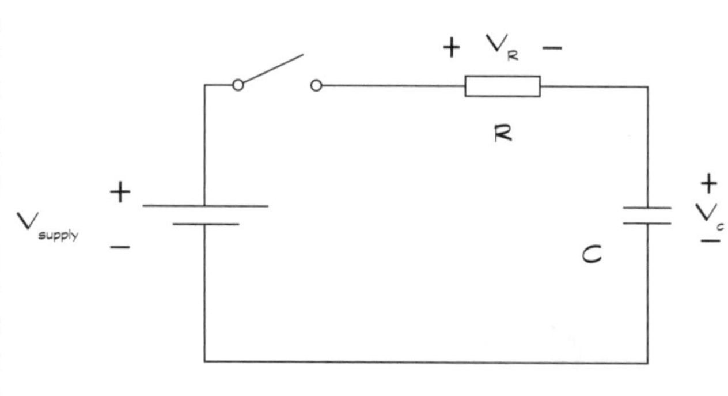

Figure 4-5: In an RC circuit, the capacitor charges up through the resistor. The values of the resistor and capacitor determine how quickly the capacitor charges.

If you remove the battery and connect the resistor across the capacitor, the capacitor discharges through the resistor. This time the voltage across the resistor is equal to the voltage across the capacitor ($V_r = V_c$), so the current is V_c/R. Here's what happens:

- **Initially:** Because the capacitor is fully charged, its voltage is initially V_{supply}. And because $V_r = V_c$, the resistor voltage is initially V_{supply}, the current jumps up immediately to V_{supply}/R. Therefore, the capacitor shuffles charges from one plate to the other pretty quickly.
- **Discharging:** The capacitor continues to discharge, but at a slower rate, and as the voltage falls so too does the current.

- **Fully discharged:** When the capacitor is fully discharged, current stops flowing, and no voltage drops across the resistor or the capacitor.

The waveform in Figure 4-6 shows how the capacitor voltage changes over time as it charges and discharges through the resistor when a constant voltage is applied and then removed from the circuit.

How fast the capacitor charges (and discharges) depends on the resistance and capacitance of the RC circuit. The larger the resistance, and the larger the capacitance, the smaller the current that flows for the same supply voltage, and the longer the capacitor takes to charge.

During the discharge cycle, a larger resistor slows down the electrons more as they move from one plate to the other, increasing the discharge time, and a larger capacitor holds more charge, taking longer to discharge.

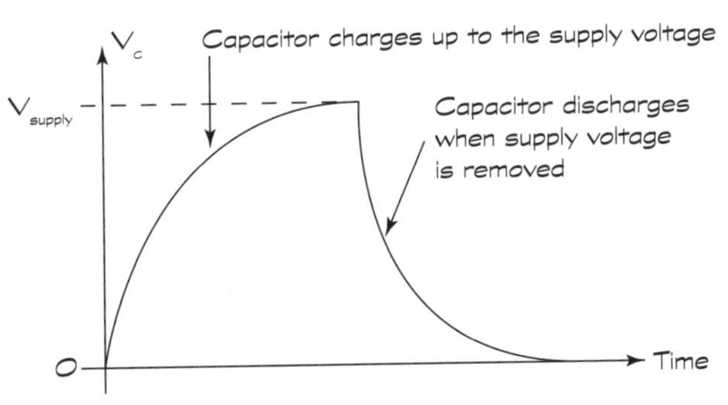

Figure 4-6: The voltage across a capacitor changes over time as the capacitor charges and discharges.

Calculating RC time constants

You can adjust a capacitor's charge and discharge time by picking the values of the capacitor and the resistor. In fact, a definitive relationship exists between your choice of resistance, R, and capacitance, C, and the time taken to charge and discharge your selected capacitor through your chosen resistor. If you multiply R (in ohms) by C (in farads), you get what's known as the *RC time constant* of your RC circuit, symbolised by T:

$$T = R \times C$$

A capacitor charges and discharges nearly completely after five times its RC time constant, or 5RC (which really means $5 \times R \times C$). After one time constant, a discharged capacitor charges to roughly two-thirds of its capacity, and a charged capacitor discharges nearly two-thirds of the way.

For instance, suppose that you chose a 2-MΩ resistor and a 15-μF capacitor for the circuit in Figure 4-5. You calculate the RC time constant as follows:

$$\text{RC time constant} = R \times C = 2{,}000{,}000 \text{ ohms} \times 0.000015 = 30 \text{ seconds}$$

Then you know that it's going to take about 150 seconds (or 2½ minutes) to fully charge or discharge the capacitor. If you want a shorter charge/discharge cycle time, you can reduce the resistor or the capacitor (or both).

Suppose that the 15-μF capacitor is the only one you have in the house, and you want to charge it in five seconds. You can figure out what resistor you need to make this happen as follows:

- **Find the RC time constant:** You know that five times the RC time constant is necessary to charge the capacitor fully, and you want that to be five seconds. So the RC time constant is one-fifth of five seconds, or one second.
- **Calculate R:** If $R \times C = 1$ second, and C is 15-μF, you know that $R = 1$ second/0.000015 farads, which is approximately 66,667 ohms or 67 kΩ.

Creating a timer

Armed with the knowledge of the RC time constant, you can use an RC circuit to create a timer. Say, for example, your freeloading flatmate keeps raiding your fridge in the middle of the night. You decide to set up a little noisemaker to scare her off whenever she opens the fridge. Just for fun, you want to allow her some time to gaze at the lovely cool beers inside before you shock the pants off her with a few choice decibels of sound from a buzzer, triggered by a switch that closes when she opens the fridge door.

If you've got a buzzer that requires a voltage of 6 volts in order to sound off, and you're using a 9-volt battery to power your little scare circuit, you can build an RC circuit like the one in Figure 4-5 and use the capacitor voltage to trigger the buzzer. The idea is to charge the capacitor to about 6 volts in the time you want your flatmate to think about whether she wants a beer or a lager and reach out for one, but then blast her with the buzzer.

Suppose that this period is about 10 seconds. You have a 15-μF capacitor handy, so you need to calculate the resistance necessary to charge the capacitor to 6 volts in 10 seconds. Because the capacitor keeps charging until it reaches the full power supply voltage of 9 volts, the trigger point (when the buzzer sounds) is when the capacitor reaches ⅔ of its capacity, or about two-thirds capacity. That happens after just one time constant. You calculate the resistance for your 10-second delay-to-buzzing interval as follows:

$$10 \text{ seconds} = R \times 0.000015 \text{ farads}$$

$$R = \frac{10 \text{ seconds}}{0.000015 \text{ farads}} \approx 667 \text{ ohms}$$

You have a 620 Ω resistor handy, so you have an actual RC time constant of about 9.3 seconds (that's 620 ohms × 15-μF). It takes about 9.3 seconds for the capacitor voltage to reach 6 volts, so your flatmate has enough time to look – but not enough time to drink – before she gets caught red-handed.

If you want to fine-tune the delay, use a resistor with a slightly smaller value than you need and add a potentiometer (pot) in series with the resistor. Because the total resistance is the sum of the value of the fixed resistor and that of the pot, you can increase or decrease the resistance by adjusting the pot. Just tweak the pot until you get the delay you want. (We cover potentiometers in Chapter 3.)

Selecting Frequencies with Simple RC Filters

Because capacitors behave differently depending on the frequency of the voltage or current in a circuit, they're often used in special circuits called *filters* to allow or reject various signals. Capacitors naturally block DC signals and allow AC signals to pass, but you can control just which AC signals to allow to pass by carefully choosing the components in filter circuits.

We discuss here very simple filters and how you can control which frequencies make their way through them. Electronic filter design, which is a field of study on its own and beyond the scope of this book, usually involves more complex circuits to control the output precisely. However, the underlying concepts are the same as for the simple filters we discuss.

Looking at low-pass filters

In the circuit in Figure 4-7, a variable voltage source, labelled V_{in}, is applied to a series RC circuit, and the output of the circuit, V_{out}, is the voltage across the capacitor.

Suppose that you apply a constant voltage (f = 0 Hz) to this circuit. No current flows, and so the full input voltage drops across the capacitor: $V_{out} = V_{in}$. At the other end of the frequency spectrum, the capacitive reactance ($\frac{1}{2} \times \Pi\ f \times C$) for a very, very high frequency is a very, very low value. This high frequency effectively short-circuits the capacitor, so the voltage drop across the capacitor is zero: $V_{out} = 0$.

As you alter the frequency of the input signal from very low to very high, the capacitive reactance varies from very high to very low. The higher the reactance, the more voltage drops across the capacitor (at the expense of voltage dropped across the resistor). The lower the reactance, the less voltage drops across the capacitor (and the more is dropped across the resistor).

This circuit tends to allow lower frequencies to pass from input to output while blocking higher frequencies from getting through and is therefore known as a *low-pass filter*.

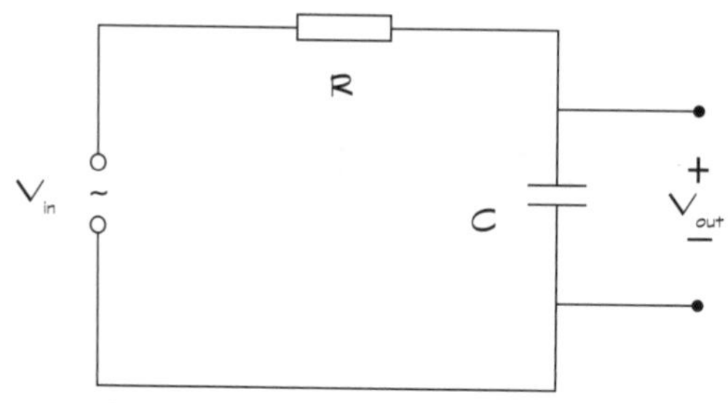

Figure 4-7: A low-pass filter allows input signals with low frequencies to pass through to the output.

Encountering high-pass filters

By reversing the roles of the resistor and capacitor in the low-pass RC circuit, you can create the opposite effect: a *high-pass filter*. In the circuit in Figure 4-8, the output voltage is the voltage across the resistor. For very low-frequency input signals, the capacitor blocks current from flowing, so no voltage drops across the resistor: $V_{out} = 0$. For very high-frequency input signals, the capacitor acts like a short, so current flows and all the input voltage drops across the resistor: $V_{out} = V_{in}$.

As you alter the frequency from low to high, the capacitive reactance varies from high to low. You can think of this situation as placing an imaginary device – a frequency-controlled potentiometer – in the circuit in place of the capacitor: as the input frequency increases, the reactance decreases, and more and more of the input voltage drops across the resistor.

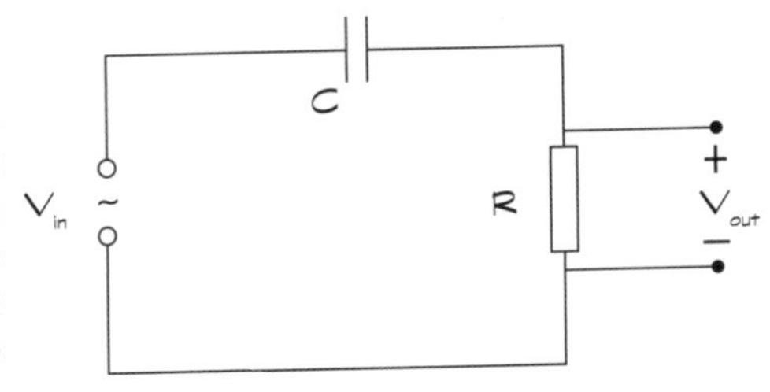

Figure 4-8: A high-pass filter allows input signals with high frequencies to pass through to the output.

Cutting off frequencies at the knees

Filter circuits are designed to pass certain frequencies while *attenuating*, or reducing the amplitude of, other frequencies. No filter is perfect; it can't pass all signals above or below a specific frequency completely, while totally blocking all other frequencies. Complex filter designs are much better than simple RC filters at distinguishing which frequencies get through and which don't, but all filters – simple or complex – share a design parameter known as the cut-off frequency.

The *cut-off frequency*, F_c, is the frequency at which the filter starts to restrict the passage of the input signal. Figure 4-9 shows a graph of the amplitude of the output signal for various input frequencies for a high-pass filter (notice that this graph is a frequency plot, not a time plot!). The graph shows that frequencies above the cut-off frequency are allowed to pass with little or no attenuation, while frequencies below the cut-off frequency are attenuated substantially. The cut-off frequency, which occurs near the knee of the curve, can be found using the following equation:

$$F_c = \frac{1}{2 \times \Pi \times T}$$

Because $T = R \times C$, you can control the cut-off frequency of your simple low- or high-pass filter by carefully selecting the values of the resistor and capacitor, according to the following equation:

$$F_c = \frac{1}{2 \times \Pi \times R \times C}$$

For instance, say you have a high-pass filter configuration with a 220-Ω resistor and a 0.1-μF capacitor. The cut-off frequency of your filter is roughly $1/6.28 \times 220$ ohms $\times$ 0.0000001 farads, or about 7,200 Hz. If you use such a filter in an audio system, don't be surprised when you don't hear much of your favourite band's voices or their instruments: the sounds they make fall well below 7,000 Hz, and your simple filter attenuates those sounds!

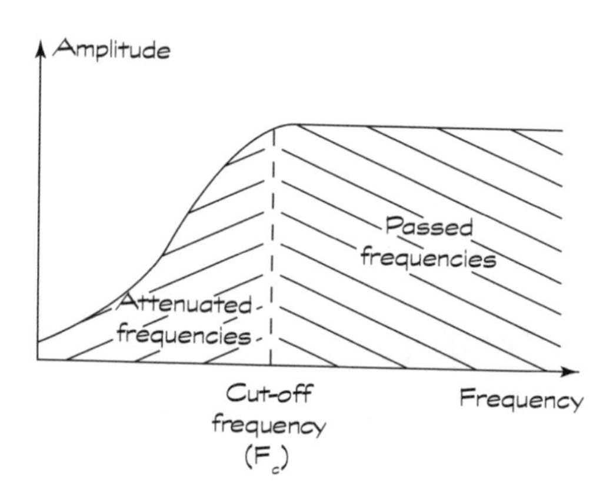

Figure 4-9: The filter cut-off frequency, F_c, is the frequency at which the filter starts to attenuate the signal.

Filtering frequency bands

You can design filters to have two cut-off frequencies – one high and one low – to allow a certain *band*, or range, of frequencies to pass through or to reject a band of frequencies. Such filters, known respectively as *band-pass* and *band-reject* filters, are constructed by combining low- and high-pass filters in just the right way.

Band-pass filters are commonly used in radio receiver systems to select just the right signal from among many signals transmitted through the air. You can use a band-reject filter to filter out unwanted hum from a 60-Hz power line – as long as you know the frequency range of the hum. Most of these complex filters employ inductors, as well as capacitors and resistors (we discuss inductors in Chapter 5).

Trying Out Simple Capacitive Circuits

If you're anxious to play around with some real capacitive circuits, take a peek at Chapter 13 in Part III. If you decide to do so, however, we ask that you please read through Part II first, so that you know how to set up your electronics shop, read schematics and – most importantly – can build circuits safely.

Chapter 5

Curling Up with Coils and Crystals

In This Chapter

- Inducing currents in coils with a changing magnetic field
- Opposing changes in current with an inductor
- Using inductors in filter circuits
- Resonating with RLC circuits
- Making frequencies crystal clear
- Coupling magnetic flux to transfer energy between circuits

Many of the best inventions in the world, including penicillin, Post-it notes, champagne and the pacemaker, were the result of pure accidental discovery (in some cases, attributable to utter carelessness or second-rate science). One such lucky discovery – the interaction between electricity and magnetism – led to the development of two amazingly useful electronic components: the induction coil and the transformer.

The induction coil, or inductor, stores electrical energy in a magnetic field and shapes electrical signals in a different way to a capacitor. Whether operating alone, in special pairs known as transformers or as part of a team along with capacitors and resistors, inductors are at the heart of many modern-day conveniences you may not want to live without, including radios, television and even the mains electricity network.

This chapter reveals the relationship between electricity and magnetism and explains how 19th-century scientists purposely exploited that relationship to create inductors and transformers. We look at what happens when you try to change the direction of current through an inductor too quickly and how Ohm's Law can be applied to inductors.

We also explore how inductors are used in circuits, why crystals do something rather special at just one frequency and discover how transformers transfer electrical energy from one circuit to another – without any direct contact between the circuits.

Working Both Ways: Magnetism and Electricity

Magnetism and electricity were once thought to be two completely separate phenomena until a 19th-century scientist named Hans Christian Ørsted discovered that a compass needle moved away from magnetic north when current supplied by a nearby battery was switched on and off.

Ørsted's keen observation led to lots of research and experimentation, ultimately confirming the fact that electricity and magnetism are closely related. After several years (and many more accidental discoveries), Michael Faraday and other scientists figured out how to capitalise on the phenomenon known as *electromagnetism*. Today's power transformers, electromagnetic generators and many industrial motors are based upon the principles of electromagnetism.

In this section, we look at how electricity and magnetism interact.

Drawing the (flux) lines with magnets

Just as electricity involves a force (voltage) between two electrical charges, so magnetism involves a force between two magnetic poles. If you've ever performed the classic school science experiment where you place a magnet on a surface and toss a bunch of iron filings near the magnet, you've seen the effects of magnetic force. Remember what happened to the filings? They settled into curved linear paths from the north pole of the magnet to its south pole.

Those filings showed you the magnetic lines of force – also known as *flux lines* – within the magnetic field created by the magnet. You may have seen more filings closer to the magnet because that's where the magnetic field is strongest. Figure 5-1 shows the pattern produced by invisible lines of flux around a magnet.

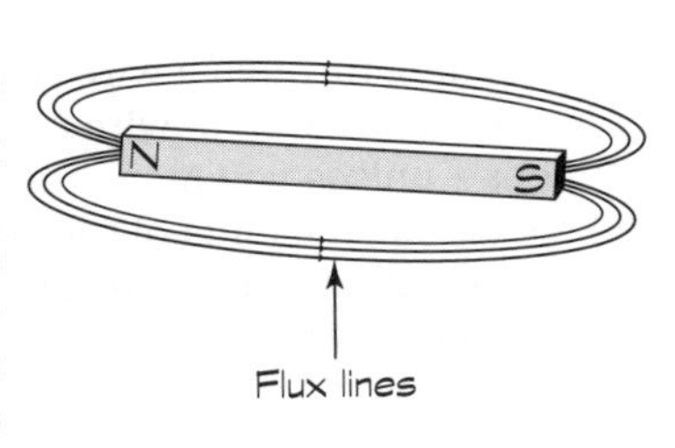

Figure 5-1: Magnetic lines of force exist in parallel flux lines from a magnet's north pole to its south pole.

Producing a magnetic field with electricity

As Ørsted discovered, electrical current running through a wire produces a weak magnetic field surrounding the wire, which is why the compass needle moved when the compass was close to Ørsted's circuit. Stop the current from flowing, and the magnetic field disappears. You control this temporary magnet, known as an *electromagnet*, electronically: you can turn the magnet on and off by switching the current on and off.

With the current on, the lines of force encircle the wire and are spaced evenly along the length of the wire, as shown in Figure 5-2. Picture a paper towel roll with a wire running through its exact centre. If you pass current through the wire, invisible flux lines wrap around the wire along the surface of the paper towel roll and along similar rings around the wire at various distances from the wire. The strength of the magnetic force decreases as the flux lines get farther away from the wire. If you wind the current-carrying wire into a uniform coil of wire, the flux lines align and reinforce each other, strengthening the magnetic field.

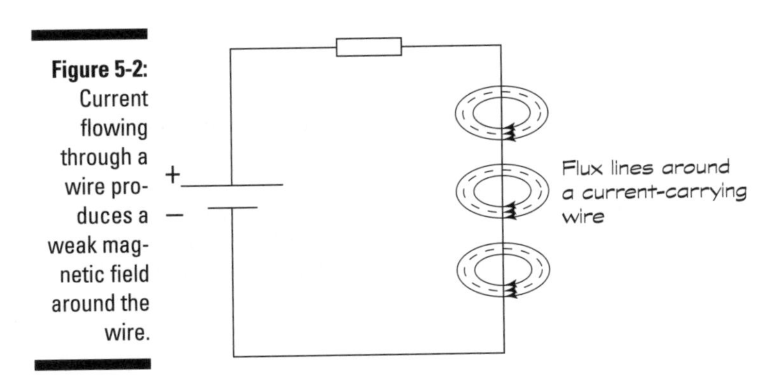

Figure 5-2: Current flowing through a wire produces a weak magnetic field around the wire.

Inducing current with a magnet

Perhaps you're wondering, if electricity running through a wire produces a magnetic field, what happens if you place a closed loop of wire near a permanent magnet? Actually, nothing happens – unless you move the magnet. A moving magnetic field *induces* a voltage across the ends of the wire, causing current to flow through the wire. *Electromagnetic induction* seems to make current magically appear – without any direct contact with the wire.

The strength of the current depends on a lot of things, such as the strength of the magnet, the number of flux lines intercepted by the wire, the angle at which the wire cuts across flux lines and the speed of the magnet motion. You can increase your chances of inducing a strong current by wrapping the wire in a coil and placing the magnet through the centre (*core*) of the coil: the more turns of wire you wrap, the stronger the current.

Suppose you place a strong permanent magnet in the centre of a coil of wire that's connected as in Figure 5-3. If you move the magnet up, current is induced in the wire and flows in one direction. If you move the magnet down, current is induced, but flows in the other direction. By moving the magnet up and down repeatedly, you can produce an alternating current (AC) in the wire. Alternatively, you can move the *wire* up and down around the magnet, and the same thing happens. As long as relative motion exists between the wire and the magnet, current is induced in the wire.

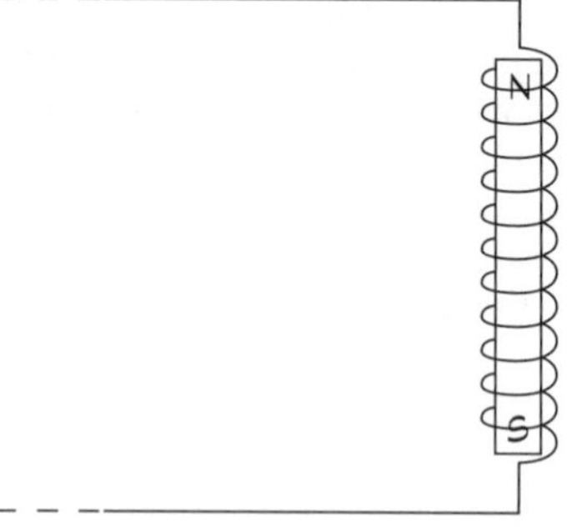

Figure 5-3: Moving a magnet inside a coil of wire induces a current in the wire.

Many power plants generate AC by rotating a conductor inside a very powerful magnet.

Moving a conductor is easier than moving a magnet, and power plants use different methods to move the conductor. Most methods involve attaching the conductor to a rotating turbine, which turns as water or steam applies pressure to its fins. As the conductor makes one full rotation inside the magnet, the magnet pulls electrons first in one direction and then in the other direction, producing alternating current.

Introducing the Inductor: A Coil with a Magnetic Personality

An *inductor* is a passive electronic component made from a coil of wire wrapped around a core, which can be air, iron or ferrite (a brittle material made from iron). Wrapping wire around iron or ferrite increases the strength of the magnetic field induced by current several hundred times. Inductors are sometimes known as coils, chokes, electromagnets and solenoids, depending on how they're used in circuits. You can see the circuit symbol for an inductor below:

If you pass current through an inductor, you create a magnetic field around the wire. If you *change* the current, increasing it or decreasing it, the magnetic flux around the coil changes, inducing a voltage across the inductor. That voltage, sometimes called *back voltage*, causes a current flow that opposes the main current. This property of inductors is known as *self-inductance* or simply *inductance*.

Measuring inductance

Inductance, symbolised by L, is measured in units called *henrys* (named after Joseph Henry, a New Yorker who liked to play with magnets and discovered the property of self-inductance).

An inductance of one henry (symbol H) induces one volt when the current changes its rate of flow by one ampere per second. Naturally, one henry is much too large for everyday electronics, so you're more likely to hear about millihenrys (mH), so-named because thousandths of a henry is more commonplace (not because Joseph's wife's name was Milly), or microhenrys (μH), which are millionths of a henry (not named after Joseph's baby).

Opposing current changes

Figure 5-4 shows the application of a DC voltage to a resistor in series with an inductor. When the DC voltage is first switched on, current starts to flow, inducing a magnetic field around the coils of the inductor. As the current increases, the strength of the magnetic field increases proportionally. The changing magnetic field induces a back voltage. The inductor seems to be trying to prevent the current from changing, and the effect is that the current doesn't increase as quickly.

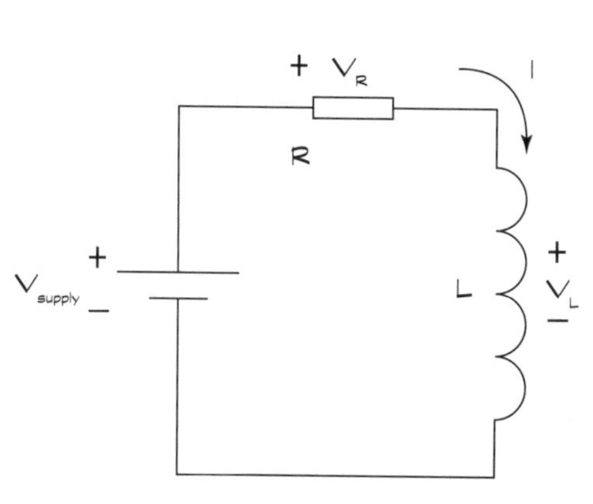

Figure 5-4: Inductors oppose changes in current, so it takes time for a direct current (DC) to flow through an inductor. When it does flow, the inductor acts like a short in a DC circuit.

The induced current reduces the strength of the expanding magnetic field a little. As the source current keeps rising, the magnetic field continues to expand (but more and more slowly), and current opposing the source current continues to be induced (but it gets smaller and smaller). The cycle continues, until finally, the current settles down to a steady DC. When current reaches a steady level, the magnetic field no longer changes and the inductor stops opposing the current.

The overall effect is that the current flowing through the inductor takes some time to reach a steady DC value. When this happens, the current flows freely through the inductor, which acts like a simple wire (or short circuit, with V_L = 0 volts), and the current is determined by the source voltage and the resistor according to Ohm's Law ($I = V_{supply}/R$). (We discuss Ohm's Law in Chapter 3.)

If you then remove the DC voltage source and connect the resistor across the inductor, current flows for a short period of time, with the inductor again opposing the sudden drop in current, until finally, the current settles down to zero and the magnetic field disappears.

From an energy perspective, when you apply a DC source to an inductor, it stores electrical energy in a magnetic field. When you remove the DC source and connect a resistor across the inductor, the energy transfers to the resistor, where it dissipates as heat. Inductors store electrical energy in magnetic fields.

To help understand inductors, think about water flowing through a pipe with a turbine in it. At first the water is slowed by the inertia of the wheel and the blades. Think of it like trying to push a car from a standstill. But after a while, the water flows at a constant rate, hardly slowed by the turbine turning at the same rate as the water is moving. If you suddenly remove the water pressure, the blades keep turning for a time, pushing water along with them until eventually both the turbine and the water stop.

Don't worry about the ins and outs of induced currents, expanding and contracting magnetic fields, and the like. Just remember a few things about inductors:

- Inductors oppose current changes.
- Inductors act like an open circuit when DC is first applied (no current flows right away, and the entire source voltage drops across the inductor).
- Inductors eventually act like a short circuit in DC circuits – when all the magnetic field magic settles down. The voltage is zero and it passes the full DC current.

Alternating current gets nowhere fast

When you apply an AC voltage to a circuit containing an inductor, the inductor fights against any changes in the source current. If you keep varying the supply voltage up and down at a very high frequency, the inductor keeps opposing the sudden changes in current. At the extreme high end of the frequency spectrum, no current flows at all because the inductor simply can't react quickly enough to the change in current.

Imagine standing between two very tempting chocolate pastries. You're torn between which to try first. You start out running towards the éclair, but quickly change your mind, turn around and run towards the brownie. Then you change your mind again, turn and start racing towards the éclair, and so on. The faster you change your mind, the more you're stuck in the middle – not getting anywhere (or any chocolate).

Those tempting pastries make you act like the electrons in an inductor when a high-frequency signal is applied to the circuit: neither you nor the electrons make any progress.

Understanding Inductive Reactance

An inductor's opposition to changing current is known as *inductive reactance*. The faster the current tries to change, the more the inductor resists the change.

In the torn-between-two-pastries example in the preceding section, if you don't change your mind quite so fast, you can run to the éclair, snatch a bite and then run to the brownie, take a bite and so on. If you change your mind a bit faster, you may find yourself running towards each pastry, getting halfway there and then turning around and running towards the other, getting half-way there and so on. How far you get depends on how quickly you change your mind. The same thing applies to current flow through an inductor: how far a current gets depends on how quickly the current is changing.

Inductive reactance, symbolised by X_L, is measured in – did you guess? – ohms. You use this formula for X_L:

$$X_L = 2 \times \Pi \times f \times L$$

In this formula, f represents the frequency in hertz (Hz) of the applied AC voltage, L is the inductance in henrys and Π is a constant that can be approx-imated as 3.14. You can approximate $2 \times \Pi$ as 6.28 and simplify the formula as follows:

$$X_L \approx 6.28 \times f \times L$$

You calculate the inductive reactance of a 0.1 mH inductor for a source with a frequency of 1 Hz as follows:

$$\begin{aligned} X_L &\approx 6.28 \times 1 \times 0.0001 \\ &\approx 0.000628 \text{ ohms} \end{aligned}$$

If you raise the frequency to 2 MHz (that's 2,000,000 Hz), the inductive reactance is:

$$X_L \approx 6.28 \times 2{,}000{,}000 \times 0.0001$$
$$\approx 1.3 \text{ kilohms}$$

Notice that for a low-frequency signal (1 Hz), the inductive reactance is nearly zero, so the inductor looks almost like a short circuit, presenting no opposition to current. For a high-frequency signal (2 MHz), the inductor puts up significant opposition to current flow (1.3 kilohms of reactance).

Using Ohm's Law for inductive reactance

You can use Ohm's Law for inductive reactance, as for capacitive reactance, as long as you're careful and apply it for a single frequency at a time. Suppose you apply a 2-MHz AC voltage with a peak value of 5 volts to your 0.1-mH inductor. You can calculate the peak current flowing through the inductor by applying Ohm's Law:

$$I_{peak} = V_{peak} / X_L$$

The inductive reactance for a 0.1-mH inductor at 2 MHz is 1.3 kΩ (as calculated above), so the peak current is:

$$I_{peak} \approx 5 \text{ volts}/1300 \text{ ohms}$$
$$\approx 0.0038 \text{ amps or } 3.8 \text{ milliamps}$$

Discovering that behaviour depends on frequency (again!)

Like capacitors, inductors in an AC circuit act differently depending on the frequency of the voltage applied to them. Because the current passing through an inductor is affected by frequency, so are the voltage drops across the inductor and other components in the circuit. This frequency-dependent behaviour forms the basis for useful functions, such as low-, high- and band-pass filters.

Inductors are sort of the alter egos of capacitors (which we discuss in Chapter 4). Capacitors oppose voltage changes; inductors oppose current changes. Capacitors block DC and pass more and more AC as frequency increases; inductors pass DC and block more and more AC as frequency increases. Both store electrical energy until you need it.

Using Inductors in Circuits

Inductors are used primarily in tuned circuits, to select or reject signals of specific frequencies, and to block (or choke) high frequency signals, such as eliminating radio frequency interference in cable transmissions. Inductors are also commonly used to remove the 50-Hz hum known as noise from nearby power lines.

You probably run into (or over) inductors more often than you think. Some traffic lights use an inductor to sense when a car is waiting for the lights to change. An inductive loop, consisting of several turns of a huge coil covering several metres, is embedded in the street just before the junction. This loop is connected to a circuit that controls the traffic signal. As you pass over the loop, the steel underbody of your car changes the magnetic flux of the loop. The circuit detects this change – and may give you a green light.

Insulating and shielding inductors

The wire that makes up an inductor must be insulated to prevent short circuits between the turns. Inductors used in most electronic circuits are also *shielded*, or encased in a metal can, to prevent the magnetic lines of flux from interfering with other components in a circuit.

You use a shielded inductor when you don't want to induce voltages or currents in other circuit elements. You use an unshielded inductor (or coil) when you do want to affect other circuit elements. We discuss the use of unshielded coils in circuits in the later section 'Calling on the Coil Next Door: Transformers'.

Reading inductance values

The value of an inductor is typically marked on its package using the same colour-coding technique used for resistors, which you can read about in Chapter 3. You can often find the value of larger inductors printed directly on the components. Smaller value inductors look a lot like low-wattage resistors, and these inductors and resistors even have similar colour-coding marks. Larger value inductors come in a variety of sizes and shapes, depending upon their application.

Inductors can be fixed or variable. With both types, a slender wire winds around an insulating coil. The number of turns of the wire, the core material, the wire's diameter and length of the coil determine the value of the inductor.

Fixed inductors have a constant value, whereas variable inductors sport a knob that you can turn to adjust the value. The core of an inductor can be made of air, iron ferrite or any number of other materials (including your car). Air and ferrite are the most common core materials.

Combining shielded inductors

Although you're unlikely to use inductors in the basic electronic circuits you set up, you may run across circuit diagrams for power supplies and other devices that include multiple inductors. Just in case you do, you need to know how to calculate the equivalent inductance of combinations of shielded inductors so that you can understand how the circuit operates.

Adding inductors in series

Inductors in series add up, just as resistors do:

$$L_{series} = L1 + L2 + L3\ldots$$

Reciprocating inductors in parallel

Like resistors, inductors in parallel combine by adding the reciprocals of each individual inductance and then taking the reciprocal of that sum:

$$L_{parallel} = \frac{1}{\frac{1}{L1} + \frac{1}{L2} + \frac{1}{L3}}$$

If you have just two inductors in parallel, you can simplify this equation as follows:

$$L_{parallel} = \frac{L1 \times L2}{L1 + L2}$$

Filtering signals with inductors

You can use inductors in filter circuits to do exactly the opposite of what capacitors do in filter circuits, which you can read about in Chapter 4.

Figure 5-5 shows an *RL circuit*, which is simply a circuit that contains both a resistor and an inductor, with the output voltage, V_{out}, taken across the inductor. The lower the frequency of the input voltage, the more the inductor is able to react to changes in current, so the more the inductor looks like a short circuit.

As a result, for low input frequencies, the output voltage is nearly zero. The higher the frequency of the input voltage, the more the inductor fights the change in current and the less successful the input voltage is in pushing current through the circuit. As a result, for high input frequencies, very little voltage drops across the resistor (because $V_R = I \times R$, and I is very low), so most of the input voltage drops across the inductor, and the output voltage is nearly the same as the input voltage.

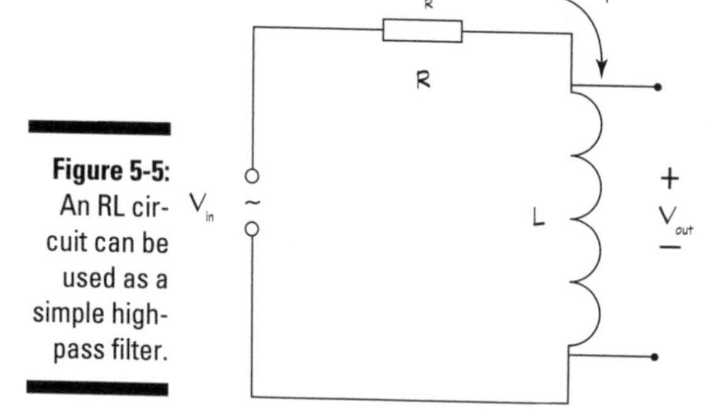

Figure 5-5: An RL circuit can be used as a simple high-pass filter.

This RL circuit is configured as a *high-pass filter*, because it allows high-frequency input signals to pass through to the output, while blocking DC and low frequencies from getting through.

If you reverse the roles of the resistor and the inductor in Figure 5-5, taking the output voltage across the resistor instead, you have a *low-pass filter*. For low frequencies, because the inductor acts like a short circuit, nearly all the input voltage drops across the resistor; at high frequencies, the inductor acts more and more like an open circuit, allowing less and less current to flow through the resistor, so the output voltage *attenuates* (decreases) dramatically.

Calculating the RL time constant

You can calculate the amount of time (in seconds) taken for an inductor's induced voltage to reach roughly two-thirds of its value using the RL time constant, T:

$$T = L/R$$

Just as the RC time constant in RC circuits (which we discuss in Chapter 4) gives you an idea of how long a capacitor takes to charge up to its full capacity, so the RL time constant helps you figure out how long an inductor takes to fully conduct a DC current: direct current settles down to a steady value after roughly five RL time constants. You can also use the RL time constant to calculate the *filter cutoff frequency* (the frequency at which a filter starts to affect an input signal) as follows:

$$F_c = \frac{1}{(2 \times \Pi \times T)} = \frac{R}{(2 \times \Pi \times L)}$$

Now Introducing Impedance!

An inductor's opposition to current is known as inductive reactance, and in Chapter 4 we state that a capacitor's opposition to current is known as capacitive reactance. In circuits that have reactance and resistance, for instance the RL circuit in Figure 5-5, you may need to know the total opposition to current in the circuit for an input signal at a certain frequency.

Both resistance and reactance are measured in ohms, so you may think that you can simply add the inductive reactance, X_L, to the resistance, R, to get the total opposition to current (like adding up series resistances) – but you can't. The reason you can't is due to the time taken for an inductor (and similarly, a capacitor) to react to changes in the circuit. The good news is that a way does exist to calculate the total opposition to current in a circuit at a particular frequency.

Impedance is the total opposition a circuit provides to varying current for a given frequency. You use the symbol Z to represent impedance, which takes into account the total resistance and total reactance of a circuit.

For a series RL circuit, the total impedance can be found using this formula:

$$Z_L = \sqrt{(R^2 + X_L^{\ 2})}$$

Similarly, for a series RC circuit, the total impedance is:

$$Z_C = \sqrt{(R^2 + X_C^{\ 2})}$$

So how do you use those scary-looking formulas? Well, if you want to use Ohm's Law to calculate the current through a series RL or RC circuit when you apply an input signal of a particular frequency, you can calculate the total impedance of the circuit at that frequency and then apply Ohm's Law, using the impedance to calculate the peak value of the current passing through the circuit for that one particular frequency:

$$I_{peak} = V_{peek} / Z_L$$

If you have a circuit with a resistor, capacitor and inductor all in series (one type of *RLC circuit*), the formula for impedance is more complicated, because the total reactance in the circuit (X_T) is the *difference* between the inductive reactance, X_L, and the capacitive reactance, X_C ($X_T = X_L - X_C$). For parallel arrangements of resistors and inductors, or resistors and capacitors, the formula for impedance is even more complicated, and we aren't going to go there in this book.

Tuning in to Radio Broadcasts

Inductors are natural low-pass filters, and capacitors are natural high-pass filters, so what happens when you put the two in the same circuit? As you may guess, inductors and capacitors are often used together in tuning circuits, to tune in a specific radio station broadcast frequency.

Resonating with RLC circuits

Look at the RLC circuit in Figure 5-6. The total impedance of this circuit can be found using the following really scary formula:

$$\begin{aligned} Z &= \sqrt{\left(R^2 + X^2\right)} \\ &= \sqrt{\left(R^2 + \left(X_L - X_C\right)^2\right)} \end{aligned}$$

Don't panic! Just notice one thing: if $X_L = X_C$, the total impedance becomes the square root of R^2, which is just R, or just the resistance.

In other words, when $X_L = X_C$, the inductive reactance and the capacitive reactance cancel each other out, almost as if only a resistor is in the circuit. This effect happens at exactly one frequency, known as the *resonant frequency*. The resonant frequency is the value of frequency, f, that makes $X_L = X_C$ for a given combination of inductance (L) and capacitance (C).

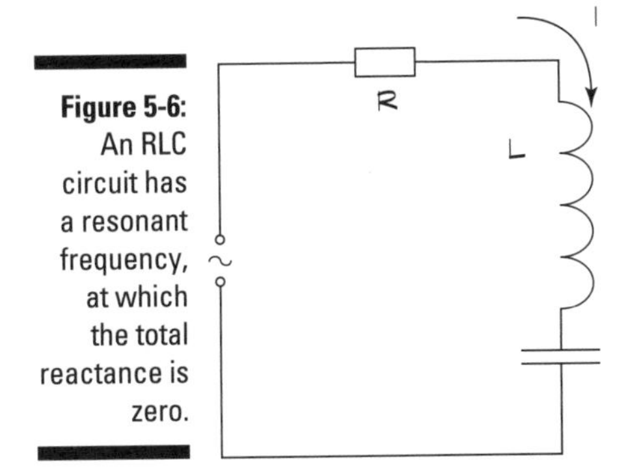

Figure 5-6: An RLC circuit has a resonant frequency, at which the total reactance is zero.

So what's the significance of the resonant frequency? Well, something special happens at that frequency that doesn't happen at any other frequency.

At frequencies above and below the resonant frequency, some overall reactance exists in the circuit, adding to the resistor's opposition to current flow. For very low frequencies, the capacitor exhibits a high reactance, whereas for very high frequencies, the inductor exhibits a high reactance. So for low frequency signals, the capacitor squashes current flow, and for high frequencies, the inductor squashes current flow. At the resonant frequency, the overall reactance is zero, and only the resistance appears to oppose current flow.

The circuit is said to *resonate* at that particular frequency, and so is known as a *resonant circuit*. In Figure 5-7, which shows a frequency plot of the current passing through the circuit, notice that the current is highest at the resonant frequency.

Radio receivers use RLC circuits to allow just one frequency to pass through the circuit. This technique is known as *tuning in* to the frequency and used this way, the circuit is known as a *tuning circuit*. Use a variable capacitor to adjust the resonant frequency, so that you can tune in different stations broadcasting at different frequencies. The knob that allows the capacitance to be changed is attached to the tuning control knob on your radio.

By shifting components around a bit, placing the inductor in parallel with the capacitor, you create a circuit that produces the *minimum* current at the resonant frequency. This sort of resonant circuit tunes out that frequency, allowing all others to pass, and is used to create *band-stop filters*. You may find such a circuit filtering out the 50-Hz hum that electronic equipment sometimes picks up from a nearby power line.

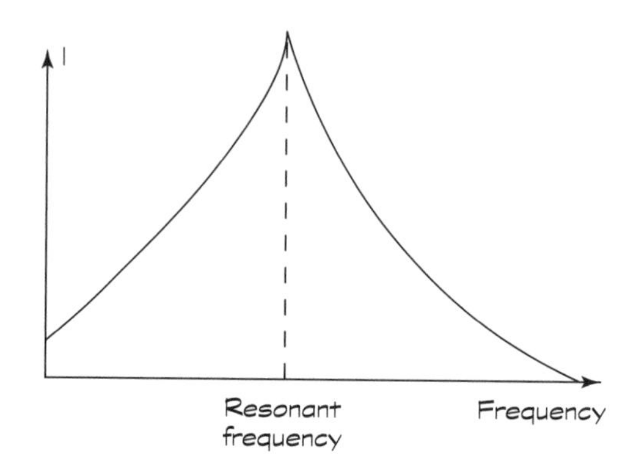

Figure 5-7: The current in a series RLC circuit is highest at the resonant frequency.

Ensuring crystal-clear resonance

If you slice a quartz crystal in just the right way, mount two leads to it and enclose it in a hermetically sealed package, you create a single component that acts like an RLC combo in an RLC circuit, resonating at a particular frequency. Quartz crystals, or simply crystals, are used in circuits to generate an electrical signal at a very precise frequency. The circuit symbol for a crystal, abbreviated to XTAL, is shown below:

Crystals work because of something called the *piezoelectric effect.* If you apply a voltage in just the right way across a quartz crystal, it vibrates at a specific frequency, known as the resonant frequency. If you then remove the applied voltage, the crystal continues to vibrate until it settles back to its previous shape. As it vibrates, the crystal generates a voltage at the resonant frequency.

The frequency at which a crystal resonates depends upon its thickness and size, and you can find crystals with resonant frequencies ranging from a few tens of kilohertz to tens of megahertz. Crystals are more precise and more reliable than combinations of capacitors and inductors, but there's a catch: they're usually more expensive. And with a crystal, you're pretty much stuck with a particular resonant frequency, so you don't find crystals in tuning circuits that allow you to adjust the resonant frequency during operation.

You do find crystals used in circuits called *oscillators* to generate electric signals at a very precise frequency. Oscillators are responsible for the ticks and tocks that control quartz wristwatches and digital integrated circuits (which we discuss in Chapter 7), and for controlling the accuracy of radio equipment.

A quartz crystal is accurate to within roughly 0.001 per cent of its stated resonant frequency. (That why they're worth paying extra for!) You may also hear of ceramic resonators, which work the same way but cost less and aren't as accurate as quartz. Ceramic resonators have a 0.5 per cent frequency tolerance and are used in many consumer electronics devices, such as TVs, cameras and toys.

Calling on the Coil Next Door: Transformers

Inductors used in tuning circuits are shielded so that the magnetic field they produce doesn't interact with other circuit components. Unshielded coils are sometimes placed close to one another so that their magnetic fields can interact. In this section, we describe how unshielded coils interact and how you can exploit their interaction to do some useful things with an electronic device known as a *transformer.*

Letting unshielded coils interact

When you place two unshielded coils near each other, the varying magnetic field created as a result of passing AC through one coil induces a voltage in that coil *as well as in the other coil.*

Mutual inductance is the term used to describe the effect of inducing a voltage in another coil, and *self-inductance* refers to the effect of inducing a voltage in the same coil that produced the varying magnetic field in the first place: the closer the coils, the stronger the interaction. Mutual inductance can add to or oppose the self-inductance of each coil, depending on how you match up the north and south poles of the inductors.

If you have an unshielded coil in one circuit and place it close to an unshielded coil in another circuit, the coils interact. By passing a current through one coil, you cause a voltage to be induced in the neighbouring coil – even though that coil is in a separate, unconnected circuit. This effect is known as *transformer action.*

A *transformer* is an electronic device that consists of two coils wound around the same core material in such a way that the mutual inductance is maximised. Current passing through one coil, known as the *primary*, induces a voltage in the other coil, known as the *secondary*. The job of a transformer is to transfer electrical energy from one circuit to another.

The circuit symbols for an air core transformer and solid core transformer respectively are shown below:

Isolating circuits from a power source

If the number of turns of wire in the primary winding of a transformer is the same as the number of turns in the secondary winding, theoretically all the voltage across the primary is induced across the secondary. This transformer is known as a *1:1 transformer*, because a 1:1 (read 'one-to-one') relationship exists between the two coils. (In reality, no transformer is perfect, or *lossless*, and some of the electrical energy gets lost in the translation.)

1:1 transformers are also known as *isolation transformers* and are commonly used to electrically separate two circuits while allowing power or an AC signal from one to feed into the other. The first circuit typically contains the power source while the second circuit contains the load. (In Chapter 1, we define the load as the destination for the electrical energy, or the thing you ultimately want to perform work on, such as a speaker diaphragm.) You may want to isolate circuits to reduce the risk of electrical shocks or to prevent one circuit from interfering with the other.

Stepping up, stepping down voltages

If the number of turns in the primary winding of a transformer isn't the same as the number of turns in the secondary winding, the voltage induced in the secondary is different from the voltage across the primary. The two voltages are proportional to each other, with the proportion determined by the ratio of the number of turns in the secondary to the number of turns in the primary, as follows:

$$V_S / V_P = N_S / N_P$$

In this equation, V_S is the voltage induced in the secondary, V_P is the voltage across the primary, N_S is the number of turns in the secondary and N_P is the number of turns in the primary.

Say, for instance, the secondary consists of 200 turns of wire, twice as many as the primary, which consists of 100 turns of wire. If you apply an AC voltage with a peak value of 50 volts to the primary, the peak voltage induced across the secondary is 100 volts, or twice the value of the peak voltage across the primary. This type of transformer is known as a *step-up transformer*, because it steps up the voltage from the primary to the secondary.

If instead, the secondary consists of 50 turns of wire, while the primary consists of 100 turns, for the same AC signal applied to the same primary, the peak voltage across the secondary is 25 volts, or half the primary's voltage. This transformer is known as a *step-down transformer*, for obvious reasons.

In each case, the power applied to the primary winding is transferred to the secondary. Because power is the product of voltage and current (the Joule's Law in Chapter 3: $P = V \times I$), the current induced in the secondary winding is inversely proportional to the voltage induced in the secondary. So a step-up transformer steps up the voltage while stepping down the current, and a step-down transformer steps down the voltage while stepping up the current.

Step-up and step-down transformers are used in the mains network to transmit power over long distances and to supply lower voltages in homes and businesses.

Chapter 6

The Wide World of Semiconductors

In This Chapter

- Looking inside a semiconductor
- Combining semiconductors to make diodes and transistors
- Letting current flow one way but not the other with diodes
- Understanding and using transistors

Semiconductors are at the heart of almost every electronic system in use today, from your washing machine to the space shuttle. These tiny components have revolutionised modern medicine, space exploration, industrial automation, home entertainment, transport, communications and just about every industry or workplace you can think of.

Semiconductor diodes and transistors can be made to conduct or block electric current – depending on how you control them electrically. They can allow current to flow in one direction, but not the other, and can amplify tiny signals – tasks that your average passive electronic component can't perform.

In this chapter, we go inside semiconductor materials, discover how to make them conduct current and explore ways to combine semiconductors to create diodes and transistors. We observe the valve-like behaviour of diodes and show you how to exploit that behaviour in circuits. In addition, we investigate how transistors work and why they're so popular, setting the stage for modern-day electronics.

Are We Conducting, or Aren't We?

Somewhere between insulators and conductors are a group of materials that can't seem to make up their minds about whether to hold on to their electrons or let them roam freely. *Semiconductors* behave like conductors under some conditions and insulators under other conditions. Silicon and germanium are two semiconductor elements commonly used to create electronic devices. The electrons encircling the nucleus in a semiconductor atom aren't quite as free to jump around as they are in a good conductor, because they're bound to the atom in a special way.

The atoms of semiconductor materials align themselves in a structured way, forming a very regular pattern, known as a crystal, as shown in Figure 6-1. Atoms within the crystal are held together by a special bond, called a *covalent bond*, with each atom sharing its outermost electrons with its neighbours. (Bonding – it's a wonderful thing; like next-door neighbours sharing a front drive, and each behaving as if the drive belongs only to her – except when it needs resurfacing!)

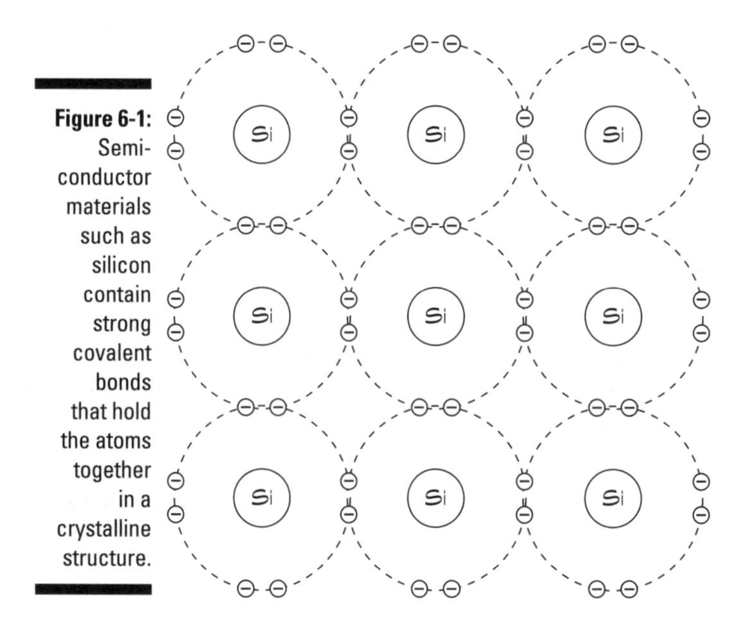

Figure 6-1: Semiconductor materials such as silicon contain strong covalent bonds that hold the atoms together in a crystalline structure.

Precisely because of this special touchy-feely bonding and sharing of electrons, the semiconductor crystal acts like an insulator most of the time. Those electrons aren't as likely to break free and wander off on their own as the free electrons in conductors. But you can do something to a pure semiconductor that changes the electrical properties of the material: dope them.

If you want to find out a little bit about the underlying theory of semiconductor physics, read the rest of this section. But you don't really need the theory to understand how semiconductor components work. If you skip ahead to 'Forming a Junction Diode', you can still read what you need to use semiconductors in circuits.

Doping semiconductors

If you add impurities (meaning different types of atoms, not dirt or dander) to a pure semiconductor material, such as silicon, you upset the bonding apple cart. This process is known as *doping,* and the impurities are called *dopants.* Arsenic and boron are two dopants that are commonly used to dope pure silicon.

Dopants are no dopes; these atoms pretend to be the crystal's atoms, attempting to bond with the other atoms, but they're just different enough to stir things up a bit. For instance, an atom of arsenic has one more outer electron than an atom of silicon. When you add a small amount of arsenic to a bunch of silicon atoms, each arsenic atom muscles its way in, bonding with the silicon atoms, but leaving its extra electron drifting around through the crystal. Even though the doped material is electrically neutral, it has more free electrons wandering around than undoped silicon – making it much more conductive. By doping the silicon, you change its electrical properties.

Another way to dope semiconductors is to use materials such as boron, which has one *less* outer electron than silicon. For every boron atom you add to a silicon crystal, you get what's known as a *hole* in the crystalline structure. Now the great thing about holes is that electrons love dropping into them. So an electron falls into the hole, which leaves another hole, and another electron drops into that hole, and so on.

Think of this process as the hole moving around inside the crystal, as illustrated in Figure 6-2. (Really, the electrons are moving, but it looks like the hole is moving.) Each hole represents a missing electron, and so the movement of holes has the same effect as a flow of positive charges.

Impurities, such as arsenic, that free up electrons (negative charges) to move through a semiconductor are called *donor dopants*, and the doped semiconductor is known as an *n-type semiconductor.* Impurities, such as boron, that free up holes (positive charges) to move through a semiconductor are called *acceptor dopants*, and the doped semiconductor is known as a *p-type semiconductor.*

Combining n-types and p-types to create components

If you apply a voltage source across an n-type or a p-type semiconductor, current flows due to the movement of electrons through the material from the negative voltage towards the positive voltage. (For p-type semiconductors, you'll hear this process described as a movement of holes from the positive voltage towards the negative voltage.)

Diving into a pn-junction

To really understand the reasons why current flows or doesn't flow when you apply voltages across a pn-junction, you need to dive a bit deeper into the physics that lies beneath the surface of the junction. Although we don't dive too deeply here, we do tell you that the important aspect is which way the holes are pushed in the p-type material, which way the electrons are pushed in the n-type material and how holes and electrons sometimes re-combine at the junction.

Even with no external voltage applied, a small voltage difference still exists across the junction. That voltage is caused by holes and electrons meeting up at the junction, crossing over to the other side and re-combining (meaning that an electron fills a hole). This process leaves the area on either side of the junction with a net charge: on the p-side of the junction, the net charge is negative; on the n-side, the net charge is positive. Therefore, a difference in charge, or a small voltage, exists across the junction.

By connecting the positive terminal of a battery to the p-type material and the negative terminal of the battery to the n-type material, you push holes through the p-type material towards the junction and push electrons through the n-type material towards the junction. If you send enough of those two types of charges toward the junction, you overcome the small voltage that exists there already, and holes cross into the n-type material and electrons cross into the p-type material.

The external voltage supply keeps pushing charges towards the junction, so charges keep crossing the junction. The net effect of holes moving one way and electrons moving the other way is a current. Conventional current (movement of positive charges) moves from the positive battery terminal through the p-type material, across the junction and through the n-type material towards the negative battery terminal.

If you connect the battery the other way, the negative terminal attracts holes from the p-type material, pulling them away from the junction, and the positive terminal attracts electrons from the n-type material, pulling them away from the junction. This process in effect strengthens the already-existing voltage across the junction, making it even harder for current to flow. If the externally applied voltage is strong enough, eventually current can be made to flow – in the reverse direction across the junction. The voltage at which this happens is called the *breakdown voltage,* and this breakdown process is what makes a Zener diode work (see the later section 'Regulating voltage with Zener diodes' for more information on these heavily doped diodes that break down at low voltages.

So far, this isn't very exciting: the doped semiconductors are simply acting like conductors, and we can just as easily use copper wire to achieve the same effect!

Things start to get interesting, however, when you fuse together an n-type and a p-type semiconductor and apply a voltage across this *pn-junction*. Whether or not a current flows depends on which way you apply the voltage. If you connect the positive terminal of a battery to the p-type material and the negative terminal to the n-type material, current flows (as long as the applied voltage exceeds a certain minimum). But if you reverse the battery, current doesn't flow (unless you apply a very large voltage).

Exactly how these n-type and p-type semiconductors are combined determines what sort of semiconductor device they become and how they allow current to flow (or not) when voltage is applied. The pn-junction is the foundation for *solid-state* electronics, which involves smaller, solid electronic devices rather than more cumbersome and fragile hollow glass parts. Semiconductors have largely replaced vacuum tubes in electronics.

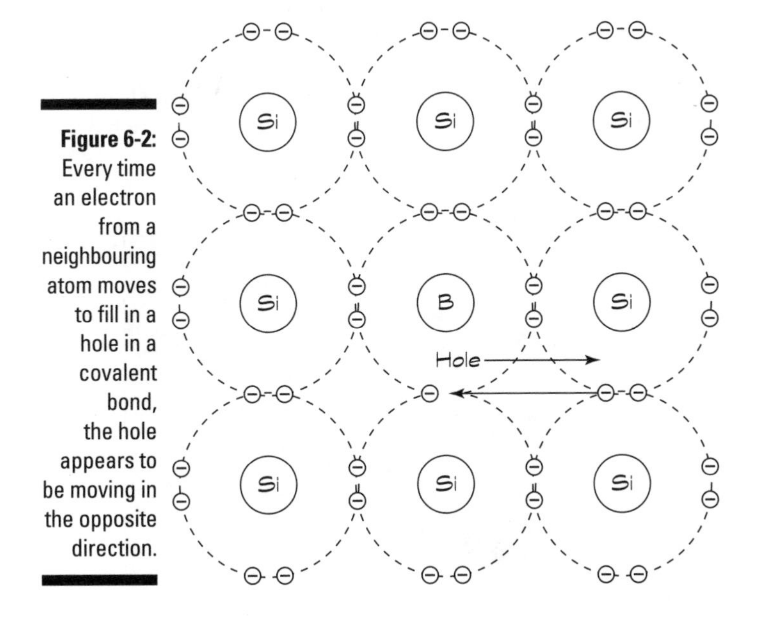

Figure 6-2: Every time an electron from a neighbouring atom moves to fill in a hole in a covalent bond, the hole appears to be moving in the opposite direction.

Forming a Junction Diode

Junctions in diodes can do some amazing things, from converting alternating current to direct current and emitting light when a current passes through it (in the case of a light-emitting diode), or even the other way around – generating a current when exposed to light (as in a solar cell).

A semiconductor *diode* is a two-terminal electronic device that consists of a single pn-junction. Diodes act like one-way valves, allowing current to flow in one direction, but not the other when a voltage is applied to them. This capability is sometimes referred to as the *rectifying* property. The circuit symbol for a diode is shown below:

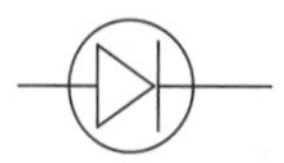

You refer to the p-side of the pn-junction in a diode as the *anode*, and the n-side as the *cathode*. In the circuit symbol above, the anode is on the left (broad end of the arrowhead) and the cathode is on the right (short vertical line segment). Most diodes allow current to flow from the anode to the cathode. (Zener diodes, which we discuss in the later section 'Regulating voltage with Zener diodes', are an exception.) You can think of the junction within a diode as a hill and the current as a ball you're trying to move from one side of the hill to the other. Pushing the ball down the hill (from anode to cathode) is easy, but pushing the ball up the hill (from cathode to anode) is very difficult.

Diodes are cylindrical, like resistors, but aren't quite as colourful. Most diodes sport a stripe or other mark at one end, signifying the cathode. Figure 6-3 shows some diodes.

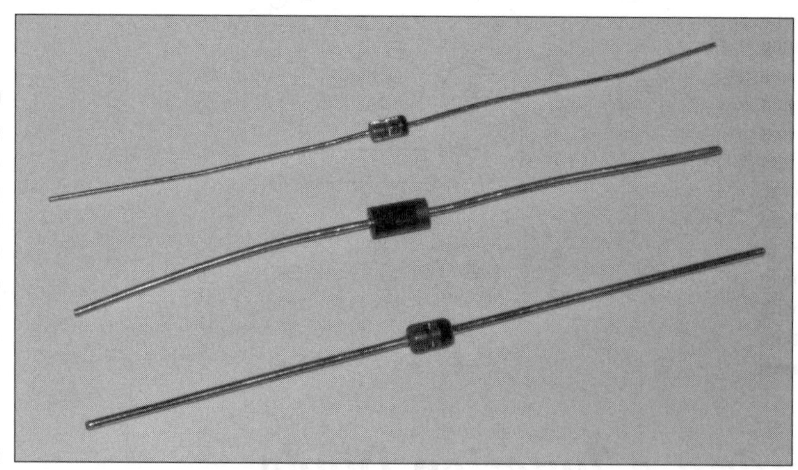

Figure 6-3: Diodes are similar in size and shape to resistors, but include just one stripe indicating the cathode.

Biasing the diode

You *forward-bias* a diode when you apply a voltage across it with the proper polarity to allow current to flow from anode to cathode. To do this, you apply a more positive voltage to the anode than to the cathode. *Forward current* flows easily across the pn-junction when the bias voltage reaches a certain level, known as the *forward voltage*. The forward voltage varies, depending on the type of diode, and can range from 0.2 to 4.0 volts. Figure 6-4 shows a forward-biased diode allowing current to flow through a lamp.

You *reverse-bias* a diode when you apply a *reverse voltage* (a more positive voltage is applied to the cathode than to the anode) across the diode, prohibiting current from flowing, as shown in Figure 6-5. If the reverse-bias voltage exceeds a certain level, the diode breaks down and *reverse current* starts flowing from cathode to anode. The reverse voltage at which the diode breaks down is known as the *peak inverse voltage* (PIV).

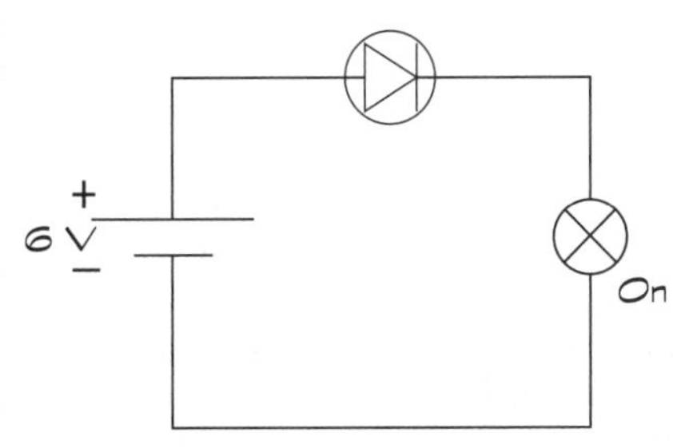

Figure 6-4: The battery forward-biases the diode in this circuit, allowing current to flow through a lamp.

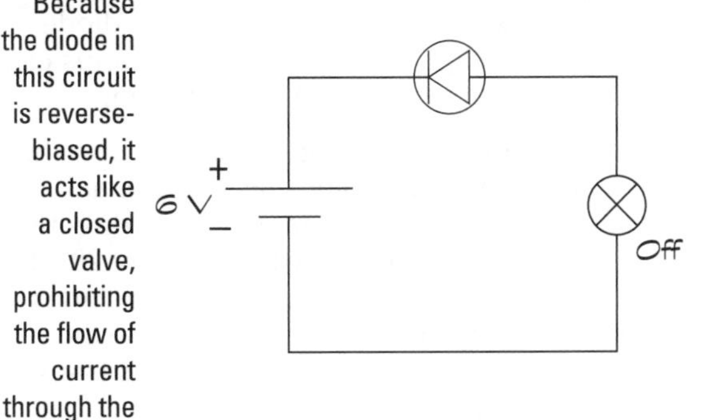

Figure 6-5: Because the diode in this circuit is reverse-biased, it acts like a closed valve, prohibiting the flow of current through the circuit.

Conducting current through a diode

When current starts flowing through a diode, the forward voltage drop across the diode remains fairly constant – even if you increase the forward current. For instance, most silicon diodes have a forward voltage of between 0.6 and 0.7 volts over a wide range of forward currents. If you're analysing a circuit that contains a silicon diode, such as the circuit in Figure 6-4, you can assume that the voltage drop across the diode is about 0.7 volts – even if you increase the source voltage from 6 to 9 volts. Increasing the source voltage increases the current through the circuit, but the diode voltage remains the same, so the increased source voltage is dropped across the lamp.

Of course, every electronic component has its limits. If you increase the current through a diode too much, you generate a lot of heat within the diode. At some point, the junction becomes damaged from all that heat, so you have to be careful not to turn up the source voltage too high.

Rating your diode

Resistors and capacitors come in a range of sizes, measured in ohms and farads. But a diode isn't rated as big or small in that way – it just does its own thing! A diode controls the on/off flow of electrons, without altering the shape or size of the electron flow. But that doesn't mean all diodes are the same. Standard diodes are rated according to two main criteria: peak inverse voltage (PIV) and current. These criteria guide you to choosing the right diode for a particular circuit:

- The **PIV rating** tells you the maximum reverse voltage the diode can handle before breaking down. For example, if the diode is rated at 100 volts, you shouldn't use it in a circuit that applies more than 100 volts to the diode.
- The **current rating** tells you the maximum forward current the diode can withstand without sustaining damage. A diode rated for 3 amps can't safely conduct more than 3 amps without overheating and failing.

  Rectifier diodes (flip to the later section 'Rectifying AC' for more on these diodes) rated to about 3 to 5 amps generally come encased in black or grey epoxy, and they're designed so that you can mount them directly on printed circuit boards. Higher-current diodes, with ratings of 20, 30 or 40 amps, are commonly contained in a metal housing that includes a heat sink or a mounting stud designed to allow you to affix the diode securely to a heat sink. A few diodes use the same packaging as transistors (which we describe in the next section).

Discovering what's in a name

Most diodes are named with an industry-standard five- or six-digit code that starts with '1N' ('N' labels it as a semiconductor, and '1' is the number of pn-junctions in the diode). The remaining three or four digits reveal the type of diode.

A classic example is the series of rectifier diodes identified as 1N40xx, where xx may be 00, 01 and so forth up to 08. They're rated at 1 amp with PIV ratings ranging from 50 to 1000 volts, depending upon the xx. For instance, the

1N4001 rectifier diode is rated at 1 amp and 50 volts, and the 1N4008 is rated at 1 amp and 1000 volts.

Diodes in the 1N54xx series have a 3 amp rating with PIV ratings from 50 to 1000 volts. You can readily find such information in any electronics component catalogue or diode data cross-reference book. (A *cross-reference book* tells you what parts can be substituted for other parts, in case you can't find a particular part specified in a circuit diagram.)

Just to make things interesting (not to mention confusing), some diodes use the same colour-coding scheme on their packaging as resistors, but instead of translating the code into a value (like resistance), the colour code simply gives you the semiconductor identification number for the diode. For instance, the colour sequence brown-orange-red indicates the numerical sequence 1-3-2, meaning that the diode is a 1N132 germanium diode.

Orientating yourself: Which way is up?

When you use a diode in a circuit, orienting it the right way round is extremely important. The stripe or other mark on the diode package corresponds to the line segment in the circuit symbol for a diode: both indicate the cathode, or negative terminal, of the diode.

You can also determine which end is what by measuring the resistance of the diode (before you insert it in your circuit) with an ohmmeter or multimeter (which we discuss in Chapter 12). The diode has a low resistance when forward-biased and a high resistance when reverse-biased. By applying the positive lead of your meter to the anode and the negative lead to the cathode, your meter is essentially forward-biasing the diode (because when used to measure resistance, a multimeter applies a small voltage across its leads). You can measure the resistance twice, applying the leads first one way and then the other way. The lower measurement result indicates the forward-bias condition.

Diodes are like one-way valves, allowing current to flow in one direction only. If you insert a diode backwards into a circuit, your circuit doesn't work at all (because no current flows) or you may damage some components (because you may exceed the PIV and allow current to flow in reverse, which can damage components such as electrolytic capacitors). Always note the orientation of the diode when you use it in a circuit, double-checking to make sure that you have it right!

Using Diodes in Circuits

We now describe several different types of semiconductor diode that are designed for various uses in electronic circuits.

Rectifying AC

Figure 6-6 shows a circuit with a silicon diode, a resistor and an AC power source. Notice the orientation of the diode in the circuit: its anode (positive end) is connected to the power source. The diode conducts current when forward-biased, but not when the diode is reverse-biased. When the AC source is positive (and at least 0.7 volts), the diode conducts current; when the AC source is negative, the diode doesn't conduct current. The output voltage is a clipped version of the input voltage, passing only the positive portion of the input signal through to the output.

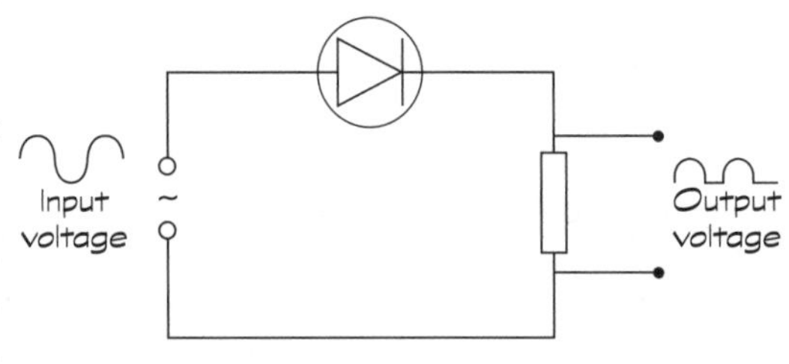

Figure 6-6: The diode in this circuit clips off the negative half of the AC source voltage.

If the diode orientation is reversed in the circuit, the opposite happens. When the input voltage is positive, the diode is reverse-biased and no current flows. When the input is sufficiently negative (at least –0.7 volts), the diode is forward-biased and current flows. Only the negative part of the input voltage is passed through to the output.

Diodes are often used in this way to convert AC current into varying DC current (the current is DC because it's flowing in one direction only, but it's not a constant current). This process is known as *rectification.* Using a single diode to clip an AC signal is known as *half-wave rectification* because it converts half of the AC signal into DC.

Diodes used for rectification are sometimes called *rectifier diodes* or just *rectifiers.* They usually handle currents ranging from several hundred milliamps to a few amps, whereas general-purpose *signal diodes* are designed to handle smaller currents (up to about 100 milliamps).

By arranging four diodes in a circuit known as a *bridge rectifier*, you can convert both the ups and the downs of an AC voltage into just ups, as shown in Figure 6-7. This process is known as *full-wave rectification*, and it's the first stage in converting AC power to a steady DC power supply. Bridge rectifiers are so popular that you can purchase them as a single four-terminal part, with two leads for the AC input and two leads for the DC output.

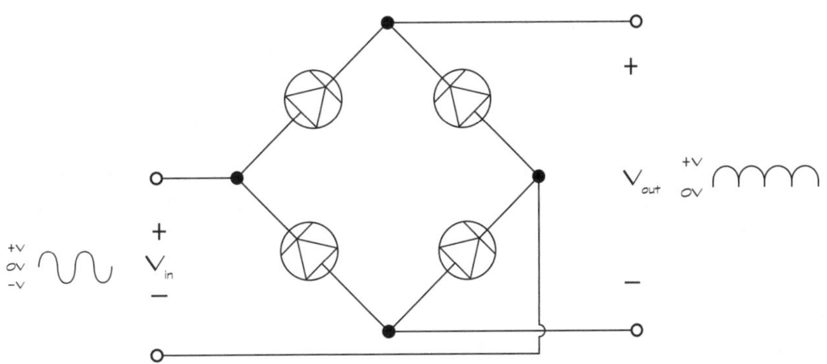

Figure 6-7: A bridge rectifier uses four diodes to transform AC current into a varying DC current.

Regulating voltage with Zener diodes

Zener diodes are special diodes that are meant to break down. They're really just heavily doped diodes that break down at much lower voltages than standard diodes. When you reverse-bias a Zener diode, and the voltage across it reaches or exceeds its breakdown voltage, the Zener diode suddenly starts conducting current backwards through the diode (from cathode to anode). As you continue to increase the reverse-bias voltage beyond the breakdown point, the Zener continues to conduct more and more current – while maintaining a steady voltage drop.

The two important ratings for Zener diodes are:

- The **breakdown voltage**, commonly called the Zener voltage, is the reverse-bias voltage that causes the diode to break down and conduct current. Breakdown voltages, which are controlled by the semiconductor doping process, range from 2.4 volts up to hundreds of volts.
- The **power rating** tells you the maximum power (voltage × current) that the Zener diode can handle. (Even diodes designed to break down can *really* break down if you exceed their power ratings.)

The circuit symbol for a Zener diode is shown below:

Zener diodes are so good at maintaining a constant reverse-bias voltage, even as current varies, that they're used to regulate voltage in a circuit. In the circuit in Figure 6-8, a 9-volt DC supply is being used to power a load, and a Zener diode is placed so that the DC supply exceeds the breakdown voltage of 6.8 volts. (Note that this voltage is reverse-biasing the diode.) Because the load is in parallel with the Zener diode, the voltage drop across the load is the same as the Zener voltage, which is 6.8 volts. The remaining supply voltage is dropped across the resistor (which is there to limit the current through the diode so that the power rating isn't exceeded).

Here's the important thing: if the supply voltage varies up or down around its nominal 9-volt value, the current in the circuit fluctuates, but the voltage across the load remains the same: a constant 6.8 volts. The Zener diode allows current fluctuations while stabilising the voltage, whereas the resistor voltage varies as the current fluctuates.

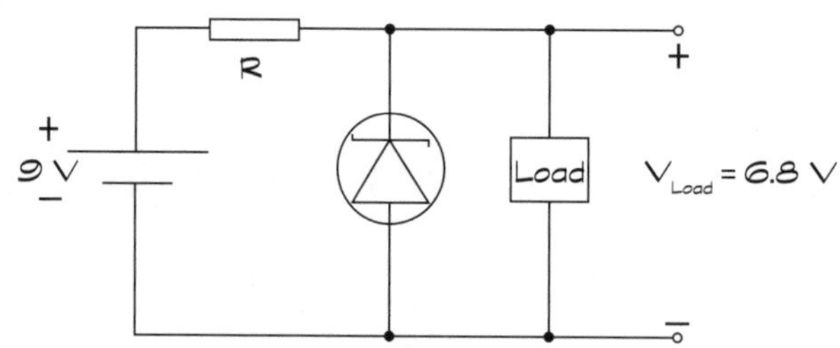

Figure 6-8: The Zener diode stabilises the voltage drop across the load in this circuit.

Seeing the light with LEDs

All diodes release energy in the form of light when forward-biased. The light that the standard silicon diodes release is in the infrared range, which isn't visible to the human eye. *Infrared light-emitting diodes* (IR LEDs) are commonly used in remote control devices to send invisible messages to other electronic devices, such as your TV or DVD player.

Diodes known as *visible LEDs* (or just LEDs) are specially made to emit copious amounts of visible light. By varying the semiconductor materials used, diodes can be engineered to emit red, yellow or green light, and you can find special-purpose LEDs that emit blue or even white light. Tri-colour LEDs contain two or three different diodes within one package.

The diode in an LED is housed in a plastic bulb designed to focus the light in a particular direction. The lead connected to the cathode is shorter than the lead connected to the anode. LEDs last longer and use power more efficiently than standard light bulbs, which generate heat in addition to light, eventually burning out the filament. Indicator lights in cars, computers and audio electronics are often made from LEDs.

You may have noticed household light bulbs, each containing an array of diodes, starting to appear in the shops. That's because they last longer and are cheaper to run (although they still cost more to buy in the first place) than conventional tungsten or halogen light bulbs. Figure 6-9 shows the circuit symbol for an LED along with some single-colour LEDs.

LEDs carry the same specifications as standard diodes, but they usually have fairly low current and PIV ratings. A typical LED has a PIV rating of about 5 volts with a maximum current rating of under 50 milliamps. If more current passes through an LED than its maximum rating specifies, the LED burns up like a marshmallow in a campfire. Forward voltages vary depending on the type of LED, ranging from 1.5 volts for IR LEDs up to 4.6 volts for blue LEDs. Red, yellow and green LEDs typically have a forward voltage of about 2 volts.

Make sure that you check the specifications of any LEDs you use in circuits.

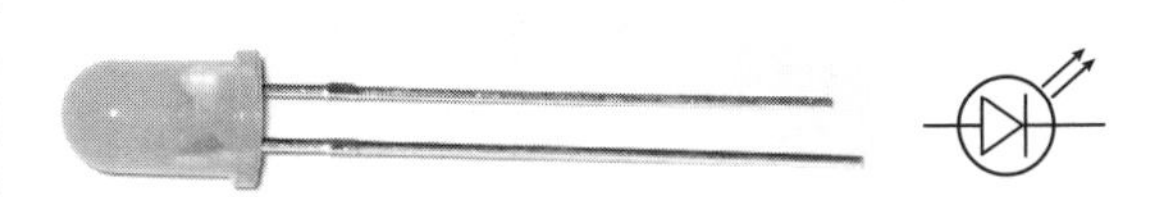

Figure 6-9: The shorter lead of a typical single-colour LED is attached to the cathode. The symbol for an LED is also shown here.

The maximum current rating for an LED is usually referred to as the maximum *forward current*, which is different from another LED rating, known as the *peak current* or *pulse current*. The peak/pulse current, which is higher than the maximum forward current, is the absolute maximum current that you can pass through the LED for a very short period of time. Here, short means short – in the order of milliseconds. If you confuse forward current with peak current, you may wreck your LED.

Never connect an LED directly to a power source because it's going to go bang! Instead, use a resistor in series with the LED to limit the forward current.

For instance, in the circuit in Figure 6-10, a 6-volt battery is used to power a red LED. The LED has a forward voltage drop of 2 volts and a maximum current rating of 30 milliamps. The voltage drop across the resistor is the difference between the source voltage and the LED forward voltage, or 6 volts – 2 volts = 4 volts. The question is, how big should the resistor be to limit the current to 30 milliamps (that's 0.030 amps) *or less* when the voltage dropped across the resistor is 4 volts? You apply Ohm's Law (which we discuss in Chapter 3) to calculate the *minimum* value of resistance required to keep the current below the maximum current rating as follows:

$$R = V_R / I_{max}$$
$$= \frac{4 \text{ volts}}{0.030 \text{ amps}}$$
$$\approx 133 \text{ ohms}$$

You're unlikely to find a resistor with the exact value you calculated, so choose a standard resistor with a *higher* value, such as 150 Ω, to limit the current a bit more. If you choose a lower value, such as 120 Ω, the current exceeds the maximum current rating.

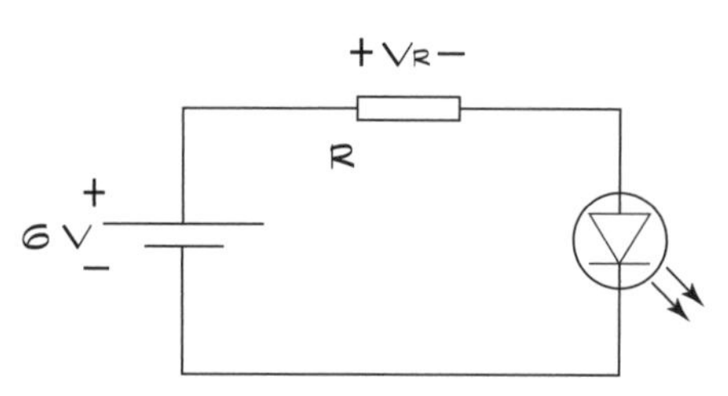

Figure 6-10: A resistor inserted in series with an LED is necessary to limit current to the LED.

Using diodes in other ways

Diodes have lots of other uses in circuits including:

- **Overvoltage protection:** Diodes placed in parallel with a piece of sensitive electronic equipment protect the equipment from large voltage spikes. The diode is placed backwards so that it's normally reverse-biased, acting like an open circuit and not playing any part in the normal operation of the circuit. However, an abnormal voltage spike forward-biases the diode, limiting the voltage across the sensitive component and drawing excess current away from it.
- **Making logic gates:** Diodes are the building blocks of specialised circuits known as *logic*, which produce a binary (0 or 1) output when given a set of two or more binary inputs. We discuss logic a bit more in Chapter 7.
- **Current steering:** Diodes are sometimes used in uninterruptible power supplies (UPSs) to prevent current from being drawn out of a backup battery under normal circumstances, while allowing current to be drawn from the battery during a power outage.

Trillions of Transistors

Imagine the world without the amazing electronics building block known as the transistor. Your phone wouldn't be mobile, and your computer wouldn't fit in your palm, or in your desk – or even in the room.

Transistors are at the heart of nearly every electronic device in the world, quietly working away without taking up much space, generating a lot of heat, or breaking down. Generally regarded as the most important technological innovation of the 20th century, transistors were developed as an alternative to the vacuum tube, which drove the development of electronic systems ranging from radio broadcasting to computers, but were big, heavy and fragile. The solid-state transistor enabled the miniaturisation of electronics, leading to the development of mobile phones, iPods, satnavs and pretty much every gadget or computer that we take for granted.

Shrinking circuits

Transistors are microscopically small, have no moving parts, are very reliable and dissipate a heck of a lot less power than their vacuum tube predecessors. They basically do just two things in electronic circuits: switch and amplify. But with those two things, they can do a lot. If you can switch electron flow on and off, you have control over the flow, and you can build very involved circuits by incorporating lots of switches in the right places. If you can amplify an electrical signal, you can store and transmit tiny signals and boost them up when you need them to make something happen (like move the diaphragm of a speaker).

Examining the anatomy of a transistor

Although many different types of transistors exist, the two most common types are bipolar junction transistors and field effect transistors.

Bipolar junction transistors

The first transistors to be invented were *bipolar junction transistors* (BJTs), and they're what most hobbyists use in home-made circuits. BJTs consist of two pn-junctions in a three-layer sandwich-like structure. Leads are attached to each section of the transistor and labelled the *base*, *collector* and *emitter*. The two types of bipolar transistors are:

- **NPN transistors:** A thin piece of p-type semiconductor is sandwiched between two thicker pieces of n-type semiconductor, and leads are attached to each of the three sections. The symbol for an NPN transistor is shown below:

- **PNP transistors:** A thin piece of n-type semiconductor is sandwiched between two thicker pieces of p-type semiconductor, and leads are attached to each section. The symbol for a PNP transistor is shown below:

Bipolar transistors essentially contain two pn-junctions: the base-emitter junction and the base-collector junction. By controlling the voltage applied to the base-emitter junction, you control how that junction is biased (forward or reverse), ultimately controlling the flow of electrical current through the transistor. We explore exactly how an NPN transistor works in the later section 'How Transistors Really Work'.

Field effect transistors

A *field effect transistor* (FET) consists of a channel of n- or p-type semiconductor material through which current can flow, and a different material, laid across a section of the channel, that controls the conductivity of the channel.

One end of the channel is known as the *source*, the other end of the channel is called the *drain*, and the control mechanism is called the *gate*. By applying a voltage to the gate, you control the flow of current from the source to the drain. Leads are attached to the source, drain and gate. Some FETs have a fourth lead, grounding part of the FET to the chassis of the circuit.

FETs come in two flavours – N-channel (similar to NPN) and P-channel (similar to PNP) – depending on the type of semiconductor material through which current flows. The two major types of FETs are *MOSFETs (metal-oxide-semiconductor FETs)* and *JFETs (junction FETs)*. Which is which depends on exactly how the gate is constructed, and that in turn results in different electrical properties and, consequently, different uses of each type.

Although you don't need to concern yourself with the details of gate construction, you should be aware of the names of the two major types of FETs: MOSFETs and JFETs.

FETs (particularly MOSFETs) have become much more popular than bipolar transistors for use in integrated circuits, which we discuss in Chapter 7, where thousands of transistors work together to perform a task. That's because they're low-power devices with a structure that allows for thousands of N- and P-channel MOSFETs to be crammed together like sardines on a single piece of silicon.

Electrostatic discharge (ESD) can damage FETs. If you purchase FETs, be sure to keep them in an antistatic bag or tube and leave them there until you're ready to use them. You can read more about the harmful effects of ESD in Chapter 9.

Operating a transistor

BJTs and FETs work in basically the same way. The voltage you apply to the input (base for a BJT or gate for an FET) determines whether or not current flows through the transistor (from collector to emitter, for a BJT, and from source to drain for an FET). Below a certain voltage level, no current flows. Above a certain voltage level, the maximum current possible flows. Between those two voltage levels, an in-between amount of current flows.

In this in-between operating mode, small fluctuations in input current produce large fluctuations in output current. By allowing only the first two possibilities of input voltage (all or nothin'), you use the transistor as an on/off switch for current flow. By allowing the third possibility, you use the transistor as an amplifier.

To understand how a transistor works (specifically, a FET), think of a pipe connecting a source of water to a drain with a controllable valve across a section of the pipe, as shown in Figure 6-11. By controlling whether the valve is fully closed, fully open or partially open, you control the flow of water from the source to the drain. You can set up the control mechanism for your valve in two different ways: it can act like an on/off switch, fully opening or fully closing with nothing in between; or it can open partially, depending on how much force you exert on it. That's how a transistor can act as a switch or as an amplifier.

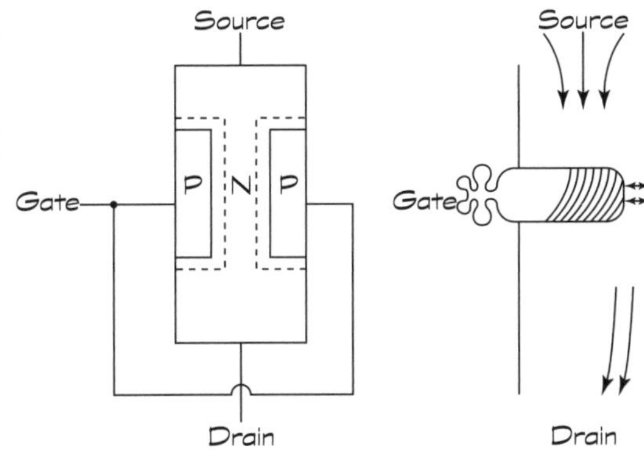

Figure 6-11: In an FET, voltage applied to the gate controls the flow of current from the source to the drain.

How Transistors Really Work

If you're curious about how applying a voltage to one pn-junction in a bipolar transistor can control the current flowing through the rest of the transistor, this section is for you. If you're not all that interested in the goings-on of free electrons and holes in doped semiconductors, you can skip this section all together and head off to the later section, 'Using a Model to Understand Transistors'.

To get an insider's view of how transistors really work, we take a look at an NPN transistor. Figure 6-12 shows a representation of the structure of an NPN transistor, which includes a narrow p-type section sandwiched between two thicker n-type sections. The p-type section forms the base of the transistor and has a lead sticking out of it.

One of the n-type sections is the emitter and the other is the collector. The emitter and collector aren't interchangeable: they're doped differently and so have different concentrations of free electrons. The base is so narrow that it contains many fewer holes than free electrons are available in the emitter and collector. This fact is important.

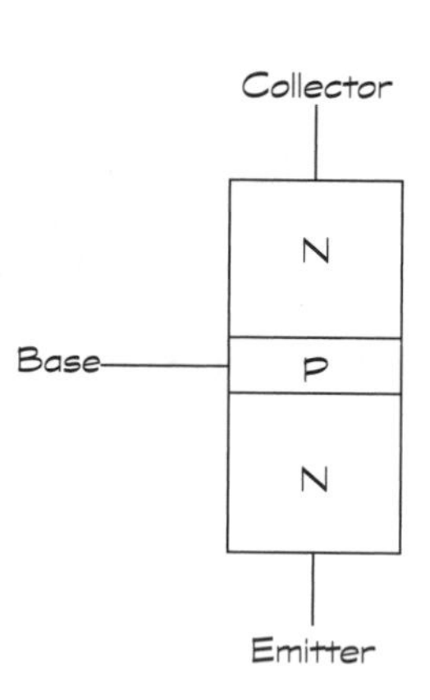

Figure 6-12: An NPN transistor has a narrow base made up of p-type semiconductor material sandwiched between two thicker pieces of n-type semiconductor material, known as the emitter and collector.

Emitting and collecting electrons

An NPN transistor has two pn-junctions: the base-to-emitter pn-junction and the base-collector pn-junction. This arrangement is like sticking two diodes together, anode to anode. The way we bias the two p-n junctions controls the operation of a transistor. Say you connect two different voltage sources across the two junctions, as shown in Figure 6-13. If $V_{CE} > V_{BE}$, the voltage at the base is more negative than the voltage at the collector, so the base-collector pn-junction is reverse-biased, and no current flows across that junction. If you raise V_{BE} to about 0.7 volts or more (for silicon transistors), you forward-bias the base-emitter junction, and current flows across the junction.

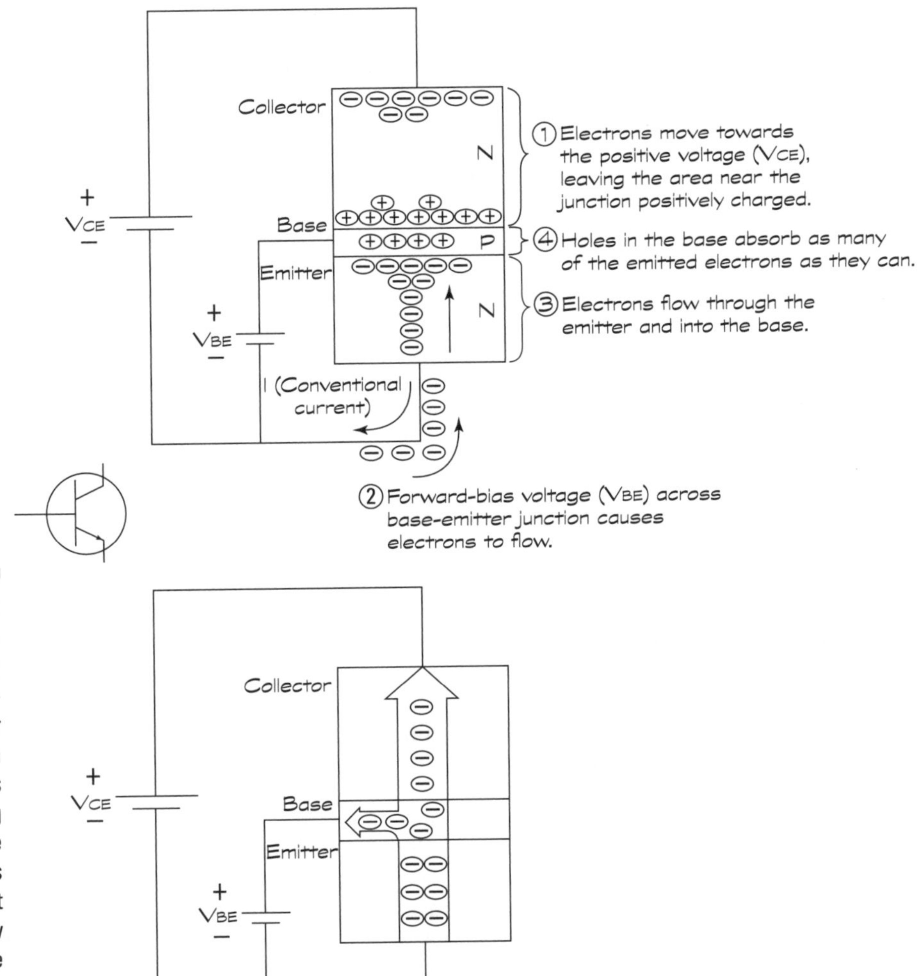

Figure 6-13: Forward-biasing the base-emitter junction in a transistor is like turning on a valve that allows current to flow through the transistor.

The current that flows across the base-emitter junction consists of the set of free electrons in the emitter. The emitter is said to emit electrons towards the base (hence the name). (Of course, in fact an external force – coming from the power supply voltage connected to the collector – is nudging the free electrons in the emitter.)

Some of the electrons that get into the base *recombine* with the holes floating around in the base (remember, the base is made of a p-type semiconductor). But because the base is so narrow, too few holes exist to absorb all the electrons crossing the junction, so the base becomes negatively charged and tries to force the extra electrons out.

All those surplus electrons can travel out of the base via two paths: through the base connection that leads to the positive voltage source or through the base-collector junction and into the collector. So which way do they go? Remember that the base-collector junction is reverse-biased, with a strong positive voltage connected to the collector lead. That positive voltage tends to attract the free electrons that normally exist in the n-type collector towards one end of the collector. This makes the region at the other end of the collector (right near the base-collector junction) look as though it's positively-charged.

Hmmm . . . so the region of the collector right near the base-collector junction looks positively charged, while loads of electrons that were emitted by the emitter are now crowding into the base looking for a way out. So what happens? *Most of those electrons (about 99 per cent) get pulled across the base-collector junction.* The electrons that were emitted by the emitter are now collected by the collector. A small percentage (less than 1 per cent) of the electrons move out of the base through the lead that's connected to the base voltage source, but the pull across the base and out isn't as strong as the other pull across the junction and into the collector. The collector wins the tug-of-war, attracting most of the emitter's electrons.

What you see as an outside observer, looking at the leads connected to the transistor, is that when the base-emitter junction is forward-biased, electrons flow from the emitter and then get split between the collector and the base, with most of the electrons (about 99 per cent) going to the collector. *By controlling the voltage at the base-emitter junction, you cause a large amount of electrons to flow through the transistor from emitter to collector.* This fact is the crux of transistor operation. Biasing the transistor is like opening a valve that controls the flow of current through the transistor.

Because electronic circuits always refer to conventional current, which is just the opposite of the real electron flow we've been talking about in this section, in circuit-speak, you say that forward-biasing the base-emitter junction of an NPN transistor causes a small (conventional) current to flow from the base to the emitter, and a large (conventional) current to flow from the collector to the emitter. That's why the circuit symbol for an NPN transistor shows an arrow pointing out of the emitter; the arrow's pointing in the direction of conventional current flow.

You forward-bias a silicon NPN transistor by applying a voltage of at least 0.7 volts to the base-emitter junction. Transistors made of germanium (much less common than silicon) have smaller forward-bias voltages (about 0.3 volts), but the same cause-and-effect transistor action occurs.

PNP transistors work the same way, except that all the polarities are reversed because the junctions are reversed. To forward-bias the base-emitter junction of a PNP transistor, for example, you apply a voltage of –0.7 volts from base to emitter. The circuit symbol for a PNP transistor shows an arrow pointing *into* the emitter, which indicates the direction of conventional current flow when the transistor is conducting.

Gaining current

If you increase the current flowing into the base, I_B, when a transistor is conducting current, an interesting thing happens: the current flowing into the collector, I_C, increases too. If you decrease the base current, the collector current decreases too. In fact, the 1–99 per cent relationship between the base current and the collector current holds true as you vary the base current (within limits, which we discuss in the next section, 'Saturating the transistor').

The pattern of current changes at the collector exactly tracks the pattern of current changes at the base – just much bigger, which is why transistors are known as current amplifiers. The amount of collector current that flows is directly proportional to the amount of base current. The *current gain* (symbolised by h_{FE}) of the transistor depends on several factors, including the specific transistor in use.

Even for a single, specific transistor, the current gain varies according to several factors. You should never design a circuit that hinges on a specific value of current gain, because your circuit is likely to act strangely.

When you're using the transistor as an on/off switch (which we discuss in the later section 'Switching Signals with a Transistor'), the exact current gain doesn't matter. If you're using the transistor as an amplifier, you can avoid a crazy circuit by configuring your transistor, along with other components such as resistors, in a way that removes any dependency on the exact value of current gain.

This clever little work-around, which we discuss in 'Amplifying Signals with a Transistor' later in this chapter, is very easy to design.

Understanding the word 'transistor'

The word 'transistor' is a combination of two word parts: *trans* and re*sistor*. The *trans* part of the name conveys the fact that by placing a forward-bias voltage on the base-emitter junction, you cause electrons to flow in another part of the component, from emitter to collector. You *transfer* the action from one part of the transistor to another, known as *transistor action*. Fluctuations in base current result in proportional fluctuations in collector/emitter current, so you can think of the transistor as a sort of variable resistor: when you turn the dial (by varying the base current), the resistance changes, producing a proportionally varying collector/emitter current. That's where the *sistor* part of the name comes from.

Saturating the transistor

In transistors, the current gain from base to collector is proportional up to a point.

The insides of a transistor consist of doped semiconductors with a limited number of free electrons or holes that can move around. As you increase the voltage supply feeding into the base, you allow more electrons to flow out of the base, which means more electrons coming from the emitter. But only so many free electrons are available in the emitter, so an upper limit exists as to how much current can flow. When the transistor maxes out, you say it's *saturated.*

You can think of saturating a transistor as opening a valve wider and wider, so that more and more water flows through a pipe, until the pipe is handling as much water as it can; you can open the valve more, but you can't get any more water to flow.

When a transistor is saturated, both of its junctions (base-emitter and base-collector) are forward-biased. The voltage drop across the output of the transistor – from collector to emitter – is nearly zero, as if a wire is connected across the output of the transistor. And because I_C is so much larger than I_B, and $I_E = I_B + I_C$, you can say that $I_C \approx I_E$.

This approximation comes in handy when analysing and designing transistor circuits.

If you operate the transistor so that current maxes out or doesn't flow at all, you're using the transistor as an on/off switch. You do this by designing your circuit so that the base-emitter junction is nonconducting (the voltage across it is less than 0.7 volts) or fully conducting – with nothing in between.

Using a Model to Understand Transistors

Free electrons, moving holes, pn-junctions and biasing as described in the preceding sections are all very nice, but you really don't need to know all that technical stuff in order to use transistors in circuits. Instead, you can familiarise yourself with a functional model of a transistor and find out enough to get going.

Figure 6-14 shows a simple model of an NPN transistor. The model shows a diode between the base and the emitter controlling a variable resistance, R_{CE}, between the collector and the emitter.

A transistor can operate in three different ways:

- **Transistor off:** If $V_{BE} < 0.7$ volts, the diode is off, so $I_B = 0$. This result makes the resistance R_{CE} infinite, which means $I_C = 0$. The output of the transistor (collector-to-emitter) is like an open switch. You call this mode of operation *cutoff.*
- **Transistor partially on:** If $V_{BE} \geq 0.7$ volts, the diode is on, so base current flows. If I_B is small, the resistance R_{CE} is reduced and some collector current, I_C, flows. I_C is directly proportional to I_B, with a *current gain*, h_{FE}, equal to I_C/I_B, and the transistor functions as a current amplifier. You say that the transistor is operating in *active* mode.
- **Transistor fully on:** If $V_{BE} \geq 0.7$ volts, and I_B is increased a lot, the resistance R_{CE} is zero and the maximum possible collector current, I_C, flows. The voltage from collector to emitter, V_{CE}, is nearly zero, so the output of the transistor (collector-to-emitter) is like a closed switch. You say that the transistor is *saturated.*

Choosing the right switch

You may wonder why you'd use a transistor as a switch when so many other types of switches and relays are available (as we describe in Chapter 8). Well, transistors have several advantages over other types of switches, and so they're used under certain circumstances.

Transistors use very little power, can switch on and off several times per second and can be made microscopically small, so integrated circuits (which we discuss in Chapter 7) use thousands of transistors to switch signals around on a single tiny chip. Mechanical switches and relays have their uses, too, in situations where transistors just can't handle the load, such as switching currents over about 5 amps, switching higher voltages (such as in electrical power systems) and switching several contacts simultaneously.

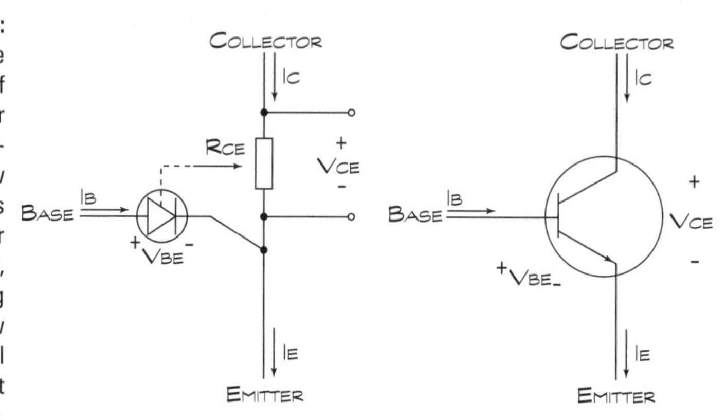

Figure 6-14: You can use a model of a transistor to understand how it works as a switch or an amplifier, depending upon how you control what's input to the base.

When you design a transistor circuit, you choose components that put the transistor into the right operating mode (cutoff, active or saturation), depending on what you want the transistor to do. For instance, if you want to use the transistor as an amplifier, you select supply voltages and resistors that forward-bias the base-emitter junction and allow just enough base current to flow, but not so much that the transistor becomes saturated. This setting is known as *biasing* the transistor.

If you want the transistor to act like a switch, you choose values that allow no base current to flow or enough base current to saturate the transistor, and you control that two-state operation with a switch or the output of a previous stage of electronic circuitry.

Amplifying Signals with a Transistor

Transistors are commonly used to amplify small signals. Say, for instance, you produce an audio signal as the output of one stage of an electronic circuit, and you want to amplify it before shipping it off to another stage of electronics, such as a speaker. You use a transistor, as shown in Figure 6-15, to amplify the small up-and-down fluctuations in the audio signal (v_{in}), which you input to the base of the transistor, into large signal fluctuations (v_{out}), which appear at the output (collector) of the transistor. Then you take the transistor output and apply it to the input of your speakers.

Turning it on

A transistor must be partially on in order to work as an amplifier. To do this, you apply a small voltage to the base to bias the transistor. In the example in Figure 6-15, resistors R1 and R2 are connected to the base of the transistor and configured as a voltage divider (we discuss how a voltage divider works in Chapter 3). The output of this voltage divider (R1/(R1 + R2) × V) supplies enough voltage to the base to turn the transistor on and allow current to flow through it, biasing the transistor so that it's in the active mode (partially on).

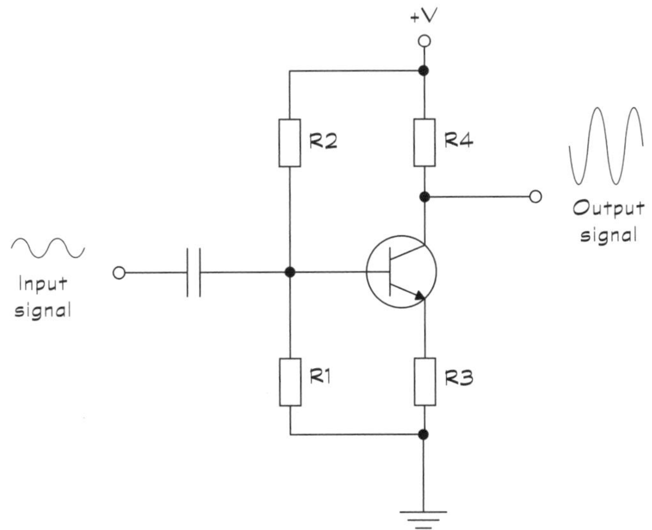

Figure 6-15: By strategically positioning a few resistors in a transistor circuit, you can properly bias a transistor and control the gain of the circuit.

The capacitor at the input allows only AC to pass through to the transistor, blocking any DC component (known as a DC offset) of the input signal, as shown in Figure 6-16. Without that capacitor, any DC offset in the input signal can upset the bias of the transistor, potentially shutting the transistor off (cutoff) or saturating it so that it no longer acts like an amplifier.

Turning it up

With the transistor in Figure 6-15 partially on, the current fluctuations caused by the AC input signal get amplified by the transistor. Because the current gain of any transistor you happen to choose can be highly unpredictable, you design your amplifier circuit in such a way so as to eliminate any dependency on that flaky current gain. You give up some of the strength of the amplification, but you get stability and predictability in return.

By placing resistors R3 and R4 in the circuit, you can control the *voltage gain*, or how much the input signal is amplified – without worrying about the exact current gain of the specific transistor at the heart of your circuit – amazing but true!

The AC voltage gain of a transistor circuit with resistors as shown in Figure 6-15 is $-R4/R3$. The negative sign just means that the input signal is *inverted*: as the input voltage varies up and then down, the output voltage varies down and then up, as shown by the input and output signal waveforms in Figure 6-15.

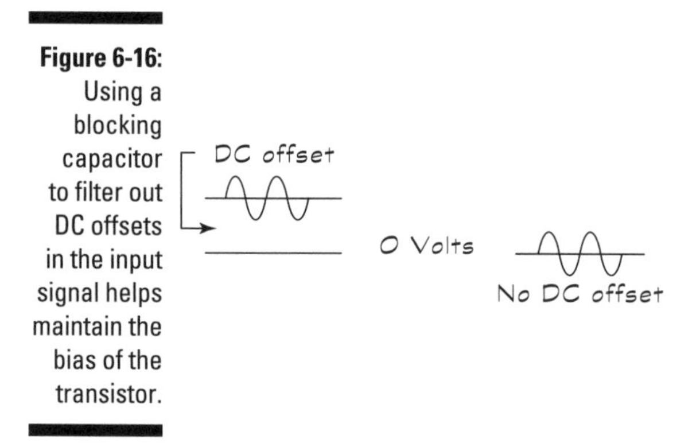

Figure 6-16: Using a blocking capacitor to filter out DC offsets in the input signal helps maintain the bias of the transistor.

Turning it to your needs

The type of transistor setup we discuss in the preceding section, known as a *common-emitter* circuit, is just one of many different ways to configure transistor circuits for use as amplifiers. You use different configurations to achieve different goals, such as high power gain versus high voltage gain.

How the circuit behaves depends on how you connect up the transistor to power supplies, where the load is and what other circuit components, such as resistors, capacitors and other transistors, you add and where you add them. For instance, you can piggy-back two bipolar transistors in a setup known as a *Darlington pair* to produce multiple stages of amplification (or you can just purchase a three-lead component called a *Darlington transistor*, which includes a Darlington pair already wired up for you).

Transistor amplifier circuit design is a field of study on its own, and many excellent books have been written on the subject. If you're interested in discovering more about transistors and how to design amplifier circuits using transistors, try getting your hands on a good electronics design book, such as *The Art of Electronics*, by Paul Horowitz and Winfield Hill (Cambridge University Press). The book's not cheap, but it's a classic.

Switching Signals with a Transistor

You can also use a transistor as an electrically operated switch. The base lead of the transistor works like the toggle on a mechanical switch. The transistor is off when no current flows into the base (in cutoff), and the transistor acts like an open circuit – even if a voltage difference exists from collector to emitter. The transistor is on when current flows into the base (in saturation), and the transistor acts like a closed switch, allowing current to flow from collector to emitter – and out to whatever load you want to turn on.

How do you get this on/off thing to work? Say that you use an electronic gadget to feed your cat at sunrise. You can use a *photodiode*, which conducts current when exposed to light, to control the input to a transistor switch that delivers current to your cat feeder (the load). At night, the photodiode doesn't generate any current, so the transistor is off. When the sun rises, the photodiode generates current, turning the transistor on. This reaction allows current to drive a motor that raises a dish cover or opens a door so moggie can get her breakfast and you can get a lie-in.

Why don't you just supply the current from the photodiode to the gadget? Because your gadget may need a larger current than can be supplied by the photodiode. The small photodiode current controls the on/off action of the transistor, which acts like a switch to allow a larger current from a battery to power your gadget.

One of the reasons transistors are so popular for switching is that they don't dissipate a lot of power. Remember that power is the product of current and voltage. When a transistor is off, no current flows, so the power dissipated is zero. When a transistor is fully on, V_{ce} is nearly zero, so the power dissipated is nearly zero.

Choosing Transistors

So, you may be wondering, how do you make sense of the thousands upon thousands of different transistors available, and how do you select the right one for your circuit? Good questions.

If you're designing a transistor circuit, you need to understand how your circuit operates under various conditions. What's the maximum amount of collector current that your transistor has to handle? What's the minimum current gain you need in order to amplify an input signal? How much power can possibly be dissipated in your transistor under extreme operating conditions (for instance, when the transistor is off and the entire power supply voltage may be dropped across the collector-emitter)?

When you understand the ins and outs of how your circuit operates, you can start looking up transistor specifications to find one that meets your needs.

Tackling transistor ratings

Transistor makers use lots of different parameters to describe their endless variety of products, but you really need to be familiar with only a few in order to choose the right transistor for your circuit. For bipolar (NPN or PNP) transistors, here's what you need to know:

- **I_c max:** The *maximum collector current* the transistor can handle. When designing a circuit, you need to make sure that you use a resistor to limit the collector current so that it doesn't exceed this value.
- **h_{FE}:** The *DC current gain* from base to collector. Because the current gain can vary, even among transistors of the same type, you need to know the guaranteed minimum value of h_{FE}, and that's what this parameter tells you. h_{FE} also varies for different values of I_c, so sometimes h_{FE} is given for a specific value of I_c, such as 20 milliamps.
- **V_{ce}:** The *maximum collector-to-emitter voltage*. Usually at least 30 volts, so for low power applications like hobby electronics circuits, you don't really need to worry about it.
- **P_{total} max:** The *maximum power dissipation*, which is roughly $V_{ce} \times I_c$ max. If you're using the transistor as a switch, you don't need to worry about this rating, because power dissipation is nearly zero anyway. If you're using the transistor as an amplifier, you need to be aware of this rating. If you think that your circuit may approach this value, you need to attach a heat sink to the transistor.

Of course, none of these ratings appears anywhere on the transistor itself – that would be too easy. Look at the retailer's production description, a specifications book or the technical documentation at the manufacturer's website. If you're building a circuit someone else designed, you can simply use the transistor specified by the designer or consult a cross-reference book to find a similar model to substitute.

Identifying transistors

Most bipolar transistors are named using an industry-standard five- or six-digit code (see the 'Discovering what's in a name' section earlier in this chapter). Transistors start '2N': the 'N' tells you that it's a semiconductor and the '2' gives the number of pn-junctions. The remaining three or four digits indicate the specific features of the transistor. However, different manufacturers may use different coding schemes, so your best bet is to consult the appropriate website, catalogue or specification sheet to make sure that you're getting what you need for your circuit.

Many suppliers categorise transistors according to the type of application in which they're used, such as low power, medium power, high power, audio (low noise) or general purpose. This system helps guide you to make the right selection of a transistor for your particular circuit.

Recognising transistors

The semiconductor material in a transistor is the size of a grain of sand or even smaller, so manufacturers put these microscopic little components in a metal or plastic case with leads sticking out so that you can see them in the first place and then connect them up in your circuits.

You can find literally dozens upon dozens of different shapes and sizes of transistors, some of which are shown in Figure 6-17. The smaller packages generally house *signal transistors*, which are rated to handle smaller currents, whereas larger packages contain *power transistors* designed to handle larger currents. Most signal transistors come in plastic cases, but some precision applications require signal transistors housed in metal cases to reduce the likelihood of stray radio frequency (RF) interference.

Figure 6-17: Signal and power transistors come in a variety of sizes and packages.

Bipolar transistors typically have three wire leads so that you can access the base, collector and emitter of the transistor. One exception is a *phototransistor* (which we discuss in Chapter 8), which is packaged in a clear case and has just two leads (collector and emitter), because light is used to bias the transistor so you don't have to apply a voltage to the base. All FETs have leads for the source, drain and gate, and some include a fourth lead so that you can ground its case to the chassis of your circuit.

To figure out which package lead is which, you need to consult the documentation for the specific transistor. Be careful how you interpret the documentation: transistor connections are often (though not always) shown from the underside of the case, as if you've turned the transistor over and are gazing at it from the bottom.

You absolutely must install transistors the proper way in your circuits. Switching the connections around can damage a transistor and may even damage other circuit components.

Integrating Components

Transistors can be combined in all sorts of different ways to make lots of incredible things happen. The small size of the semiconductor material making up a transistor allows the creation of a single circuit containing hundreds or thousands of transistors (along with resistors and other components) in a single component that fits easily into the palm of your hand.

These amazing creations, known as *integrated circuits* (ICs), enable you to build really complex circuits with just a couple of parts. To find out more, take a look at Chapter 7, where we discuss some of the ICs that are available today as a result of the semiconductor revolution.

Chapter 7

Cramming Components into Chips

In This Chapter

- Packing parts onto an integrated circuit (IC)
- Thinking logically about digital circuits
- Pondering packages and pinouts
- Boosting signals with op amps
- Timing, counting and controlling everything in sight

This chapter is about chips – apologies if that makes you feel hungry, but you can't eat these ones. We're going to tell you all about silicon chips, also known as *integrated circuits* (ICs). Space exploration, desktop computers, consumer electronics and much more would be impossible without this incredible innovation – really, a series of incredible innovations – that drives the modern information age. The silicon chip has completely revolutionised the world in the past half century – and it's still changing our lives faster than ever before. Just don't smother one in tomato sauce!

An IC incorporates anywhere from a few dozen to many millions of circuit components into a single device that can sit on the tip of your finger. Like the web of neurons that makes up your central nervous system, each IC contains an intricate mesh of tiny transistor-based taskmasters, with access to the outside world provided via a finite number of inputs and outputs akin to the nerve endings of your five sense organs.

In this chapter, we explore how integrated circuits came into being, identify the three major IC flavours and dissect the inner workings of one variety – digital ICs. We look at how computers and other digital devices manipulate low and high voltages to process information using special rules known as logic. In addition, we explain how to read an IC to understand what the heck it does (because its cover doesn't tell you) and how to connect it up for use in real circuits. Finally, we introduce you to three best-selling ICs, show you what they do and how you can use them to create your own innovative circuits.

The birth of the IC

With the invention of the transistor in 1947, the focus of electronic design shifted away from bulky vacuum tubes to this new, smaller, more reliable device. This innovation stirred up tremendous excitement, as engineers worked to build more and more advanced circuits, because size was no longer an obstacle. The success in creating advanced designs led to some practical problems: interconnecting hundreds of components inevitably resulted in errors that were extremely difficult to isolate. Additionally, complex circuits often failed to meet speed requirements (because electrons take some time to travel through a bunch of wires and components). Throughout the 1950s, a major focus within the electronics industry was figuring out how to make circuits smaller and more reliable.

Then in the summer of 1958, an engineer named Jack Kirby working alone in a lab at Texas Instruments (while his colleagues were on vacation) came up with a brilliant idea: why not build multiple circuit components out of a single monolithic piece of semiconductor material and lay metal connectors in patterns on top of it? He reasoned that this arrangement would eliminate faulty wiring and the cumbersome manual assembly of discrete components. Luckily, he got the go-ahead from management to try his idea out, and the first IC was born. (Kirby was awarded the Nobel Prize in Physics for his invention – but not until 42 years later.)

A lot has happened since 1958. All those really smart, creative people continued to plug away at their work, and many more innovations took place. As a result, the electronics industry exploded as *chip densities* (a measure of how closely packed the transistors are) increased exponentially. Semiconductor manufacturers routinely carve millions of transistors into a piece of silicon smaller than a postage stamp, and still ICs continue to shrink.

Discovering Integrated Circuits

The IC was invented in 1958 (see the following sidebar 'The birth of the IC') to solve the problems inherent in manually assembling mass quantities of tiny transistors. These amazing creations are miniature circuit boards etched onto a single piece of semiconductor, which is usually silicon (made from sand).

A typical IC contains hundreds of transistors, resistors, diodes and capacitors, and the most advanced contain several hundred million components. Because of this circuit efficiency, you can build really complex circuits with just a couple of parts. ICs are the building blocks of larger circuits. You string them together to form just about any electronic device you can think up.

Linear, Digital or Both?

Over the years, chip makers have produced lots of different integrated circuits, each of which performs a specific function depending on how the components inside are wired up.

Many of the ICs you encounter are so popular that chip makers have standardised them. Manufacturers and hobbyists the world over buy and use these standardised ICs in various projects, and you can find a wealth of information about them online and in books. In contrast to standardised ICs, some companies also design so-called special-purpose ICs to accomplish unique tasks.

Whether standardised or special-purpose, you can separate ICs into three main categories: *linear (analogue)*, *digital* and *mixed signal*. These terms relate to the kinds of electrical signals (we discuss electrical signals in Chapter 2) that work within the circuit:

- **Linear (analogue) ICs:** These ICs contain circuits that process *analogue signals*, which consist of continuously varying voltages and currents. Such circuits are known as analogue circuits, and you can see simple examples of those in Chapters 3, 4 and 5. Examples of analogue ICs are power management circuits, sensors, amplifiers and filters.
- **Digital ICs:** These ICs contain circuits that process *digital signals*, which are patterns consisting of just two voltage (or current) levels representing binary digital data, for instance, on/off, high/low or 1/0. (We discuss digital data a bit more in the next section, 'Making Decisions with Logic'.) Such circuits are known as digital circuits. Some digital ICs, such as microprocessors, contain millions of tiny circuits within just a few square millimetres.
- **Mixed signal ICs:** These ICs contain a combination of analogue and digital circuits.

The majority of standardised ICs fall into the linear or digital category, and most mail-order businesses that sell ICs separate them into linear and digital lists. We present projects that use both linear and digital ICs in Chapter 14. In this chapter, however, we focus more on how digital ICs work than on linear ICs because the special way in which they process digital, or binary, data has made them the building blocks of computers. When you understand how they think, you too can use them in some ingenious ways.

Making Decisions with Logic

When you first find out how to add numbers, you memorise facts such as '2 + 2 = 4', '3 + 6 = 9' and so forth. Then, when you discover how to add multi-digit numbers, you use those simple facts as well as a new rule – carrying to us older folks, regrouping to the younger generation. By applying a few simple addition facts and one simple rule, you can add two large numbers together fairly easily.

The digital microprocessor in your computer works in much the same way. It uses lots of microscopically small digital circuits, known as *logic*, to process simple functions similar to '2 + 2 = 4', and then combines the outputs of those functions by applying rules similar to carrying/regrouping to get an answer. By piggybacking lots of these answers together in a complex web of circuitry, the microprocessor is able to perform some pretty complicated mathematical tasks. Deep down inside, though, is just a bunch of logic applying simple little rules.

In this section, we take a look at how digital logic circuits work.

Beginning with bits

When you add two digits together, you have ten choices for each digit (0 to 9) because that's how the generally used numbering system (known as a *base 10* or *decimal system*) works. When a computer adds two digits together, it uses only two possible digits: 0 and 1 (which is known as a *base 2* or *binary system*). Because only two digits are used, they're known as binary digits, or *bits*. Bits can be strung together to represent letters or numbers. The sidebar 'Adding up bits of numbers' tells you more about how the binary system works, if you want to know. You don't need this information, however, to work with digit circuits

In addition to representing numbers and letters, bits can also be used to carry information. As information carriers, data bits can represent many *two-state* (binary) things: pixel on or off; CTRL key up or down; DVD laser pit or absence; ATM transaction authorised or not – and much more. By assigning logical values of 1 and 0 to a particular on/off choice, you can use bits to carry information about real physical events and allow that information to control other things by processing the bits in a digital circuit.

Logical 1 and logical 0 are also referred to as true and false, or high and low. But what exactly are these ones and zeros in a digital circuit? They're simply high or low currents or voltages that are controlled and processed by transistors. (In Chapter 6, we discuss how transistors work and how they can be used as on/off switches.) Common voltage levels used to represent digital data are 0 volts for logical 0 (low) and (often) 5 volts for logical 1 (high).

Adding up bits of numbers

In the decimal (*base 10*) system, if you want to express a number greater than 9, you need to use more than one digit. Each position, or place, in a decimal number represents a *power of ten* (10^0, 10^1, 10^2, 10^3 and so forth) and the value of the digit (0–9) sitting in that position is a multiplier for that power of ten. With powers of ten, the *exponent* (that tiny number raised up next to the 10) tells you how many times to multiply 10 times itself, so 10^1 equals 10, 10^2 equals 10×10, which is 100, 10^3 equals $10 \times 10 \times 10$, which is 1000 and so on. As for 10^0, it just equals 1 because *any* number raised to the power of zero equals 1. So the positions in a decimal number, starting from the rightmost position, represent 1, 10, 100, 1000 and so forth. These positions are also known as *place values* (ones or units, tens, hundreds, thousands and so forth). The digit (0–9) sitting in that position (or place) tells you how many ones, tens, hundreds, thousands and so forth are contained in that decimal number.

For example, the number 9452 can be written in *expanded notation* as:

$$(9 \times 1000) + (4 \times 100) + (5 \times 10) + (2 \times 1)$$

The entire mathematics system is based on the number 10 (if humans had only eight fingers, we may be using a *base 8* system), and so your brain has been trained to automatically think in decimal format (it's like a maths language). When you add two digits together, such as 6 and 7, you automatically interpret the result, 13, as '1 group of 10 plus 3 groups of 1'. The system is as ingrained in your brain as your native language.

The binary system is like another language: it uses the exact same methodology, but it's based on the number 2. If you want to represent a number greater than 1, you need more than one digit and each position in your number represents a *power of two*: 2^0, 2^1, 2^2, 2^3, 2^4 and so forth, which is the same as 1, 2, 4, 8, 16 and so on. The *bit* (a bit is a binary digit, just 0 or 1) that sits in that position in your number is a multiplier for that power of two. For example, the *binary number* 1101 can be written in expanded notation as:

$$(1 \times 2^3) + (1 \times 2^2) + (0 \times 2^1) + (1 \times 2^0)$$

By translating this into decimal format, you can see what numerical quantity the bit string 1101 represents:

$$(1 \times 8) + (1 \times 4) + (0 \times 2) + (1 \times 1)$$

$$= 8 \quad + \quad 4 \quad + \quad 0 \quad + \quad 1$$

$$= 13 \text{ (in decimal format)}$$

So the binary number 1101 is the same as the decimal number 13. They're just two different ways of representing the same physical quantity, just as you may say *'bonjour'* or *'buenos días'* rather than 'hello'. They're just different words for the same greeting.

When you add two binary numbers together, you use the same methodology that's used in the decimal system, but using 2 as a base. In the decimal system, $1 + 1 = 2$, but in the binary system, $1 + 1 = 10$ (remember, the binary number 10 represents the same quantity as the decimal number 2). Computers use the binary system for arithmetic operations because the electronic circuits inside computers can work easily with bits, which are just high or low voltages (or currents) to them. The circuit that performs addition inside a computer contains several transistors arranged in just the right way so that when high or low signals representing the bits of two numbers are applied to the transistor inputs, the circuit produces the right combination of high or low outputs to represent the bits of the numerical sum. Exactly how this is done is beyond the scope of this book, but hopefully, you now have an idea of how this sort of thing works.

A *byte*, which you've probably heard about quite a bit, is a grouping of eight bits used as a basic unit of information for storage in computer systems. Computer memory, such as Random Access Memory (RAM), and storage devices such as CDs and memory sticks, use bytes to organise chunks of data. Just as banks pack coins into bags of £10, £20 or other nice round values to simplify the job of counting out cash, so computer systems pack data bits together in bytes to simplify the storage of information.

Processing data with gates

Logic gates, or simply *gates*, are tiny digital circuits that accept one or more data bits as inputs and produce a single output bit whose value (1 or 0) is based on a specific rule. Just as different arithmetic operators produce different outputs for the same two inputs (for instance, three *plus* two produces five, whereas three *minus* two produces one), so different types of logic gates produce different outputs for the same inputs:

- **AND gate:** The output is high (1) only if both inputs (one input AND the other input) are high. If either input is low (0), the output is low. A standard AND gate has two inputs, but you can also find three-, four- and eight-input AND gates. For those gates, the output is high only if *all* inputs are high.
- **NOT gate (inverter):** This single-input gate produces an output that inverts the input: a high input generates a low output, and a low input generates a high output. A more common name for a NOT gate is an *inverter.*
- **NAND gate:** This function behaves like an AND gate followed by an inverter (hence the NAND, which means NOT AND). It produces a low output only if all its inputs are high. If any input is low, the output is high.
- **OR gate:** The output is high when one, OR the other, OR both of its inputs are high. It produces a low output only if both inputs are low. A standard OR gate has two inputs, but three- and four-input OR gates are also available. For these gates, a low output is generated only when all inputs are low; if one or more inputs is high, the output is high.
- **NOR gate:** This behaves like an OR gate followed by a NOT gate. It produces a low output if one or more of its inputs are high, and generates a high output only if all inputs are low.
- **XOR gate:** The exclusive OR gate produces a high output if one input OR the other is high, but not both; otherwise, it produces a low output. All XOR gates have two inputs, but multiple XOR gates can be cascaded together to create the effect of XORing multiple inputs.
- **XNOR gate:** The exclusive NOR gate produces a low output if either of its inputs – but not both – is high. All XNOR gates have two inputs.

Figure 7-1 shows the circuit symbols for these common logic gates.

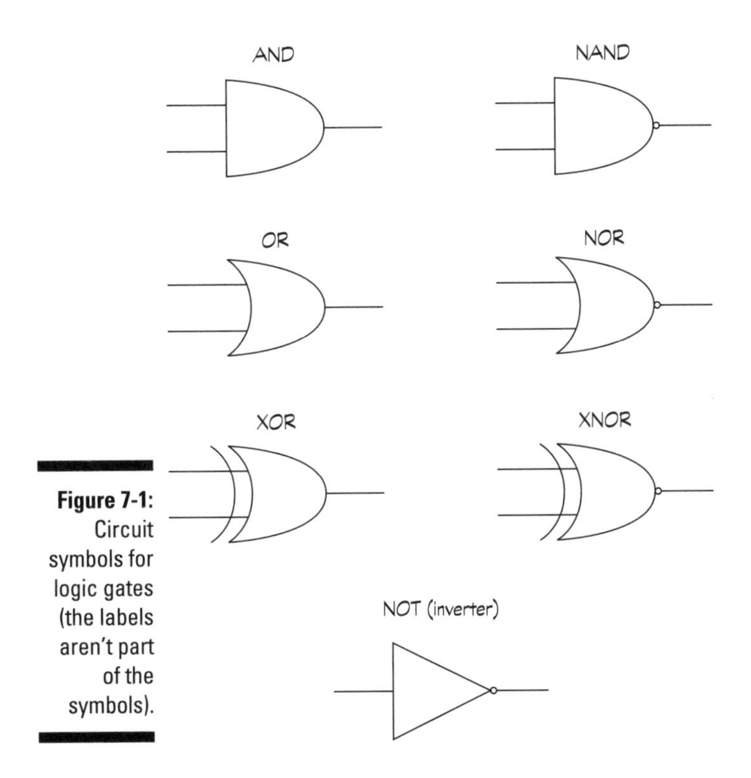

Figure 7-1: Circuit symbols for logic gates (the labels aren't part of the symbols).

Most logic gates are built using diodes and transistors, which we discuss in Chapter 6. Inside each logic gate is a circuit that arranges these components in just the right way, so that when you apply input voltages (or currents) representing a specific combination of input bits, you get an output voltage (or current) that represents the appropriate output bit. The circuitry is built into a single chip with leads, known as *pins*, providing access to the inputs, outputs and power connections in the circuit.

You usually find multiple logic gates sold in integrated circuits, such as an IC containing four two-input AND gates (called a quad 2-input AND gate), as shown in Figure 7-2. The package sports pins leading to each gate's inputs and output and pins you use to connect a power supply to the circuitry.

IC manufacturers produce datasheets for their products (see the later section 'Relying on IC datasheets' for more about these useful products). A *datasheet* is like a user's manual, providing technical specifications and performance information about the chip. Look on the website of the IC's manufacturer for a datasheet that tells you which pins are inputs, outputs, V+ (voltage) and ground.

Make sure that the part you buy has the number of inputs that you need for your project. You can buy logic gates with more than two inputs. For example, you can get a 3-input AND gate from most electronics suppliers.

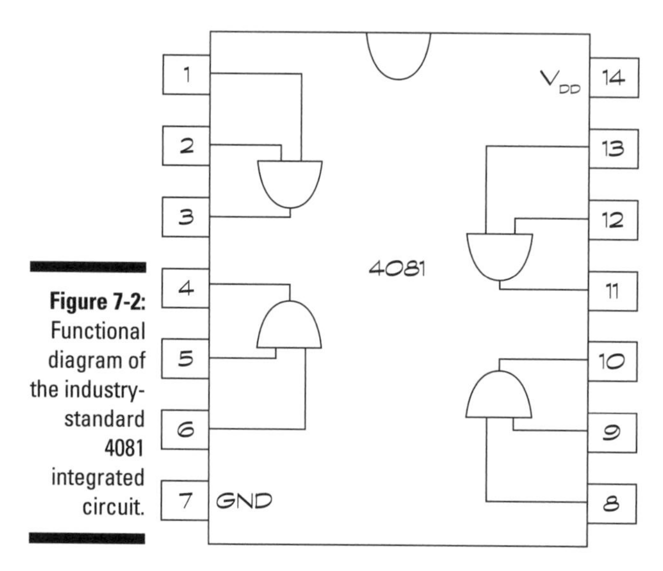

Figure 7-2: Functional diagram of the industry-standard 4081 integrated circuit.

You can create any of the other logical functions by combining just NAND gates or just NOR gates with inverters in the right way. Chip makers typically build digital circuits using these so-called *universal gates* almost exclusively, so that they can focus their research and development efforts on making process and design improvements to just two basic logic gates.

Telling the truth

Tracking all the high and low inputs to logic gates and the outputs they produce can get a bit confusing – especially for gates with more than two inputs – so designers use a tool called a truth table to keep things organised. A *truth table* lists all the possible combinations of inputs and corresponding outputs for a given logical function. Truth tables for the AND and NAND logic gates are shown below to show you how they work. In these tables, A and B represent input bits and the rightmost column shows the output of the gate.

A	B	A AND B
0	0	0
0	1	0
1	0	0
1	1	1

A	B	A NAND B
0	0	1
0	1	1
1	0	1
1	1	0

Understanding How to Use ICs

Integrated circuits are nothing like discrete components, such as individual resistors, capacitors and transistors, which have two or more leads connected directly to the component inside the package. The miniature pre-built components inside an IC are already interconnected into one big happy circuit, ready to perform a specific task. You just have to add a few ingredients, like power and one or more input signals, and the IC does its thing.

Sounds simple, right? Well, it is. You just have to know how to read IC packages – they all look like squashed black bugs with lots of legs – so that you know how to make the right connections.

Identifying ICs with part numbers

Every IC has a unique code, such as 7400 or 4017, to identify the type of device – really, the circuit that's inside. You can use this code, also known as a *part number*, to look up specifications and parameters about an IC in a reference book or online resource. The code is printed on the top of the chip.

Many ICs also contain other information, including manufacturing catalogue number and maybe even a code that represents when the chip was made. Don't confuse the date code or catalogue number with the part number used to identify the device. Manufacturers don't have any standards for how they stamp the date code on their integrated circuits, so you may have to do some detective work to pick out the actual part number of the IC.

Packaging is everything

Great things really do come in small packages. Many ICs that can balance on your fingertip contain incredibly complex circuitry, such as an entire AM/FM radio circuit (minus the battery and antenna) in an IC package the size of a stamp. The actual circuit is so small that manufacturers have to mount it onto a reasonably sized plastic or ceramic package so that humans can use it. They also attach little wires leading out of the package so that the chip inside can connect to your circuit outside.

ICs used in hobby electronics projects are assembled into *dual in-line packages* (DIPs), such as the ones in Figure 7-3. DIPs (sometimes called DILs) are rectangular-shaped plastic or ceramic packages with two parallel rows of leads, called *pins*, down either side. DIPs contain between 8 and 52 pins, but the most common sizes are 8-, 14- and 16-pin. You can solder DIP pins directly or insert them into *sockets* designed to hold the chip without bending the pins, and then solder the socket connections into your circuit.

Figure 7-3: The dual in-line pin (DIP) package is one of the most popular styles for ICs.

Some ICs are very sensitive to static electricity (which we discuss in Chapter 1), so when storing ICs, be sure to enclose them in special conductive foam (sold by most electronics suppliers). And before handling an IC, make sure that you discharge yourself by touching a conductive material, such as a metal pipe, so you don't zap your IC and wonder why it's not working.

ICs used in mass-produced products are generally more complex and require a higher number of pins than DIPs can provide, so manufacturers have developed (and continue to develop) clever ways of packaging ICs and connecting them to printed circuit boards (PCBs). To save space on the board, most ICs today are mounted directly to metal connections built onto the PCBs. This technique is known as *surface mount technology* (SMT), and many IC packages are specially designed to be used this way.

Understanding IC pinouts

The pins on a DIP or other IC package provide connections to the tiny integrated circuits inside, but alas, the pins aren't labelled on the package so you have to rely on the datasheet for the particular IC in order to make the proper connections. Among other things, the datasheet provides you with the IC's *pinout*, which describes the function of each pin.

You can find datasheets for most common (and many uncommon) ICs on the Internet. Try using a search engine, such as Google or Yahoo!, to help you locate them.

To determine which pin is which, you look down onto the top of the IC (not up at the little bug's belly) and look for the clocking mark on the package. The *clocking mark* is usually a small notch in the packaging, but it can also be a little dimple, or a white or coloured stripe.

Conventionally, the pins on an IC are numbered counter-clockwise, starting with the upper-left pin closest to the clocking mark. So, for example, the pins of a 14-pin IC are numbered 1 through 7 down the left side and 8 through 14 up the right side, as shown in Figure 7-4.

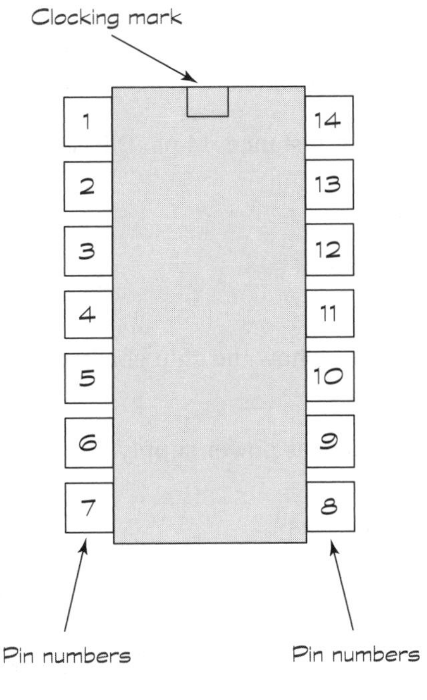

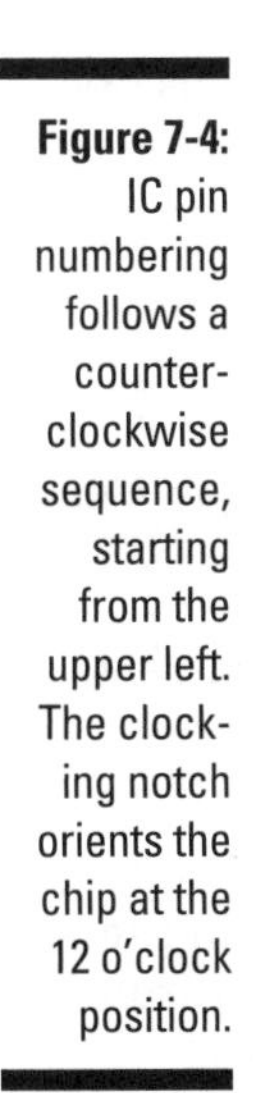

Figure 7-4: IC pin numbering follows a counter-clockwise sequence, starting from the upper left. The clocking notch orients the chip at the 12 o'clock position.

Don't assume that all ICs with the same number of pins have the same pinouts, or even that they use the same pins for power connections. And never – *never!* – make random connections to IC pins, under the misguided notion that you can explore different connections until you get the IC to work. That's a sure-fire way to destroy a poor defenceless circuit.

Many circuit diagrams (schematics) indicate the connections to integrated circuits by showing an outline of the IC with numbers beside each pin. The numbers correspond to the clocked pinout of the device. (Remember, you start with 1 in the upper left and count up as you go counter-clockwise around the chip.) You can easily wire up an IC with these kinds of diagrams because you don't need to look up the device in a book or datasheet. Just make sure that you follow the schematic and that you count the pins properly.

If a schematic lacks pin numbers, you need to find a copy of the pinout diagram. For standard ICs, you can find these diagrams in reference books and online; for non-standard ICs, you have to visit the manufacturer's website to get the datasheet.

Relying on IC datasheets

IC datasheets are like owner's manuals, providing detailed information about the insides, outsides and recommended use of an integrated circuit. They're created by the IC manufacturer and are usually several pages long. Typical information contained in a datasheet includes:

- Available packaging formats (for instance, 14-pin DIP) and photos of each format
- Brief functional description
- Dimensions and pinout diagrams
- IC name and part number
- Input/output waveforms (showing how the chip changes an input signal)
- Manufacturer's name
- Minimum/maximum ratings (such as power supply voltages, currents, power and temperature)
- Recommended operating conditions

Many datasheets include sample circuit diagrams, illustrating how to use the IC in a complete circuit. You can get lots of guidance and good ideas from IC datasheets. Sometimes it really does pay to read the owner's manual!

Sourcing and sinking current

The insides of integrated circuits are hidden from view, which means that knowing exactly how current flows when you connect a load or other circuitry to the IC's output pin or pins is difficult. Typically, datasheets specify how much current an IC output can *source* or *sink.* An output is said to *source current* when current flows out of the output pin and *sink current* when current flows into the output pin. If you connect a device, say a resistor, between an output pin and the positive terminal of a power supply and the output goes low (0 volts), current flows through the resistor into the IC (the IC sinks the current). If you connect a resistor between the output pin and the negative supply (ground) and the output goes high, current flows out of the IC and through the resistor (the IC sources the current). Refer to the datasheet for the maximum source or sink current (which are usually the same) of an IC output.

Manufacturers often publish application notes for their ICs. An *application note* (often called an *app note*) is a multi-page document that explains in greater detail than the datasheet how to use the IC in an application. An app note may showcase one or more unusual uses of a particular IC.

Meeting Some Top Chips

You can find a seemingly endless supply of integrated circuits on the market today, but two are famous for their versatility and ease of use: the operational amplifier (really, a class of ICs) and the 555 timer. If you intend to get even remotely serious about developing your electronics habit, you need to get to know these two circuits fairly well.

In this section, we describe these two popular ICs as well as one additional IC, the 4017 CMOS decade counter. We use all three of these ICs in the projects in Chapter 14.

Sounding out operational amplifiers

The most popular type of linear (analogue) IC is undoubtedly the *operational amplifier*, nicknamed the *op amp,* which is designed to add muscle to (amplify) a weak signal. An op amp contains several transistors, resistors and capacitors, and offers more robust performance than a single transistor. For example, an op amp can provide uniform amplification over a much wider range of frequencies (*bandwidth*) than a single-transistor amplifier.

Most op amps come in 8-pin DIPs, as shown in Figure 7-5, and include two input pins (Pin 2, known as the *inverting input*, and Pin 3, known as the *non-inverting input*) and one output pin (Pin 6).

An op amp is one type of *differential amplifier*: the circuitry inside the op amp produces an output signal that's a multiple of the *difference* between the signals applied to the two inputs. Used a certain way, this setup can help eliminate noise (unwanted voltages) in the input signal by subtracting it out of what's amplified.

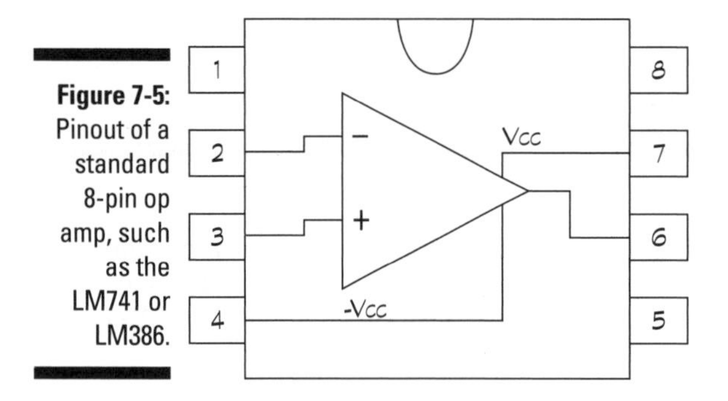

Figure 7-5: Pinout of a standard 8-pin op amp, such as the LM741 or LM386.

You can configure an op amp to multiply an input signal by a known gain factor that's determined by external resistors. One such configuration, known as an *inverting amplifier*, is shown in Figure 7-6. The values of the resistors connected to the op amp determine the gain of the inverting amplifier circuit:

$$\text{Gain} = \frac{R2}{R1}$$

For instance, if the value of R2 is 10 kΩ and that of R1 is 1 kΩ, the gain is 10. With a gain of 10, a 1-volt input signal (peak value) produces a 10-volt (peak) output signal.

To use the inverting amplifier, you just apply a signal (for instance, the output of a microphone) between the input pins; the signal, amplified several times, then appears at the output, where it can drive a component, such as a speaker. Because of the way the op amp in Figure 7-6 is configured, the input signal is flipped, or inverted, to produce the output signal.

An op amp requires both positive and negative supply voltages. A positive supply voltage in the range of 8 to 12 volts (connected to Pin 7) and a negative supply voltage in the range of –8 to –12 volts (connected to Pin 4) works.

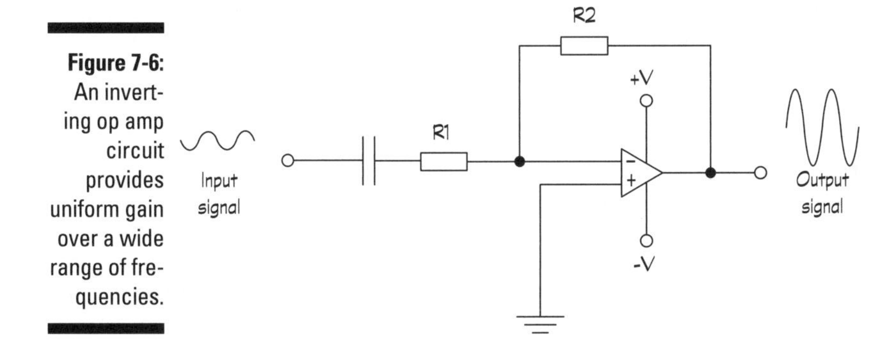

Figure 7-6: An inverting op amp circuit provides uniform gain over a wide range of frequencies.

Plenty of different op amps are available at prices ranging from under a pound for standard op amp ICs, such as the LM386 power amplifier and the LM741 general-purpose op amp, to over £100 for high-performance op amps. The 'Lil but Loud Amp project in Chapter 14 uses an LM386 op amp.

Playing with the IC time machine: the 555 timer

Although it can't take you back to the 1970s so that you can rescue all those flares from the back of the wardrobe, the versatile 555 timer is one of the most popular and easy-to-use integrated circuits. First introduced in 1971 and still in wide use today, more than one billion units are produced every year. This little workhorse can be used for a variety of functions in both analogue and digital circuits, most commonly for precision timing ranging from microseconds to hours. The 555 timer is the cornerstone of many projects you can build (including several in Chapter 14).

Figure 7-7 illustrates the pin assignments for the 555 timer. Among the pin functions are:

- **Trigger input:** When you apply a low voltage to Pin 2, you trigger the internal timing circuit to start working. This trigger is known as an *active low* trigger.
- **Output pin:** The output waveform appears on Pin 3.
- **Reset:** If you apply a low voltage to Pin 4, you reset the timing function, and the output pin (Pin 3) goes low. (Some circuits don't use the reset function, and this pin is tied to the positive supply.)
- **Control voltage input:** If you want to override the internal trigger circuit (which you normally don't), you apply a voltage to Pin 5. Otherwise, you connect Pin 5 to ground, preferably through a 0.01 μF capacitor.

- **Threshold input:** When the voltage applied to Pin 6 reaches a certain level (usually two-thirds the positive power supply voltage), the timing cycle ends. You connect a resistor between Pin 6 and the positive supply. The value of this *timing resistor* influences the length of the timing cycle.
- **Discharge pin:** You connect a capacitor to Pin 7. The discharge time of this *timing capacitor* influences the length of the timing intervals.

You can find various models of the 555 timer IC, including the 556 timer, a dual version of the 555 timer, packaged in a 14-pin DIP. The two timers inside share the same power supply pins.

By connecting a few resistors, capacitors and switches to the various pins of the 555 timer, you can get this little gem to perform loads of different functions – and doing so is remarkably easy. You can find detailed, easy-to-read information about its various applications on the manufacturers' datasheets. We discuss three popular ways to configure a timing circuit using a 555 here.

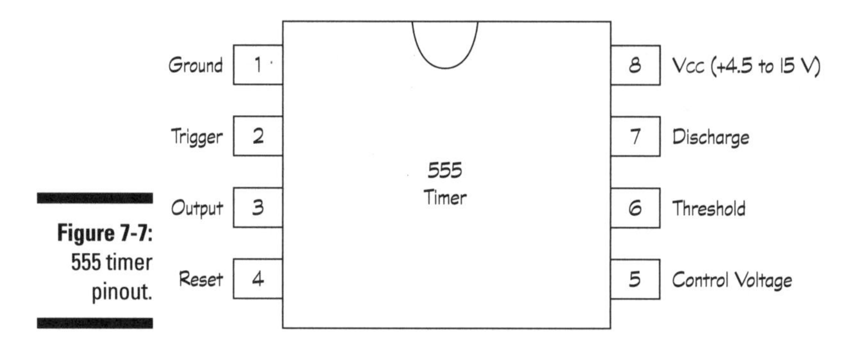

Figure 7-7: 555 timer pinout.

Astable multivibrator (oscillator)

The 555 can behave as an *astable multivibrator*, which is just a fancy term to describe a sort of electronic metronome. By connecting components to the chip as shown in Figure 7-8, you configure the 555 to produce a continuous series of voltage pulses that automatically alternate between low (0 volts) and high (the positive supply voltage, V_s), as shown in Figure 7-9. (The term *astable* refers to the fact that this circuit doesn't settle down into a stable state, but keeps changing on its own.) This self-triggering circuit is also known as an *oscillator*.

You can use the 555 astable multivibrator for lots of fun things:

- *Flashing lights:* A low-frequency (< 10 Hz) pulse train can control the on/off operation of an LED or lamp (see the Building Brilliant, Blinking, Bright Lights project in Chapter 14).

- *Electronic metronome:* Use a low-frequency (< 20 Hz) pulse train as the input to a speaker or piezoelectric transducer to generate a periodic clicking sound.
- *Sounding an alarm:* By setting the frequency to the audio range (20 Hz – 20 kHz) and feeding the output into a speaker or piezoelectric transducer, you can produce a loud, annoying tone (see the Siren and Light Alarm projects in Chapter 14).
- *Clocking a logic chip:* You can adjust the pulse widths to match the specifications for the signal that clocks the logic inside a chip (see the Generating Cool Lighting Effects project in Chapter 14).

The *frequency* (F, in Hertz), which is the number of complete up-and-down cycles per second, of the square wave produced is determined by your choice of three external components, according to this equation:

$$F = \frac{1.4}{\left(R_1 + 2R_2\right) \times C_1}$$

If you flip the numerator and dominator in this equation, you get the *time period* (T), which is the length of time (in seconds) of one complete up-and-down pulse:

$$T = 0.7 \times \left(R_1 + 2R_2\right) \times C_1$$

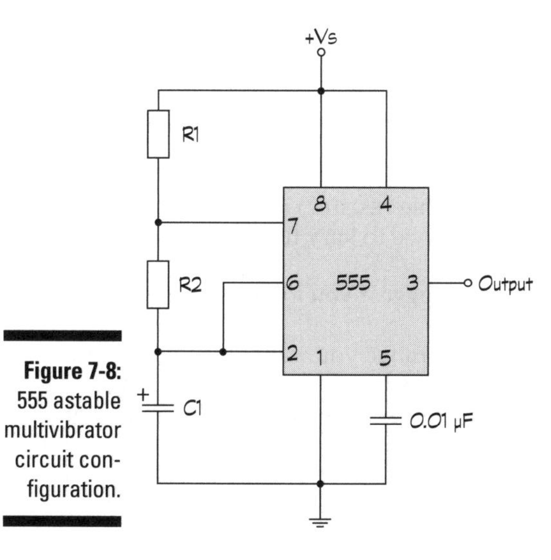

Figure 7-8: 555 astable multivibrator circuit configuration.

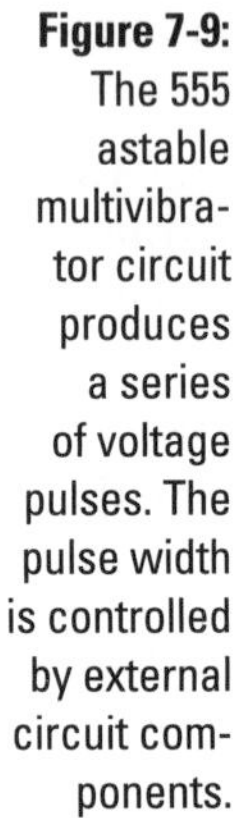

Figure 7-9: The 555 astable multivibrator circuit produces a series of voltage pulses. The pulse width is controlled by external circuit components.

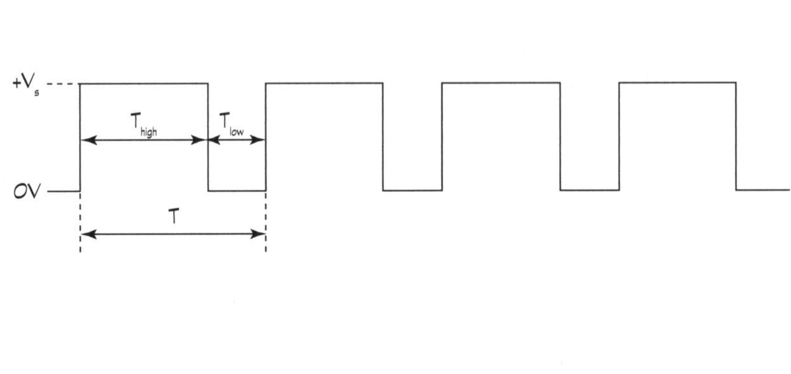

You can set up your circuit so that the width of the high part of the pulse is different from the width of the low part of the pulse. The width of the high part of the pulse, T_{high}, is:

$$T_{high} = 0.7 \times (R_1 + R_2) \times C_1$$

The width of the low part of the pulse, T_{low}, is:

$$T_{low} = 0.7 \times R_2 \times C_1$$

If R_2 is much, much bigger than R_1, the high and low pulse widths are fairly equal. If $R_2 = R_1$, the high portion of the pulse is twice as wide as the low portion. You get the idea.

You can also use a potentiometer (variable resistor) in series with a small resistor as R_1 or R_2 and adjust its resistance to vary the pulses.

To choose values for R_1, R_2 and C_1, we suggest you follow these steps:

1. **Choose C_1.** Decide what frequency range you want to generate and choose an appropriate capacitor. The lower the frequency range, the higher the capacitor you should choose. (Assume that R1 and R2 are somewhere in the 10 kΩ to 1 MΩ range.) For many low frequency applications, capacitor values of between 0.1 μF and 10 μF work well. For higher frequency applications, choose a capacitor in the range 0.01–0.001 μF.

2. **Choose R_2.** Decide how wide the low part of the pulse should be and choose the value of R_2 that produces that width, given the value of C_1 you've already determined.
3. **Choose R_1.** Decide how wide the high part of the pulse should be. Using the values of C_1 and R_2 already selected, calculate the value of R_1 that produces the desired high pulse width.

Monostable multivibrator (one-shot)

By configuring the 555 timer as shown in Figure 7-10, you can use it as a *monostable multivibrator* that generates a single pulse when triggered (sometimes called a *one-shot*). Without a trigger, this circuit produces a low (zero) voltage, which is its stable state. When triggered by closing the switch between Pin 2 and ground, an output pulse is generated at the supply voltage, V_s. The width of the pulse, T, is determined by the values of R_1 and C_1, as follows:

$$T = 1.1 \times R_1 \times C_1$$

Capacitor values can often vary by as much as 20 per cent, so you may need to choose a resistor with a somewhat different value than the formula suggests in order to produce the pulse width you desire.

You can use a one-shot to safely trigger a digital logic device, such as that in the later section 'Counting on the 4017 decade counter'. Mechanical switches tend to bounce when closed, producing multiple voltage spikes that a digital IC can misinterpret as multiple trigger signals. Instead, if you trigger a one-shot with a mechanical switch and use the output of the one-shot to trigger the digital IC, you can effectively debounce the switch.

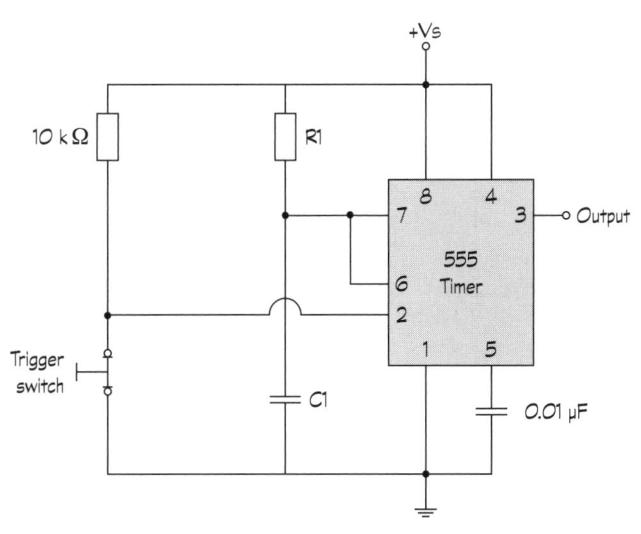

Figure 7-10: When triggered by closing the switch at Pin 2, the 555 monostable circuit produces a single pulse whose width is determined by the values of R_1 and C_1.

Bistable multivibrator (flip-flop)

If an astable circuit has no stable state, and a monostable circuit has one stable state, what's a bistable circuit? Yup, it's a circuit with two stable states. The 555 *bistable multivibrator* shown in Figure 7-11 produces alternating high and low voltages, switching from one state to the other only when triggered. Such a circuit is commonly known as a *flip-flop*. You don't need to calculate resistor values, because the timing of the pulses generated is controlled by activating the trigger switch.

Because it stays low or high until triggered, a flip-flop can be used to store a data bit. (Remember, a bit is a 0 or a 1, which is a low or a high voltage.) The registers used to store temporary outputs between stages of logic consist of multiple flip-flops. Flip-flops are also used in certain digital counter circuits, holding bits in a series of interconnected registers that form an array, the outputs of which make up a bit string representing the count.

You can use various types of 555 timer circuits to trigger other 555 timer circuits. For instance, you can use an oscillator to trigger a flip-flop (useful for clocking registers).

Try using a one-shot to produce a temporary low-volume tone, and then when it ends, change the state of a flip-flop, whose output triggers an oscillator that pulses the speaker on and off. This circuit can be used in a home alarm system: upon entering the home, the homeowner (or intruder) has 10 seconds or so to deactivate the system (when hearing a low-volume warning tone) – or the siren wakes up the neighbours.

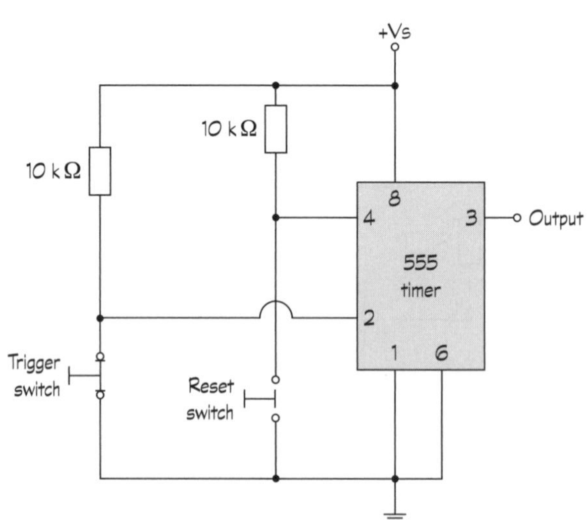

Figure 7-11: The 555 bistable circuit (or flip-flop) produces a high output when triggered by the switch at Pin 2 and a low output when reset by the switch at Pin 4.

Meet the logic families

Manufacturers use many different techniques to build digital integrated circuits. A single gate can be constructed using a resistor and a transistor, or just bipolar transistors, or just MOSFETs (another kind of transistor), or other combinations of components. Certain design approaches make cramming lots of tiny gates together in a chip easier, and other design approaches result in faster circuits or lower power consumption.

Digital ICs are classified according to the design approach, and therefore the processing technology, used to build the tiny circuits. These classifications are called *logic families*. Dozens of logic families exist, but the two most famous families are TTL and CMOS.

TTL, or *transistor-transistor logic*, uses bipolar transistors to construct both gates and amplifiers. Manufacturing TTL ICs is relatively inexpensive, but they generally draw a lot of power and require a specific (5-volt) power supply. The TTL family has several branches, but the most notable is the *Low Power Schottky* series, which draws roughly one-fifth the power of conventional TTL technology. Most TTL ICs use the 74xx and 74xxx part number format, where xx or xxx specifies the particular type of logic device. For instance, the 7400 is a quad 2-input NAND gate. The Low Power Schottky version of this part is coded 74LS00.

CMOS, which stands for *complementary metal oxide semiconductor*, is one type of technology used to make MOSFETs (metal oxide semiconductor field effect transistors). (You can see why this family shortened its name to CMOS!) CMOS chips are a little more expensive than their TTL equivalents, but they draw a lot less power and operate over a wider range of supply voltages (3 to 15 volts). They're very sensitive to static electricity, and so require special handling. Some CMOS chips are pin-for-pin equivalents for TTL chips and are identified by a 'C' in the middle of the part number. For instance, the 74C00 is a CMOS quad 2-input NAND gate with the same pinout as its cousin, the TTL 7400 IC. Chips in the 40xx series, for instance the 4017 decade counter and 4511 7-segment display driver, are also members of the CMOS family.

Counting on the 4017 decade counter

The 4017 CMOS decade counter shown in Figure 7-12 is a 16-pin IC that counts from 0 to 9 when triggered. Pins 1–7 and 9–11 go from low to high one at a time when a trigger signal is applied to Pin 14. (They don't go high in order; you have to check the pinout to determine the order.) You can use the count outputs to light up LEDs (as in the 'Building Brilliant, Blinking, Bright Lights' section and project in Chapter 14) or trigger a one-shot that controls another circuit.

Counting can take place only when the *disable* pin (Pin 13) is low; you can disable counting by applying a high signal to Pin 13. You can also force the counter to reset to zero (meaning that the 'zero' count output, which is Pin 3, goes high) by applying a high signal (+V) to Pin 14.

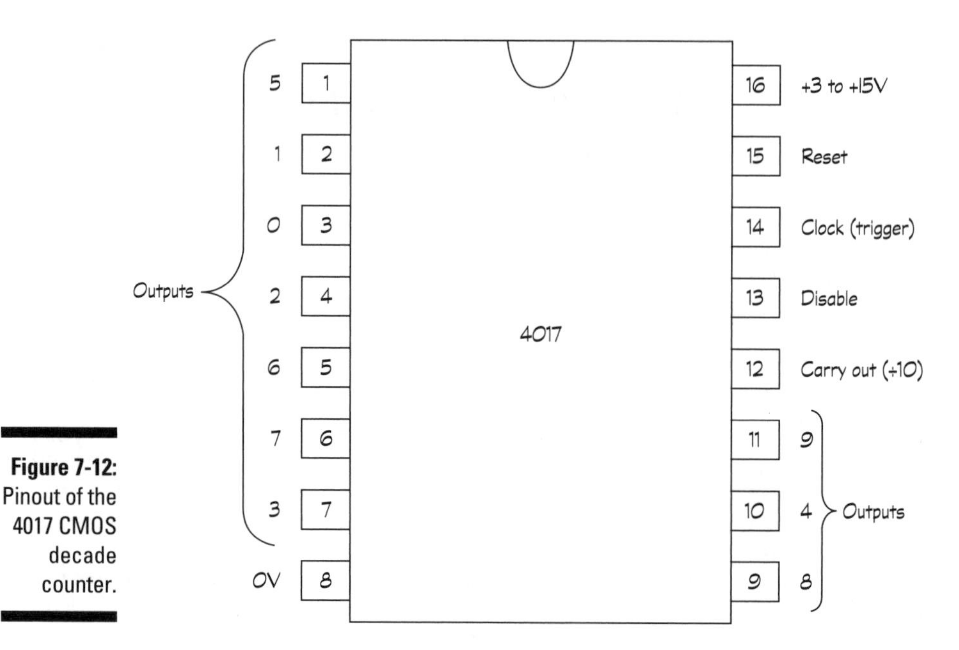

Figure 7-12: Pinout of the 4017 CMOS decade counter.

By piggybacking multiple 4017 ICs together, you can count up tens, hundreds, thousands and so forth. Pin 12 is high when the count is 0–4 and low when the count is 5–9, so it looks like a trigger signal that changes at one-tenth the rate of the count. If you feed the output of Pin 12 into the trigger input (Pin 14) of another decade counter, that second counter counts up tens. By feeding the second counter's Pin 12 output into Pin 14 of a third counter, you can count up hundreds. With enough 4017 ICs, you may even be able to tally the national debt!

You can also connect two or more of the counter's outputs using diodes to produce a variable timing sequence. To do this, connect each anode (positive side of a diode) to an output pin and connect all the cathodes (negative sides of the diodes) together and then through a resistor. With this arrangement, when any one of the outputs is high, current flows through the resistor. For instance, you can simulate the sequence of a traffic light by tying outputs 0–4 together and feeding the result (through a resistor) into a red LED, connecting output 5 to a yellow LED and tying outputs 6–9 together to control a green LED.

Expanding Your IC Horizons

In addition to the uses described in the earlier sections, ICs can also provide the following common functions:

- Mathematical operations (addition, subtraction, multiplication and division).
- Multiplexing (selecting a single output from among several inputs).
- Analogue-to-digital (A/D) converters (to convert a real-world analogue signal into a digital signal so that you can process it with a computer or other digital electronics system).
- Digital-to-analogue (D/A) converters (to convert a processed digital signal back into an analogue signal).

Of course, the microprocessor that runs your personal computer (and maybe even your personal life) is also quite popular as ICs go!

Among the most versatile integrated circuits you can find is the small, complete computer on a chip, known as a *microcontroller.* To program a microcontroller, you place it on a development board that allows the IC to interface with your personal computer.

When it's programmed, you mount the microcontroller into a socket on your electronic device. You add a few other components in circuits that interface the microcontroller to your computer, motors or switches, and *voilà*! Your little programmed IC makes things happen (for instance, it can control the motion of a robot).

The great thing about a microcontroller is that you can simply alter a few lines or code or reprogram it completely to change what it does; you don't need to swap out wires, resistors and other components in order to get this flexible IC to take on a new personality.

Integrated circuits have much more to them than we can possibly cover in this book. Really smart circuit designers are always coming up with new ideas and improvements on some of the old ideas, so you're spoilt for choice in the world of integrated circuits.

If you're interested in finding out more, refer to the Appendix for some interesting websites that provide useful how-to information for using various ICs in working projects.

Chapter 8

Rounding Out Your Parts List

In This Chapter

- Picking the perfect type of wire
- Powering up with batteries and solar cells
- Controlling connections with switches
- Triggering circuits with sensors
- Turning electricity into light, sound and motion

Many individual components and integrated circuits (such as those we discuss in Chapters 3–7) are the foundations of shaping the flow of electrons in electronic circuits. But these indispensable parts rely on a bunch of other contributing parts to help get the job done.

Some of these contributing parts, such as wires, connectors and batteries, are themselves essential ingredients in any electronic circuit. After all, you'd be hard pressed to build an electronic circuit without wires to connect things together or a source of power to make things run. You may use the other parts we discuss in this chapter only occasionally and for certain circuits – for example, when you need to make some noise, a buzzer certainly comes in handy.

In this chapter, we discuss a mixed bag of components, some of which you need to keep in stock (just like toilet paper and toothpaste – we hope) and others that you can pick up whenever you feel like it.

Making Connections

Making a circuit requires that you connect components so that electric current can flow between them. The following sections describe wires, cables and connectors that allow you to do just that.

Choosing wires wisely

Wire that you use in electronics projects is just a long strand of metal, usually made of copper. The wire has only one job: to allow electrons to travel through it. However, various types of wire are available, and in this section we help you choose the right one for the situation.

Selecting stranded or solid

Unplug and cut open the cord of an old lamp you're about to dump and you can see two or three small bundles of very fine wires, each bundle wrapped in insulation; this sort of wire is called *stranded* wire. Another type of wire, known as *solid* wire, consists of a single wire on its own. You can see examples of each in Figure 8-1.

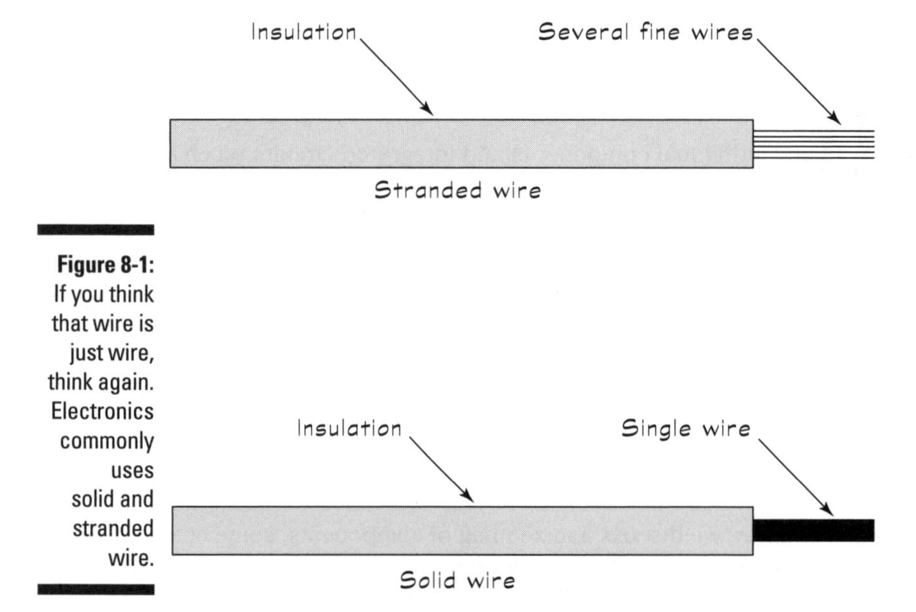

Figure 8-1: If you think that wire is just wire, think again. Electronics commonly uses solid and stranded wire.

Stranded wire is much more flexible than solid wire, and you use it in situations in which the wire is going to be moved or bent a lot (such as in cords for lamps or the cables you hook up to your home entertainment system).

You use solid wire in places where you don't plan to move the wire around, and to connect components on breadboards (check out Chapter 11 for more on breadboards). Inserting solid wire into holes in the breadboard is easy, but if you try to use a stranded wire, you have to twist the strands to get them all in the hole, and you may break a strand or two in the process (trust us, it happens!), which can short out the circuit.

Sizing up your wire gauge

You refer to the diameter of wire as the *wire gauge*. This gauge is sometimes given in millimetres for diameter or cross-sectional surface area. But you can also see it expressed as a two-digit number, such as in the American Wire Gauge (AWG) system, in which the smaller the number, the thicker the wire. Table 8-1 shows the common wire gauges.

Table 8-1 Wires Commonly Used in Electronics Projects

AWG	*Wire Diameter (mm)*	*Uses*
16	1.3	Heavy-duty electronics applications
18	1.0	Heavy-duty electronics applications
20	0.8	Most electronics projects
22	0.7	Most electronics projects
30	0.3	Wire-wrap connections on circuit boards

For most electronics projects, including the ones in Chapters 13 and 14 in this book, you use 20- or 22-gauge wire. If you're hooking up a motor to a power supply, you need to use 16- or 18-gauge wire. As you progress in your electronics dabbling, you may find yourself connecting components on special circuit boards using a technique known as *wire wrapping*, which calls for smaller 30-gauge wire.

The AWG system is most popular in America, but you see it in the UK and elsewhere too, abbreviated in many different ways or just given as a two-digit figure on its own. You may also come across a little-used British Standard Wire Gauge (SWG) and a metric gauge. So you may see the same wire described as 22 AWG, 23 SWG or the metric 7/0.25. Confused? Things are clearer when you see what's on the shelf, we promise!

If you start working on projects involving higher voltage or current than the ones we describe in this book, consult the instructions for your project or an authoritative reference to determine the appropriate wire gauge. For example, the *IEE Wiring Regulations* lists the recommended wire gauges for each type of wiring that you use in a house – to meddle with that you need planning permission, a suitable approved electrician or both. Mains wiring is outside the scope of this book but, needless to say, messing with it isn't a good idea.

Viewing the colourful world of wires

Like the colourful bands that unlock the secrets of resistor values (check out Chapter 3 for all about resistors), the colourful insulation around wire can help you keep track of connections in a circuit.

A common practice when wiring up a DC circuit (for example, when you work with a breadboard) is to use red wire for all connections to positive voltage (+V) and black wire for all connections to ground. For AC circuits, you may want to use green wire instead of black wire for ground connections. Yellow or orange wire is often used for input signals, such as the signal from a microphone into a circuit.

If you keep lots of different colours of wire handy, you can colour-code your component connections so that you can more easily tell what's going on in a circuit just by glancing at it.

Collecting wires into cables or cords

Cables are actually groups of two or more wires, each protected by an outer layer of insulation and usually with another layer of insulation around them all. Line cords that bring AC power from a wall socket to an electrical device such as a lamp are cables, as are the mishmash of connections in your home entertainment system.

Plugging in to connectors

If you look at a cable, say the one that goes from your computer to your printer, you see that it has metal or plastic bits on each end. These bits are called *plugs*, and they're one kind of *connector*. Your computer and printer also have metal or plastic receptacles into which these cable ends fit. These receptacles are another kind of connector, called a socket or *jack*. The various pins and holes in connectors connect the appropriate wire in the cable to the corresponding wire in the device.

Many different types of connectors are used for various purposes. Here are some of the connectors you're likely to hook up with in your electronics adventures:

- A *terminal* and *terminal block* work together as the simplest type of connector. A terminal block contains sets of screws in pairs. You attach the block to the case or chassis of your project. Then, for each wire you want to connect, you solder (or crimp) a wire to a terminal. Next, you connect each terminal to a screw on the block. When you want to connect two wires to each other, simply select a pair of screws and connect the terminal on each wire to one of those screws.

- *Plugs* and *jacks* that carry audio signals between pieces of equipment, such as a guitar and an amplifier, have cables like the one you see in Figure 8-2. Plugs on each end of the cable connect to jacks on the equipment being connected. These cables contain one or two signal wires (they carry the actual audio signal) and a metal *shield* surrounding the wires. The metal shield protects the signal wires from electrical interference, known as *noise*, by minimising the introduction of current into the wires.
- *Pin headers* bring signals to and from circuit boards, which are thin boards designed to house a permanent circuit. Pin headers consist of two parts: the socket, which you mount on the circuit board, and the plug, which you attach to a ribbon cable. A *ribbon cable* consists of a series of insulated wires stuck together side-by-side to form a flat, flexible cable. The rectangular shape of the connector allows for easy routing of signals from each wire in the cable to the correct part of the circuit board via the socket.

  You refer to these connectors by the number of pins; for example, you may talk about a 40-pin header. Pin headers come in handy for complex electronics projects that involve multiple circuit boards.

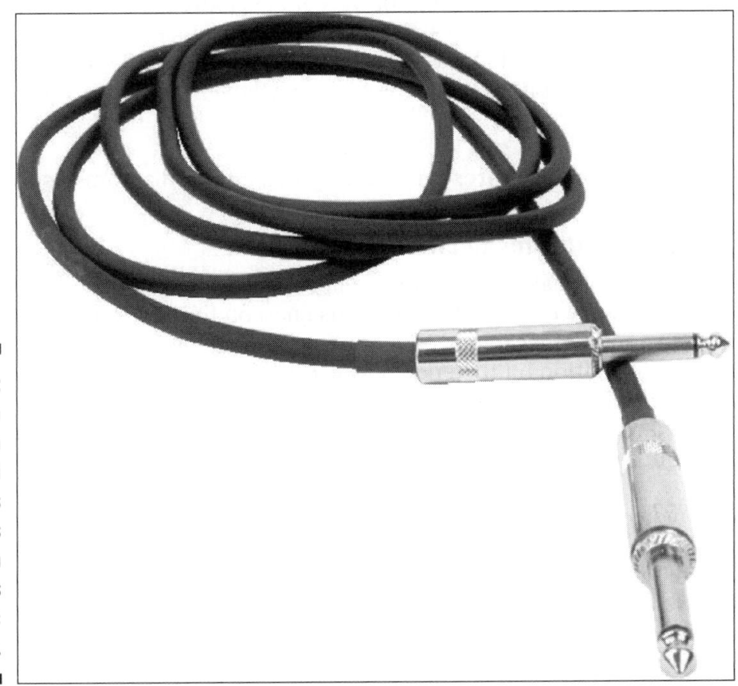

Figure 8-2: A cable with a plug on each end connects to jacks mounted on two pieces of electronic equipment.

Electronics uses many connectors that you don't need to know about until you get into more complex projects. If you want to find out more about the broad array of connectors, take a look at some of the catalogues or websites of the electronics suppliers listed in Chapter 16.

Powering Up

All the wires and connectors in the world are useless when you don't have a power source. In Chapter 2, we discuss sources of electricity, including AC power from wall sockets and DC power from batteries and solar cells (also known as photovoltaic cells). Here we discuss how to choose a power source and how to feed the power into your circuits.

Turning on the juice with batteries

For most hobby electronics projects, batteries are the way to go. They're relatively lightweight and portable, and by combining multiple batteries in series, you can create a variety of DC voltage sources. Everyday batteries, such as AAA-, AA-, C- and D-size batteries, all produce about 1.5 volts. A *transistor battery* is shaped like a 3-D rectangle; it produces 9 volts and is commonly called a 9-volt battery. The big, boxy *lantern battery* produces about 6 volts.

The symbol used to represent a battery in a circuit diagram is shown below:

Connecting batteries to circuits

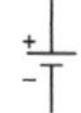

You use a 9-volt battery clip, shown in Figure 8-3, to connect an individual 9-volt battery to a circuit. Battery clips snap on to the terminals of the battery and contain black and red leads that you connect to your circuit. You strip the insulation off the ends of the wires and then connect the leads to your circuit. You can connect them to terminals, insert them into holes in a breadboard or solder the leads directly to components. We discuss all these techniques in Chapter 11.

When you connect the positive terminal of one battery to the negative terminal of another battery, the total voltage across this series connection is the sum of the individual battery voltages. Battery holders, such as the one shown in Figure 8-4, make series connections between batteries for you while holding multiple batteries in place. Red and black leads from the battery holder provide access to the total voltage.

Figure 8-3: A battery clip allows you to connect a 9-volt battery easily to your circuit.

Figure 8-4: Four 1.5-volt batteries tucked into a battery holder produce about 6 volts across the red and black leads.

Rating the life of the everyday battery

The amp-hour or milliamp-hour rating for a battery gives you an idea of how much current a battery can conduct for a given length of time. For example, a 9-volt transistor battery usually has about a 500 milliamp-hour rating. Such a battery can power a circuit using 25 milliamps for approximately 20 hours before its voltage begins to drop. (We checked a 9-volt battery that we'd used for a few days and found that it was producing only 7 volts.) An AA battery may have a 1500 milliamp-hour rating, so a battery pack containing AA batteries can power a circuit drawing 25 milliamps of current for approximately 60 hours.

Six AA batteries in series, which produce about 9 volts, last longer than a single 9-volt battery. That's because the six series batteries contain more chemicals than the single battery and can produce more current over time before becoming depleted. (In Chapter 2, we discuss how batteries are made and why they eventually run out of juice.) If you have a project that uses a lot of current or you plan to run your circuit all the time, consider using larger C- or D-size batteries, which last longer than smaller batteries or rechargeable batteries.

See the section 'Sorting batteries by what's inside' for more about different types of batteries and how long you can expect them to last.

Sorting batteries by what's inside

Batteries are classified by the chemicals they contain, and the type of chemical determines if a battery is rechargeable or not. The following types of batteries are readily available:

- **Non-rechargeable batteries**:
  - **Zinc-carbon** batteries are at the low end of the battery food chain. They may not cost very much, but they also don't last very long.
  - **Alkaline** batteries also come in a variety of sizes and last about three times as long as zinc-carbon batteries. We suggest starting with this type of battery for your projects. If you find yourself replacing them often, you can step up to rechargeable batteries.
  - **Lithium** batteries are lightweight and generate higher voltages – about 3 volts – than other types; they also have a higher current capacity than alkaline batteries. They cost more, and you can't recharge most lithium batteries, but when your project calls for a lightweight battery (for instance, if you're moving a small robot around the house), you can't beat them.

- **Rechargeable batteries:**
  - **Nickel-cadmium (Ni-Cad or Nicad)** batteries generate about 1.2 volts and are the most popular type of rechargeable batteries. Some Ni-Cad batteries still exhibit a flaw known as the memory effect, requiring you to discharge the battery fully before recharging it, to ensure that it recharges to its full capacity.
  - **Nickel-metal hydride (Ni-MH)** batteries also generate about 1.2 volts, but don't suffer from the memory effect. We suggest you use Ni-MH batteries for your projects that need rechargeable batteries.

Be careful not to mix battery types in the same circuit, and *never* attempt to recharge non-rechargeable batteries. These batteries can rupture and leak acid, or even explode.

Buying a recharger and a supply of rechargeable batteries can save you a considerable amount of money over time. Just make sure that the battery charger you use is designed for the type of rechargeable battery you select.

Getting power from the sun

If you're building a circuit designed to operate outside – or you just want to use a clean, green source of energy – you may want to purchase one or more solar panels. A solar panel consists of an array of solar cells, which are large diodes known as *photodiodes* that generate current when exposed to a light source, such as the sun. (We discuss diodes in Chapter 6, and photodiodes in the section 'Using Your Sensors', later in this chapter.)

A solar panel measuring about 12.5 x 12.5 centimetres (5 x 5 inches) may be able to generate 100 milliamps at 5 volts in bright sunlight. If you need 10 amps, you can certainly get it, but you may find the size of the panel problematic on a small or portable project.

Some solar panels contain output leads that you can connect into your circuit, much like the leads from a battery clip or battery holder. Other solar panels have no leads, so you have to solder your own leads to the two terminals.

Here are some criteria to consider to help you determine whether or not a solar panel is right for your project:

- Do you plan to have the solar panel in sunlight when you want your circuit to be on? If not, look for another power source.
- Does the solar panel fit on the gadget you're building? To answer this question, you need to know how much power your gadget needs and the size of the solar panel that can deliver enough power. If the panel is too large for your gadget, you need to redesign the gadget to use less power or look for another power source.

Working off your wall power (not recommended!)

The AC power supplied by your utility company can cause injury or death if used improperly, so we don't recommend you run circuits directly off mains electricity. And because the vast majority of hobby electronics projects run on batteries, you may never be tempted to work with AC. However, some projects need more current or higher voltages than batteries can easily provide. In those cases, you can use a wall transformer, such as the one shown in Figure 8-5, to convert AC to DC. All the working parts are self-contained in the wall transformer, so you aren't exposed to high AC currents.

Wall transformers supply currents ranging from hundreds of milliamps to a few amps at voltages ranging from 5 to 20 volts DC. Some provide both a positive DC voltage and a negative DC voltage. Different models use different types of connectors to deliver power. If you purchase a wall transformer, be sure to read the specification sheet (specs) carefully to determine how to connect it to your circuit.

Figure 8-5: A plug-in wall transformer shields you from AC house current.

Acquiring wall warts

Wall transformers are sometimes called *wall warts* because they stick out of the wall like an ugly wart. You can purchase wall warts new or surplus. Check out Chapter 16 for some good leads on suppliers. And, of course, you may already have some old wall warts saved from a discarded cordless phone or other electronic device. If so, be sure to check the voltage and current rating, usually printed on the transformer, to see whether it's suitable for your next project.

Switching Electricity On and Off

If you think that a switch is just a simple on/off mechanism, think again. You can use lots of different kinds of switches in your electronics projects, and they're categorised by how they're controlled, the type and number of connections they make and, as with all components, how much current they can handle.

A *switch* is a device that makes or breaks one or more electrical connections. When a switch is in the *open position*, the electrical connection is broken and you have an open circuit with no current flowing. When a switch is in the *closed position*, an electrical connection is made and current flows.

Controlling the action of a switch

Switches are referred to by names that indicate how the switching action is controlled. We list some of the many different types of switches below and show several in Figure 8-6:

- **Slide switch:** You slide a knob back and forth to open and close this switch, which you can find on many torches.
- **Toggle switch:** You flip a lever one way to close the switch and the other way to open the switch.
- **Rocker switch:** You press one side of the switch down to open the switch, and the other side of the switch down to close the switch. You find rocker switches on many power strips.
- **Pushbutton switch:** You push a button to change the state of the switch, but how it changes depends on the type of pushbutton switch you have:
  - **Push on/push off buttons:** Each press of the button reverses the position of the switch.

   - **Push-to-make (for example, a leaf switch):** This momentary switch is normally open (off), but if you hold the button down, the switch is closed (on). When you release the button, the switch becomes open again.
   - **Push-to-break:** This momentary switch is normally closed (on), but if you hold the button down, the switch is open (off). When you release the button, the switch becomes closed again.

- **Relay:** A relay is an electrically controlled switch. If you apply a certain voltage to a relay, an electromagnet within pulls the switch lever (known as the *armature*) closed. You may hear talk of closing or opening the *contacts* of a relay. That's just the term used to describe a relay's switch.

Figure 8-6: From top to bottom: two toggle switches, a rocker switch and a leaf switch.

Making the right contacts

Switches are categorised by how many connections they make when you flip the switch and exactly how those connections are made. A switch can have one or more *poles*, or sets of input contacts, which can be connected to an output: a *single-pole switch* can connect one input wire to an output, whereas a *double-pole switch* can connect two input wires to an output.

A switch can have one or more conducting positions, or *throws*. A *single-throw switch* connects or disconnects one input to one output, whereas a *double-throw switch* has twice as many output contacts as input contacts and connects or disconnects each input to either of two outputs.

Here's the low-down on some common switch varieties:

- **Single-pole single-throw (SPST):** This type is your basic on/off switch, with one input and one output. The symbol for an SPST switch is shown below:

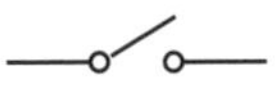

- **Single-pole double-throw (SPDT):** This on/on switch contains one input contact and two output contacts. An SPDT switch is always on; it just switches the input between two options of output. You use an SPDT switch, or *changeover switch*, when you want to have a circuit turn one device or another on (for example, a green light to let people know that they can enter a room, or a red light to tell them to stay out). The symbol for an SPDT switch is shown below:

- **Double-pole single-throw (DPST):** This dual on/off switch contains two input contacts and two output contacts and behaves like two separate SPST switches operating in sync. In the off position, both switches are open and no connections are made. In the on position, both switches are closed and connections are made between each input contact and its corresponding output contact. The symbol for a DPST switch is shown below:

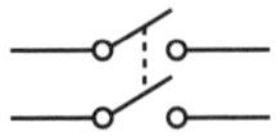

- **Double-pole double-throw (DPDT):** This dual on/on switch contains two input contacts and four output contacts, and behaves like two SPDT switches operating in sync. In one position, the two inputs are connected to one set of outputs, and in the other position, the two inputs are connected to the other set of outputs. Some DPDT switches have a third position, which disconnects all contacts.

You can use a DPDT switch as a *reversing switch* for a motor, connecting the motor to positive voltage to turn one way or negative voltage to turn the other way, and, if a third switch position is available, zero voltage to stop turning. The symbol for a DPDT switch is shown below:

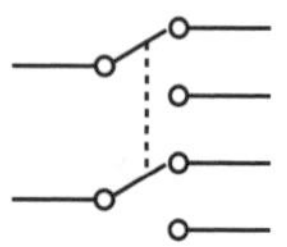

Using Your Sensors

When you want to trigger the operation of a circuit based on something physical happening, such as a change in temperature, you use electronic components known as *sensors*. Sensors take advantage of the fact that various forms of energy, including light, heat and motion, can be converted into electrical energy.

Sensors are a type of *transducer*, which is an electronic device that converts energy from one form to another. In this section, we describe some of the more common input transducers, or sensors, used in electronic circuits.

Seeing the light

Many electronic components behave differently depending on the light to which they're exposed. Manufacturers make certain versions of components to exploit this light dependency, enclosing them in clear cases so that you can use them as sensors in equipment such as burglar alarms, smoke detectors, automatic dusk-to-dawn lighting and safety devices that stop your electric garage door from closing when a cat runs under it.

You can also use these light-sensitive devices for communications between your TV remote control, which sends coded instructions via infrared light using a light-emitting diode (or LED, which we discuss in Chapter 6), and your TV, which contains a light-sensitive diode or transistor to receive the coded instructions.

Examples of light-sensitive devices used as sensors include the following:

- **Photoresistors (or photocells)** are light-dependent resistors (LDRs) made from semiconductor material. They typically exhibit a high resistance (about 1 MΩ) in darkness and a fairly low resistance (about 100 Ω) in bright light, but you can use a multimeter (as we describe in Chapter 12) to determine the actual resistances that a specific photoresistor exhibits. The typical photoresistor is most sensitive to visible light, especially in the green-yellow spectrum. The symbol for a photoresistor, which can be installed either way in your circuits, is shown below:

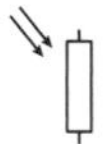

- **Photodiodes** are sort of the opposite of the LEDs that we discuss in Chapter 6. They conduct current or drop voltage only when exposed to sufficient light, usually in the invisible infrared range. Like standard diodes, photodiodes contain two leads: the shorter lead is the cathode (negative end) and the longer lead is the anode (positive end). The circuit symbol for a photo diode is shown below:

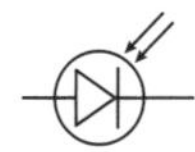

- **Phototransistors** are simply bipolar junction transistors (as we discuss in Chapter 6) encased in a clear package so that light biases the base-emitter junction. These devices have only two leads (whereas standard transistors contain three leads), because you don't need access to the base of the transistor in order to bias it – light does the job for you. Phototransistors amplify differences in the light that strikes them, but from the outside, they look just like photodiodes, so you really have to keep track of which is which. The circuit symbol for a phototransistor is shown below:

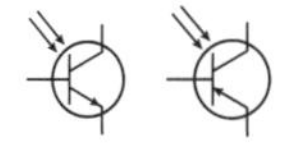

Take a look at Chapter 14 for some projects that involve light-sensitive components.

Capturing sound with microphones

Microphones are input transducers that convert acoustic energy, otherwise known as sound, into electrical energy. Most use a thin membrane, or *diaphragm*, that vibrates in response to air pressure changes from sound.

The vibrations of the membrane are translated into an AC electrical signal via any number of ways, depending on the type of microphone:

- In a *condenser microphone*, the vibrating membrane plays the role of a movable capacitor plate in a variable capacitor, so that variations in sound produce corresponding variations in capacitance. The variable capacitor is one component in the electronic circuit that carries the sound signal within the audio system. We describe variable capacitors in Chapter 4.
- In a *dynamic microphone*, the diaphragm is attached to a movable induction coil located inside a permanent magnet. As sound moves the diaphragm, the coil moves inside the magnet field, changing the current in the coil. We describe this phenomenon, known as *electromagnetic induction,* in Chapter 5.
- In a *crystal microphone*, a special *piezoelectric crystal* is used to convert sound into electrical energy, taking advantage of the *piezoelectric effect* in which certain substances produce a voltage when pressure is applied to them.
- In a *fibre-optic microphone*, a laser source directs a light beam towards the surface of a tiny reflective diaphragm. As the diaphragm moves, a detector picks up changes in light reflected off the diaphragm, which transforms the differences in light into an electrical signal.

The circuit symbol for a microphone is shown below:

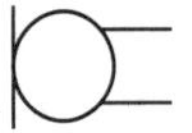

Feeling the heat

A *thermistor* is a resistor whose resistance value changes with changes in temperature. Thermistors have two leads and no polarity, so you don't need to worry about which way you insert a thermistor into your circuit. The circuit symbol for a thermistor is shown below:

Using light sensors to detect motion

If you've ever walked up to a dark doorway and seen the outdoor light going on automatically, you've seen a motion detector at work. These devices commonly use light sensors to detect the *presence* of infrared light emitted from a warm object (such as a person or animal) or the *absence* of infrared light when an object interrupts a beam emitted by another part of the device.

Many homes, schools and shops use *passive infrared detectors* to turn on lights or detect intruders. These motion detectors contain a pyroelectric infrared (PIR) sensor with at least one crystal in it, a lens and a small electronic circuit. When infrared light hits the crystal, it generates an electric charge. Because warm bodies, such as (most) humans, emit infrared light at different wavelengths than cooler objects, such as a wall, differences in the output of the PIR sensor can be used to detect the presence of a warm body. The electronic circuit interprets differences in the PIR sensor output to determine whether or not a moving warm object is nearby.

Industrial passive infrared motion detectors use or control mains-voltage circuits. For hobby projects using a battery pack, you need a compact motion detector that works with about 5 volts. A typical compact motion detector has three leads: ground, positive voltage supply and the detector output. If you supply +5 volts to the detector, the voltage on the output lead reads about 0 (zero) volts when no motion is detected or about 5 volts when motion is detected. You can find compact motion detectors through online security system vendors, but be sure to buy a motion detector, rather than just a PIR sensor. The lens included in a motion detector helps the device detect the *motion* of an object rather than just the *presence* of an object.

An *active infrared motion detector* uses its own infrared LEDs to switch a phototransistor (or other detector) on and off. The detector is an *active* infrared device, meaning it generates infrared light, as opposed to the *passive* infrared sensor that doesn't generate, but only receives, light. Such electric eye devices are used in automatic garage-door opener systems to detect moving objects.

Two types of thermistor are available:

- **Negative temperature coefficient (NTC) thermistor:** The resistance of an NTC thermistor (the most common type of thermistor) decreases with a rise in temperature.
- **Positive temperature coefficient (PTC) thermistor:** The resistance of a PTC thermistor increases with a rise in temperature.

Suppliers' catalogues usually list the resistance of thermistors as measured at 25 degrees Celsius (77 degrees Fahrenheit). You can measure the resistance of the thermistor yourself with a multimeter (see Chapter 12 for more about using multimeters) at a few temperatures. These measurements enable you to *calibrate* the thermistor, or get the exact relationship between temperature and resistance.

Other ways to take your temperature

Thermistors aren't the only type of temperature sensors. Here, we briefly summarise the characteristics of several other types for the curious among you:

- **Bimetallic strip:** The thermostat in your house uses a coiled metal strip, which shrinks as the temperature cools, to trip a switch and turn your boiler on.
- **Semiconductor temperature sensor:** The most common type of this sensor, whose output voltage depends on the temperature, contains two transistors (which we discuss in Chapter 6).
- **Thermocouple:** A thermocouple contains two wires made of different metals (for example, a copper wire and a wire made of a nickel/copper alloy) that are welded or soldered together at one point. These sensors generate a voltage that changes with temperature. The metals used determine how the voltage changes with temperature. You use a thermocouple to measure high temperatures, such as several hundred degrees, or even over a thousand degrees, Celsius.
- **Infrared temperature sensor:** This sensor measures the infrared light given off by an object. You use it when your sensor must be located at a distance from the object you plan to measure; for example, when a corrosive gas surrounds the object. Industrial plants and scientific labs typically use thermocouples and infrared temperature sensors.

If you're planning to use the thermistor to trigger an action at a particular temperature, make sure that you measure the resistance of the thermistor at that temperature.

Trying other energising input transducers

Many, many other types of input transducers are used in electronic circuits. Some examples are:

- **Antennas:** An antenna senses electromagnetic waves and transforms the energy into an electrical signal. (It also functions as an *output transducer*, converting electrical signals into electromagnetic waves.)
- **Pressure or position sensors:** These sensors take advantage of the variable resistance properties of certain materials when they undergo a deformation. Piezoelectric crystals are one such set of materials.
- **Magnetic tape heads:** These devices read magnetic field fluctuations on audio and video cassette tapes (as well as the computer floppy disks used by the ancients) and convert them into electrical signals.

Transducers are often categorised by the type of energy conversion they perform, for instance, electroacoustic, electromagnetic, photoelectric and electromechanical transducers. These amazing devices open up tremendous opportunities for electronic circuits to perform countless useful tasks.

Sensing Something's Going On

Sensors, or input transducers, take one form of energy and convert it into electrical energy, which is fed into the input of an electronic circuit. *Output transducers* do the opposite: they take the electronic signal at the output of a circuit and convert it into another form of energy, for instance sound, light or motion (which is mechanical energy).

You may not realise, but you're probably very familiar with many devices that really are output transducers. Light bulbs, LEDs, motors, speakers and cathode ray tubes (CRTs) and other displays all convert electrical energy into another form of energy. You can create, shape and send electrical signals around wires and through components all day long, but when you transform the electrical energy into a form of energy that you can experience personally, you really begin to enjoy the fruits of your labour.

Speaking of speakers

Speakers convert electrical signals into sound energy. Most speakers consist simply of a permanent magnet, an electromagnet (which is a temporary, electrically controlled magnet) and a cone. Figure 8-7 shows how the components of a speaker are arranged.

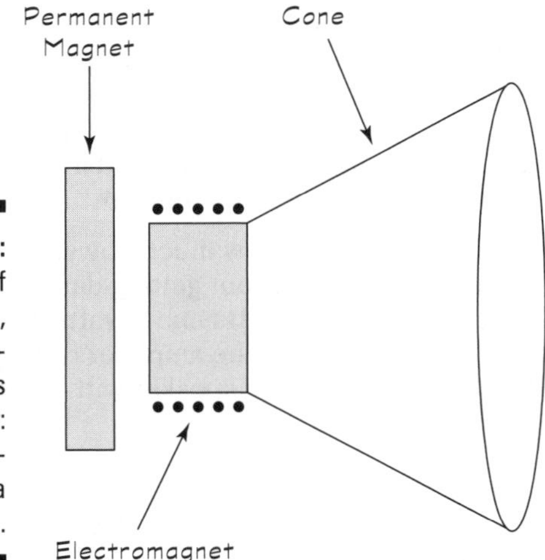

Figure 8-7: The parts of your typical, cheap-as-chips speaker: two magnets and a cone.

The electromagnet, which consists of a coil wrapped around an iron core, is attached to the cone. As electrical current alternates back and forth through the coil, the electromagnet gets pulled towards and then pushed away from the permanent magnet. (We describe the ups and downs of electromagnets in Chapter 5.) The motion of the electromagnet causes the cone to vibrate, which creates sound waves.

Most speakers come with two leads that can be used interchangeably. For more serious projects, such as speakers in stereo systems, you must pay attention to the polarity markings on the speakers because of the way they're used in electronic circuits inside the stereo system.

The circuit symbol for a speaker is shown below:

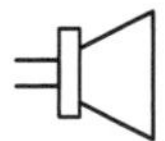

Speakers are rated according to the following criteria:

- **Frequency range:** Speakers can generate sound over different ranges of frequencies, depending on the size and design of the speakers, within the *audible frequency range,* about 20 Hz (hertz) to 20 kHz (kilohertz). For example, one speaker in a stereo system may generate sound in the bass (low audible frequency) range while another generates sound in a higher range. You need to pay close attention to speaker frequency range only if you're building a high-end audio system, when you get to know your woofers from your tweeters.
- **Impedance:** Impedance is a measure of the speaker's resistance to AC current (as we discuss in Chapter 5). You can easily find 4 Ω, 8 Ω, 16 Ω and 32 Ω speakers. You need to select a speaker that matches the minimum impedance rating of the amplifier you're using to drive the speaker. (You can find that rating in the datasheet for the amplifier on your supplier's website.) If the speaker impedance is too high, you don't get as much volume out of the speaker as possible, and if the speaker impedance is too low, you may overheat your amplifier.
- **Power rating:** The power rating tells you how much power (power = current × voltage) the speaker can handle without getting damaged. Typical power ratings are 0.25 watts, 0.5 watts, 1 watts and 2 watts. Be sure that you look up the maximum power output of the amplifier driving your speaker (check the datasheet) and choose a speaker with a power rating of at least that value.

For hobby electronics projects, miniature speakers – roughly 5–7.5 centimetres (2–3 inches) in diameter – with an input impedance of 8 Ω are often just what you need. Be careful, however, not to overpower these little noisemakers, which typically handle only 0.25 or 0.5 watts.

Sounding off with buzzers

Like speakers, buzzers generate sound, but unlike speakers, buzzers indiscriminately produce the same obnoxious sound no matter what voltage you apply (within reason). With speakers, 'Mozart in' creates 'Mozart out'; with buzzers, 'Mozart in' creates nothing but noise.

One type of buzzer, a *piezoelectric buzzer*, contains a diaphragm attached to a piezoelectric crystal. When a voltage is applied to the crystal, the crystal expands or contracts (known as the piezoelectric effect); this effect, in turn, makes the diaphragm vibrate, generating sound waves. (Note that this process is pretty much exactly the opposite of the way a crystal microphone works, as we described earlier section 'Capturing sound with microphones'.)

Buzzers have two leads and come in a variety of packages. Figure 8-8 shows a couple of typical buzzers. To connect the leads the correct way, remember that the red lead connects to a positive DC voltage. The circuit symbol for a piezoelectric buzzer is shown below:

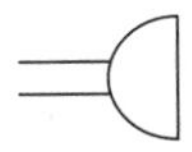

Figure 8-8: These noisy little buzzers are very simple to operate.

When shopping for a buzzer, you need to consider three specifications:

- **The frequency of sound it emits:** Most buzzers give off sound at one frequency somewhere in the range of 2–4 kHz.
- **The operating voltage and voltage range:** Make sure that you get a buzzer that works with the DC voltage that your project supplies.
- **The level of sound it produces in decibels (db):** The higher the decibel rating, the louder (and more obnoxious) the sound emitted. Higher DC voltage provides a higher sound level.

Be careful that the sound doesn't get so loud that it damages your hearing. You can start to get an annoying ringing in your ears at levels of around 85 db and above.

Creating good vibrations with DC motors

Have you ever wondered what causes a mobile phone to vibrate? No, it's not Mexican jumping beans: these devices usually use a *DC motor*. DC motors change electrical energy, such as the energy stored in a battery, into motion. That motion may involve turning the wheels of a robot that you build or vibrating your mobile. In fact, you can use a DC motor in any project where you need motion.

Electromagnets make up an important part of DC motors because these motors consist of, essentially, an electromagnet on an axle rotating between two permanent magnets, as you can see in Figure 8-9.

The positive and negative terminals of the battery connect so that each end of the electromagnet has the same polarity as the permanent magnet next to it. Alike poles of magnets repel each other. This repelling action moves the electromagnet and causes the axle to spin. As the axle spins, the positive and negative connections to the electromagnet swap places, so the magnets continue to push the axle around.

A simple mechanism consisting of a *commutator* (a segmented wheel with each segment connected to a different end of the electromagnet) and brushes that touch the commutator cause the connections to change. The commutator turns with the axle and the brushes are stationary, with one brush connected to the positive battery terminal and the other brush to the negative battery terminal. As the axle, and therefore the commutator, rotates, the segment in contact with each brush changes. This effect in turn changes which end of the electromagnet is connected to negative or positive voltage.

If you want to get a feel for the mechanism inside a DC motor, buy a cheap one for a few pounds and take it apart.

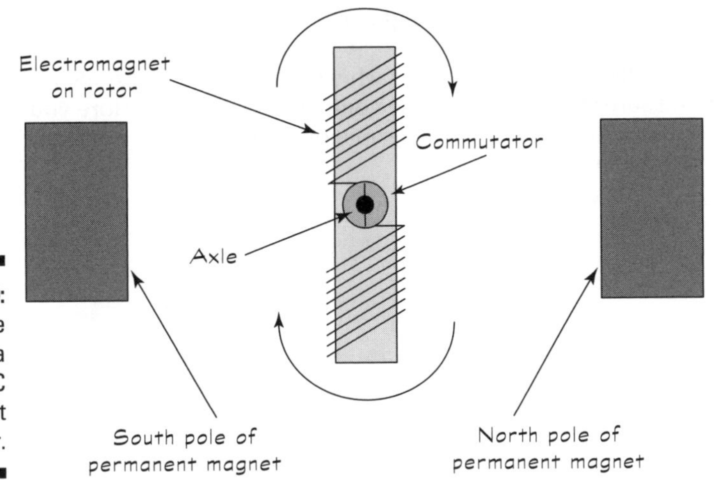

Figure 8-9: How the parts of a simple DC motor fit together.

The axle in a DC motor rotates a few thousand times per minute – a bit fast for most applications. Therefore, suppliers sell DC motors with something called a gear head pre-mounted; this device reduces the speed of the output shaft to under a hundred revolutions per minute (rpm). The result is similar to the way that changing gears in your car changes the speed of the car.

Suppliers' catalogues typically list several specifications for the motors they carry. You need to consider two key things:

- **Speed:** The speed (in rpm) that you need depends on your project. For example, when turning the wheels of a model car, you may aim for 60 rpm, with the motor rotating the wheels once per second.
- **Operating voltage:** The operating voltage is given as a range. Hobby electronics projects typically use a motor that works in the 4.5–12-volt range. Also notice the manufacturer's nominal voltage and stated rpm for the motor. The motor runs at this rpm when you supply the nominal voltage. If you supply less than the nominal voltage, the motor runs slower than the stated rpm.

DC motors have two wires (or terminals to which you solder wires), one each for the positive and negative supply voltage. You run the motor by simply supplying a DC voltage that generates the speed that you want and switching off the voltage when you want the motor to stop.

You can use a more efficient method of controlling the speed of the motor called *pulse width modulation.* This method turns voltage on and off in quick pulses. The longer the 'on' intervals, the faster the motor goes. If you're building a kit for something motor-controlled, such as a robot, the electronics for the kit supplies this speed control.

If you're attaching things such as wheels, fan blades and so on to the motor shaft, be sure that you attach the component securely before you apply power to the motor. If not, the item may spin off and hit you, or someone near and dear to you, in the face.

Part II
Getting Your Hands Dirty

'It's an exciting and rewarding career with the occasional but not <u>too</u> serious accident.'

In this part . . .

Ready to get down to business and build an electronic circuit or two (or more)? This part gives you the low-down on how to set up your own electronics workbench, stock up on tools and supplies, and stay safe. We also run through the essential 'how-tos' you need to know so that you can build – and troubleshoot – your circuits.

You find out where (and where not) to locate your electronics workbench, what shiny new tools you absolutely must have and what electronic components to buy to get you started. We reveal the language of *schematics* – circuit diagrams – and how to transfer a circuit design from paper to real live current-carrying components.

We also show you how to use a solderless breadboard to build prototype (test) circuits, and how to solder components together to create permanent circuits. You also meet the amazing multimeter, which enables you to 'see' inside your circuits and sniff out problems. And we introduce the basics of two useful but optional test tools: the logic probe and the oscilloscope.

Chapter 9

Setting Up Shop and Staying Safe

In This Chapter

- Designing a workspace that works for you
- Getting hold of tools and other supplies
- Starting a stock of electronics components
- Realising that Ohm's Law also applies to the human body
- Keeping yourself safe

Finding out all about amazing resistors, transistors and other electronic components is great, but if all you ever do is sketch out circuit diagrams and dream about how your cool circuit is going to manipulate electron flow, you never make anything buzz, beep or move. You need to start tinkering with real components, add a little power and watch your circuits perform. But before you go running out to your local electronics shop, take time to prepare for this circuit-building stage of your life.

In this chapter, we give you guidelines for setting up a little electronics laboratory in your own home. We outline the tools and supplies you need to get the circuit-building job done, and we give you a shopping list of electronic components to purchase so that you can build a bunch of different projects.

Building electronics circuits isn't for the faint-hearted (because even small currents can affect your heart), and therefore we run you through the safety information that you must know to remain a living and healthy electronics engineer.

A word to the wise right here in the chapter intro: very little electrical current is needed to seriously hurt or even kill you. Even the most seasoned professionals take appropriate precautions to stay safe. We strongly suggest (insist, even) that you thoroughly read the safety facts in this chapter and, before you start each project, review the 'Staying Safe: Safety Checklist' section at the end of this chapter.

Promise?

Picking a Place to Practise Electronics

Where you put your workshop is just as important as the projects you make and the tools you use. As with property, the motto is location, location, location. Stake out just the right spot in your house or flat and you're better organised and able to enjoy your experiments much more. Nothing is worse than working with a messy workbench in dim lighting while breathing stale air.

Creating a great lab

The prime ingredients for the well set-up electronics laboratory are:

- A comfortable, dry climate.
- A comfortable place to work, with a workbench and chair.
- Ample electrical sockets.
- A solid, flat work surface.
- Good lighting.
- Peace and quiet.
- Tools and toolboxes on nearby shelves or racks.

The workbench and work area must be inaccessible to your children. Keep your projects, tools and supplies out of reach or behind lockable doors and integrated circuits and other sharp parts off the floor – they're more painful to step on than Lego bricks!

If your work area is accessible to other family members, find ways to make the area off-limits to those with less knowledge about electronics safety, especially young children. Curious youngsters and electronics simply don't mix! (We cover safety later in this chapter in the section 'Protecting You and Your Electronics'.)

The ideal workspace is one that doesn't get disturbed when you have to leave it for hours or days. The garage is an ideal setting because it gives you the freedom to work with solder and other messy materials without worrying about soiling the carpet or nearby furniture.

You don't need much space: about 1-x-1.5 metres square ought to do it. If you can't clear that much space in your garage (or you don't have a garage), you can use a room in the house, but try to keep a corner of the room for your electronics work. When working in a carpeted room, you can prevent static electricity by spreading a protective cover, such as an anti-static mat, over the floor. We discuss this aspect in detail in the 'Minimising static electricity' section, later in this chapter.

No matter where you set up shop, consider the climate. Extremes in heat, cold or humidity can have a profound effect on your electronics circuits. If you find your work area chilly, warm or damp, take steps to control the climate in that area, or don't use that area for electronics work.

You may need to add insulation and/or use an air conditioner or dehumidifier to control the temperature and humidity. Locate your workbench away from open doors and windows that can allow moisture and extreme temperatures in. And for safety reasons, never – repeat, *never* – work in an area where the floor is wet or even slightly damp.

Choosing a workbench

The kinds of projects you do determine the size of the workbench you need, but for most applications, a table or other flat surface spanning about a metre is fine. You may even have a spare desk or table that you can use for your electronics bench.

As you work on projects, you crouch over your workbench for hours at a time. You can skimp and buy or build an inexpensive worktable, but if you don't already own a good chair, put one on the top of your shopping list. Be sure to adjust the seat for the height of the worktable. A poor-fitting chair can cause backaches and fatigue.

Tooling Up

Every hobby has its special assortment of tools and supplies, and electronics is no exception. From the lowly screwdriver to the high-speed drill, you enjoy messing with electronics much more if you have the right tools and an assortment of supplies organised and stored so that you can put your hands on them when you need them without cluttering your work area.

This section tells you exactly what tools and supplies you need to have in order to complete basic-to-intermediate electronics projects.

If you have a permanent place in your house to work on electronics, you can hang some of the hand tools mentioned in this section on the wall or a pegboard. Reserve this special treatment for the tools that you use the most. You can stash other small tools and supplies in a small toolbox and store it on your workbench. A plastic fishing tackle box with lots of small compartments and one large section can help you keep your things organised.

Shopping for soldering stuff

Soldering is the method you use to make semi-permanent connections between components as you build a circuit. Instead of using glue to hold things together, you use small globs of molten metal called *solder*, which are applied with a *soldering iron*. The metal provides a conductive physical joint, known as a *solder joint*, between the wires and component leads of your circuit.

You'll be glad to know that you need only some pretty simple tools for soldering. You can purchase a basic, no-frills soldering setup for around £10, although better soldering tools cost a bit more.

At a minimum, you need the following basic items for soldering:

- **Soldering iron:** A pencil-like tool with an insulating handle, a heating element and a polished metal tip (see Figure 9-1). Choose a soldering iron that's rated at 25–30 watts, sports a replaceable tip and has a three-pin plug so that it's grounded. Certain models allow you to use different size tips for different types of projects, and some include variable controls that allow you to change the wattage (both features are nice, but not absolutely necessary).
- **Soldering stand:** The stand holds the soldering iron and keeps the (very hot) tip from coming into contact with anything on your workbench. Some soldering irons come with a stand (a set often called a *soldering station*). The stand should have a weighted base or clamp to your workbench, so that the stand doesn't tip over. A stand isn't a fancy accessory but a must-have if you're aiming not to melt your project, singe your desk and/or burn yourself!
- **Solder:** *Solder* is a soft metal that's heated by a soldering iron, and then allowed to cool, forming a conductive joint. Solder used to contain a lot of lead and still does in North America, but solder sold in Europe is now lead-free, which means it's now mostly tin with a tiny bit of copper and sometimes a little bit of silver. Solder is sold in little spools or tubes.

We recommend that you also get your hands on these additional soldering tools and accessories:

- **Extra soldering tips:** For most electronics work, a small (a few millimetres across) pointed or chiselled tip works fine, but you can also find larger or smaller tips used for different types of projects. Be sure to purchase the correct tip for your make and model of soldering iron. Replace your tip when it shows signs of corrosion or pitting, or when the plating is peeling off; a worn tip doesn't pass as much heat.

- **Solder removal tools:** A *solder sucker* or *desoldering pump* is a spring-loaded vacuum that you can use to remove a solder joint or excess solder in your circuit. To use it, melt the solder that you want to remove, quickly position the pump over the molten blob and activate it to suck up the solder. Or you can use a *desoldering wick* or *braid*, which is a flat, woven copper wire that you place over unwanted solder and apply heat to. When the solder reaches its melting point, it adheres to the copper wire, which you then remove and dispose of.
- **Tip cleaner:** This tool gives your soldering tip a good cleaning.
- **Wetted sponge:** You use this item to wipe off excess solder and flux from the hot tip of the soldering iron. Some soldering stands include a small sponge and a built-in space to hold it, but a clean household sponge also works fine.

In Chapter 11, we explain in detail how to use a soldering iron.

Figure 9-1: Some soldering iron models are temperature-adjustable and come with their own stand.

Measuring with a multimeter

Another essential tool is a *multimeter*, which you use to measure AC and DC voltages, resistance and current when you want to explore what's going on in a circuit. Most multimeters you find today are of the digital variety (see Figure 9-2), which just means that they use a numeric display, like a digital clock or watch (you can use them to explore analogue, as well as digital, circuits). Older-style analogue multimeters use a needle to point to a set of graduated scales.

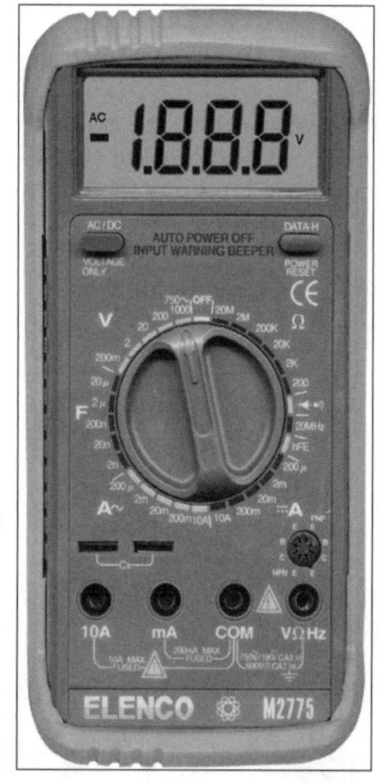

Figure 9-2: Multimeters measure voltages, resistance and current.

All multimeters come with a pair of test leads: one black (for the ground connection) and one red (for the positive connection). On small pocket units, the test leads are permanently attached to the meter, whereas on larger models, you can unplug the leads. All test leads have a cone-shaped metal tip used for probing circuits. You can also purchase test clips that slip over the tips, making testing much easier because you can attach these clips onto wires or component leads.

Prices for new multimeters range from a tenner to hundreds of pounds. Higher-priced meters include additional features, such as built-in testing capabilities for capacitors, diodes and transistors. Think of a multimeter as a window into your circuits and consider purchasing the best model you can afford. That way, as your projects grow more complex, you still get a magnificent view of what's going on inside.

We give you the low-down on how to use a multimeter and other test equipment in Chapter 12.

Getting hands-on with hand tools

Hand tools are the mainstay of any toolbox. These tools tighten screws, snip off wires, bend little pieces of metal and do all those other mundane tasks. Make sure that you have the following tools available at your workbench:

- **Wire cutter:** You can find general-purpose wire cutters at hardware and home improvement stores, but it pays to invest a fiver or so in cutters that can snip wires right up close to the work – called *nippy* or *flush cutters* and shown in Figure 9-3. You'll find them easier to use in tight places, such as above a solder joint.
- **Wire stripper:** You often need to expose 10 millimetres or so of bare wire so that you can solder a connection or insert the wire into the holes of a solderless breadboard (which we discuss in the later section, 'Practising with solderless breadboards'). A good wire stripper contains notches allowing you to neatly and easily strip just the plastic insulation from wires of various sizes (known as gauges, as we describe in Chapter 8), without nicking the copper wire inside. You can also find a combination wire cutter and stripper, but you have to perform your own gauge control.
- **Needlenose pliers (two sets):** These pliers help you bend wires, insert leads into breadboard holes and hold parts in place. Get two pairs: a mini (12.5-centimetre long – 5-inch) set for intricate work and a standard size set to use when you need to apply a bit more pressure.
- **Precision screwdrivers:** Make sure that you have both straight and *Phillips head* (cross-shaped tip) screwdrivers that are small enough for your electronics needs. Use the right size for the job to avoid damaging the head of the screw. A magnetised screwdriver can make working with small screws easier, or place a small amount of Blu-Tack into the head of the screw before inserting the screwdriver tip. Works wonders.
- **Magnifying glass:** A 3X (or more) magnifying glass can help you check solder joints and read tiny part numbers. You can buy versions on stands with built-in lights, too.
- **Third hand:** No, not the result of a mad scientist's gory experiment. A third hand is a tool that clamps onto your workbench and has adjustable clips that hold small parts (or a magnifying glass) while you're working. This makes tasks such as soldering a heck of a lot easier. See Figure 9-4 for an example of how to use it.

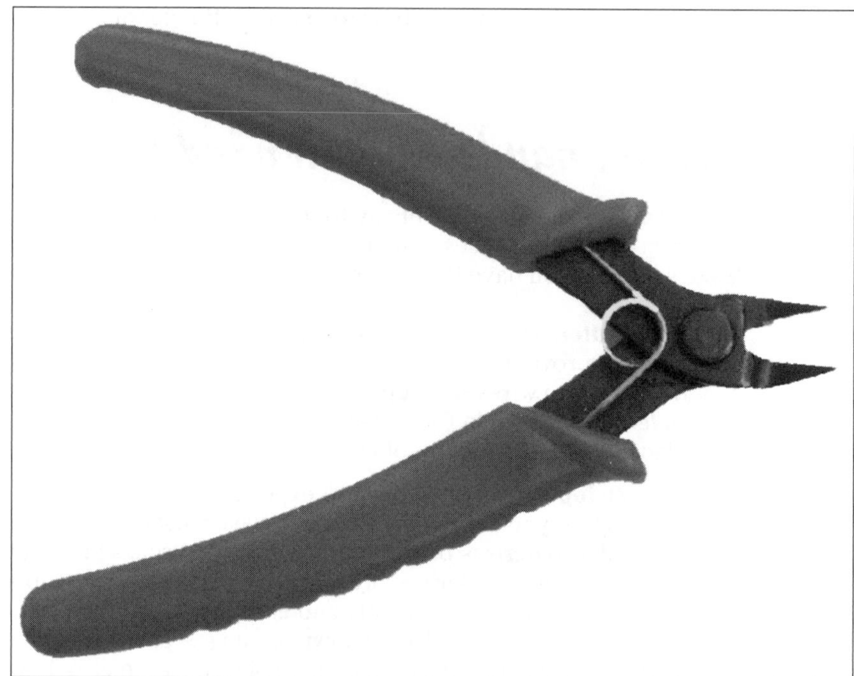

Figure 9-3: Nippy cutters trim wire ends flush to the surface.

Figure 9-4: These helping hands combine crocodile clips with a magnifying glass.

Collecting cloths and cleaners

If the circuitry, components and other parts of your electronics projects aren't kept as clean as a whistle, they may not work properly. Starting with a clean slate is especially important when you're soldering parts together or onto a circuit board. Dirt makes for bad solder joints, and bad solder joints make for faulty circuits.

Here's a list of items that can help you keep your projects spick and span:

- **Artist brushes:** Get both a small brush and a wide brush to dust away dirt, but avoid cheap brushes that shed bristles. A dry, clean toothbrush works well, too.
- **Blower brush:** Available at any photographic shop, these brushes combine the whisking action of a soft brush with the cleaning action of a strong puff of air.
- **Compressed air:** A shot of compressed air, available in cans, can remove dust from delicate electronic innards. But keep it locked away when you're not using it; if misused as an inhalant, compressed air can kill.
- **Contact cleaner:** Spray the cleaner onto a brush and then whisk the brush over electrical contacts.
- **Cotton swabs:** Soak up excess oil, lubricant and cleaner with these swabs.
- **Cuticle sticks and nail files:** Scrape junk off circuit boards and electrical contacts, and then give yourself a manicure!
- **Electronics cleaner/degreaser:** Use only a cleaner/degreaser specifically made for use on electronic components.
- **Pink pencil rubber:** Great for rubbing electrical contacts clean, especially those that have been contaminated by the acid from a leaky battery. Must be pink; other erasers can leave a hard-to-remove residue. Avoid rubbing against a circuit board, because it may create static electricity.
- **Soft cloth or bandage:** Keep your stuff dust-free by using a soft cloth or sterilised lint-free bandage. Don't use household dusting sprays because some generate a static charge that can damage electronics.
- **Water-based household cleaner:** Lightly spray to remove stubborn dirt and excess grease from tools, work surfaces and the exterior surfaces of your projects. Don't use them around powered circuits or you may short something out.

Oiling the wheels

Motors and other mechanical parts used in electronics projects require a certain amount of grease or oil to operate and need to be re-lubricated periodically. Two types of lubricants are commonly used in electronics projects:

- **Light machine oil:** Use this type of oil for parts that spin. Avoid using oil with anti-rust ingredients, which may react with plastic parts, causing them to melt. A syringe oiler with a long, thin spout is ideal for hard-to-reach places.
- **Synthetic grease:** Use lithium grease or another synthetic grease for parts that mesh or slide.

You should avoid using one type of lubricant with electronics projects: spray-on synthetic lubricants such as WD-40 and LPS. Because you can't control the width of the spray, you're bound to get some on parts that shouldn't be oiled. Also, some synthetic lubricants are non-conductive, and their fine mist can get in the way, interrupting electrical contacts.

You can find light machine oil and synthetic grease at electronics parts shops as well as many music, sewing machine, hobby and hardware stores.

Don't apply a lubricant unless you know for sure that a mechanical part needs it. Certain self-lubricating plastics used for mechanical components can break down when exposed to a petroleum-based lubricant. If you're fixing a CD player or other piece of electronic equipment, check with the manufacturer for instructions regarding the use of lubrication.

Sticking with it

Many electronics projects require that you use an adhesive of some type. For example, you may need to secure a small printed circuit board to the inside of a pocket-sized project box. Depending on the application, you can use one or more of the following adhesives:

- *White household glue* is best used for projects that involve wood or other porous materials. Allow 20–30 minutes for the glue to dry and about 12 hours to cure.
- *Epoxy cement* creates strong, moisture-resistant bonds and can be used for any material. Allow 5–30 minutes for the epoxy to set and 12 hours for it to cure.
- Super glue, or *Cyanoacrylate (CA) glue*, bonds almost anything (including fingers, so be careful), almost instantly. Use ordinary CA glue when bonding smooth and perfectly matching parts; use the heavier-bodied gap-filling CA glue if the parts don't match 100 per cent.

- *Double-sided foam tape* is a quick way to secure circuit boards to enclosures or to make sure that loosely fitting components remain in place.
- A *hot-melt glue gun* allows you to glue things with a drying time of only about 30 seconds. The waterproof, gap-sealing glue comes in a stick that you slide into a slot in the gun, which heats the glue to about 120–170° C – hot enough to hurt you, but not hot enough to melt solder.

Selecting other tools and supplies

We highly recommend that you acquire the following three other items before you begin any electronics work:

- **Anti-static wrist strap:** This inexpensive strap prevents electrostatic discharge from damaging sensitive electronic components. We discuss this aspect in the later section 'Minimising static electricity'.
- **First-aid kit and guide:** Burns (or worse) can happen when working with electronic circuits. Keeping a first-aid kit at your workbench is a good idea. Make sure that you include guidelines for applying first aid.
- **Safety glasses:** Stylish plastic safety glasses never go out of fashion. They're a must-have to protect your eyes from flying bits of wire, sputtering solder, exploding electronics parts and many other small objects. If you wear prescription glasses, place safety glasses over them to ensure complete protection all around your eyes.

A time will come when you want to enclose an electronics project in some sort of container with wires or knobs sticking out. For instance, say you build Christmas lights with a controllable blink rate. You may want to place the main circuit in a box, cut a hole through the front of the box and insert a potentiometer (variable resistor) through the hole so that you (or someone else) can control how fast the lights blink. Or you may want to build a circuit that detects intruders opening your fridge: you can disguise the circuit as a bread box and place it next to the fridge.

Whatever the project, you need some additional tools and supplies to enclose your project.

Here's a list of supplies and associated tools you may need in order to box-up your project:

- **Cable ties:** Use cable ties to attach wires to non-flat surfaces, such as a wooden dowel.
- **Electric drill:** A drill comes in handy for making holes in your box for knobs and switches. You can also use it to attach wheels or other external parts to your box.

- **Hand saws:** You can use a hack saw to cut wood or plastic to make your box and a coping saw to cut broad openings in the box.
- **Ready-made box:** You can find simple, unfinished wooden boxes at craft shops and ABS plastic boxes at most electronics suppliers. Or you can make your own box out of plywood or PVC plastic, using contact cement or another adhesive to keep it together.
- **Wire clips:** Adhesive-backed plastic clips hold wires in place along the insides of your box.

Picking up Parts

Okay, so you've got your workbench set up, complete with screwdrivers, pliers and hand saws, you've donned your anti-static wrist strap and safety glasses (along with your everyday clothes, please!), and you've got your soldering iron plugged in and ready to go. So what's missing? Oh yeah, circuit components!

When you shop for circuit components, you usually don't go out and purchase just the exact parts listed for a particular circuit diagram or schematic. You purchase an assortment of parts so that you can build several different projects without having to run out for parts each time you try something new. Think of this process as being like gathering ingredients for cooking and baking. You maintain a store of many basic ingredients, such as flour, sugar, oil, rice and spices, and you purchase enough other ingredients to enable you to cook the sorts of things you like for a week or two. The same thing applies with stocking up on electronics parts and components.

In this section, we tell you what parts and how many you should keep on hand in order to build some basic electronics projects.

Practising with solderless breadboards

A *solderless breadboard* is similar in a way to a Lego table: it's a surface that enables you to build temporary circuits simply by plugging components into holes arranged in rows and columns across the surface of the board. Breadboards allow you to take one circuit apart very easily and build another completely different circuit on the same surface.

The holes in a solderless breadboard aren't just ordinary holes; they're *contact holes* with copper lines running underneath so that anything plugged into two or more holes within a particular row is connected underneath the surface of the breadboard. You plug in your *discrete components* (resistors, capacitors, diodes and transistors) and *integrated circuits* (ICs) in just the right way, and – *voilà* – you've got a connected circuit without any soldering necessary. When you're tired of the circuit, you can simply remove the parts and build something else using the same breadboard.

Figure 9-5 shows a small solderless breadboard with a battery-powered circuit connected up. The breadboard in the figure has sections of rows and columns connected in a certain way underneath the board. We discuss just how the various contact holes are connected in Chapter 11, where we also discuss how to build circuits using breadboards. For now, just know that different sizes of breadboards are available with different numbers of contact holes.

A typical small breadboard has 400 contact holes and is useful for building smaller circuits with no more than two ICs (plus other discrete components). A typical larger breadboard contains 830 contacts and can be used to build somewhat more complex circuits. You can join some designs of boards together to make larger boards. Or you can simply link any number of breadboards together by connecting one or more wires between contact holes on one board and contact holes on the other board.

We recommend that you purchase at least two solderless breadboards and that at least one of them is a larger (830 contacts) breadboard. Also, get some stick-on Velcro strips to help hold the breadboards in place on your workbench.

Figure 9-5: You can build a circuit on a small solderless breadboard in a few minutes.

Solderless breadboards are commonly used to test your circuit design ideas or allow you to explore circuits as you discover how things work (as you can with the simple circuits in Chapter 13).

If you've created and tested a circuit using a breadboard and you want to use it on a long-term basis, you can re-create the circuit on a soldered or printed circuit board (PCB).

A PCB is a kind of breadboard, but rather than contact holes, it has ordinary holes with copper pads surrounding each hole and lines of metal connecting the holes within each row. You make connections by soldering component leads to the copper pads and ensuring that the components to be connected are located in the same row. In this book, we focus exclusively on circuit construction using solderless breadboards.

Building a circuit-building starter kit

You need an assortment of discrete electronic components (those with two or three individual leads), a few ICs, several batteries and lots of wire to connect things up. Some components, such as resistors and capacitors, come in packages of ten or more individual pieces. You'll be happy to know that all these components are really inexpensive (cheap, even), but when you add everything up the amount may reach as much as £50, depending on what you can find and where.

Discrete components come in assortment packs of different ratings for each type of component. See what's available when you start shopping, but below we list the kind of parts to start with:

- **Capacitors:** 10 each (1 package) of 0.01 μF and 0.1 μF non-polarised (polyester or ceramic disc); 10 each (1 package) of 1 μF, 10 μF, 100 μF electrolytic; 3–5 each of 220 μF and 470 μF electrolytic.
- **Diodes:** One each: 1N4001 (or any 1N400x) rectifier diode, 1N4148 small signal diode, 1 4.3-volt Zener diode (or other Zener breakdown voltage between 3 and 7 volts).
- **Fixed resistors (¼-watt or ½-watt carbon film):** 10–20 (1 or 2 packages) of each of these resistances: 1 kΩ, 10 kΩ, 100 kΩ, 1 MΩ, 2.2 kΩ, 22 kΩ, 220 kΩ, 33 kΩ, 470 kΩ, 4.7 kΩ, 47 kΩ, 470 kΩ.
- **LEDs (light emitting diodes):** 10 each (1 package): red, yellow and green 5-millimetre diffused LEDs.
- **Potentiometers:** Two each: 10 KΩ, 100 KΩ, 1 MΩ.
- **Transistors:** 3–5 general-purpose, low-power bipolar transistors (such as the 2N3904 NPN or the 2N3906 PNP) and 3–5 medium-power bipolar transistors (such as the NTE123A NPN or NTE159M PNP). (We use the 2N3906 in a project in Chapter 14.)

We suggest you obtain a few popular ICs:

- **Op amp ICs:** Get one or two op amps, such as the LM386 power amplifier (which we use in a project in Chapter 14) or the LM741 general-purpose amplifier.
- **555 timer IC:** Get 3–5 of these; you'll use them!
- **4017 CMOS decade counter IC:** One is sufficient. (We use one in a project in Chapter 14.) Get two if you want to make a decade counter, too, as we discuss in Chapter 7, or if you think that you may accidentally zap the first one with electrostatic discharge.

Don't forget these essential power and wire components:

- **Batteries:** Pick up an assortment of 9-volt batteries as well as some 1.5-volt batteries (size depends on how long you think that you're going to run your circuit).
- **Battery clips and holders:** These devices connect to batteries and provide wire leads to make it easy for you to connect battery power to your circuit. Get 3–5 clips for the size batteries you plan to use.
- **Wire:** Plenty of solid core wire is necessary for connecting components together on the breadboard. You can buy reels in a variety of colours for a few pounds each. Cut bits to various lengths and strip the insulation off each end so that you can solder them to component leads or poke them into contact holes on your solderless breadboard.

  Some electronics suppliers sell kits containing dozens of pre-cut, pre-stripped *jumper wires* of various lengths and colours, which are ideal for use in solderless breadboards. A kit with 350 jumper wires may cost you £10, but saves you the time (and trouble) of cutting and stripping your own wire (plus you get rainbow colours!). We also recommend the little packs of jumper wires with hardened ends for poking into breadboards.

You can use a jumper wire as a makeshift on/off switch in your circuit, connecting or disconnecting power or components. Just place one end of the jumper wire in your solderless breadboard and place and remove the other end to operate the switch.

Adding up the extras

Lots of other parts and components are available that can enrich your circuits. We recommend you get a few of the ones listed here:

- **Crocodile or alligator clips:** So-named because they look like the jaws of a fierce reptile, these insulated clips can help you connect test equipment to component leads, and they can double as a heat sink to take heat away from the circuit too! Get a bunch (ten or so).

- **Speaker:** You're sure to want to build a circuit that makes noise, so purchase one or two miniature 8-ohm speakers.
- **Switches:** If you think that you may enclose one or more projects in a box and you'd like a front-panel on/off control, pick up a couple of SPST (single-pole single-throw) switches, such as an SPST mini rocker switch, for about a pound each. For a few pence more, you can get one with a built-in LED that lights up when the switch is in the 'on' position.

Organising all your parts

Keeping all these parts and components organised is essential – unless you're the type who enjoys sorting through junk drawers looking for some tiny, yet important, item. An easy way to get things together is to buy a few cheap sets of clear plastic drawers. Label each drawer for a particular component (or group of components, such as LEDs, 10–99-Ω resistors and so forth). You then know at a glance where everything is, and you can see when your stock is getting low.

Protecting You and Your Electronics

You probably know the story about Benjamin Franklin flying his kite during a thunder storm in 1752 to show that lightning is actually an electric current. But he already knew it was dangerous and must have taken care to insulate himself from the conductive materials attached to the kite (the key and a metal wire), and to stay dry by taking cover in a barn. Had he not, someone else's face may be on the US $100 bill!

You must respect the power of electricity when working with electronics. In this section, we take a look at keeping yourself – and your electronic projects – safe. This section is the one that you really must read from start to finish, even if you already have some experience in electronics.

As you read this section, remember that you can describe electrical current as being one of the following:

- **Direct current (DC):** The electrons flow one way through a wire or circuit.
- **Alternating current (AC):** The electrons flow one way, and then another, in a continuing cycle.

Refer to Chapter 2 for more about these two types of electrical current.

Accepting that electricity can really hurt

The greatest danger by far in working with electronics is the risk of electrocution. Electrical shock – the body's reaction to an electrical current – can include an intense contraction of the muscles (namely, the heart) and extremely high heat where the current enters your skin. The burns can (and do) disfigure and kill, but even small currents can disrupt your heartbeat.

How much harm an electrical shock does to you depends on a lot of factors, including your age and your general health. But no matter how young and healthy you may be, voltage and current can pack a wallop, so understanding how much they can harm you is vitally important.

The two most dangerous electrical paths through the human body are hand-to-hand and left hand to either foot. If electrical current passes from one hand to the other, on its way, it passes through the heart. If current passes from the left hand to either foot, it passes through the heart as well as several major organs.

Seeing yourself as a giant resistor

Your body exhibits some resistance to electrical current, mostly due to the poor conductive qualities of dry skin. The amount of resistance can vary tremendously, depending on body chemistry, level of moisture in the skin, the total path across which resistance is measured and other factors. You see figures ranging anywhere from 50,000–1,000,000 ohms of resistance for an average human being. (In Chapter 3, we discuss what resistance is and how you measure it.)

If your skin is moist (say you have sweaty hands), you're wearing a metal ring or you're standing in a puddle, you can bet you've lowered your resistance. Industry figures indicate that such activity can result in resistances as low as 100–300 ohms from one hand to the other, or from one hand to one foot. That's not a whole lot of resistance.

To make matters worse, if you're handling high AC voltages (which you shouldn't be), your skin's resistance – wet or dry – doesn't help you at all. When you're in contact with a metal, your body and the metal form a capacitor: the tissue underneath your skin is one plate, the metal is the other plate and your skin is the dielectric. (See Chapter 4 for the low-down on capacitors.) If that metal wire you're holding is carrying an AC current, the capacitor that's your body acts like a short circuit, allowing current to bypass your skin's resistance. Voltage shocks of more than 240 volts burn right through your skin, leaving deep third-degree burns at the entry points.

Understanding how voltage and current can harm you

You've seen the signs: WARNING! HIGH VOLTAGE. So you may think that voltage is what causes harm to the human body, but actually current is what inflicts the damage. So why the warning signs? That's because the higher the voltage, the more current can flow for an equal amount of resistance. And because your body is like a giant resistor, you should avoid high voltages.

So how much current is necessary to hurt the average human being? Not much. Table 9-1 summarises some estimates of just how much – or how little – DC and AC current it takes to affect the human body. Remember that a milliamp (mA) is one one-thousandth of an amp (or 0.001 A). Please note that these amounts are *estimates* (no one has performed experiments on real humans to derive these figures), and that each person is affected differently depending on age, body chemistry, health status and other factors.

Table 9-1 Effects of Current on Average Human Body

Effect	*DC current*	*AC current*
Slight tingling sensation	0.6–1.0 mA	0.3–0.4 mA
Noticeable sensation	3.5–5.2 mA	0.7–1.1 mA
Pain felt, but muscle control maintained	41–62 mA	6–9 mA
Pain felt, and unable to let go of wires	51–76 mA	10–16 mA
Difficulty breathing (paralysis of chest muscles)	60–90 mA	15–23 mA
Heart fibrillation (within 3 seconds)	500 mA	65–100 mA

So what does all this mean to you as you pursue your electronics hobby? You probably know enough to stay away from high voltages, but what about getting up close and personal with low voltages? Well, even low voltages can be dangerous – depending on your resistance.

Remember that Ohm's Law (which we cover in Chapter 3) states that voltage is equal to current times resistance:

$$V(\text{voltage}) = I(\text{current}) \times R(\text{resistance})$$

Imagine that your hands are dry and you aren't wearing a metal ring or standing in a puddle; your hand-to-hand resistance is about 50,000 ohms (keep in mind that your resistance under these conditions may be lower). You can calculate an estimate (repeat: *estimate*) of the voltage levels that may hurt you by multiplying your resistance by the different current levels in Table 9-1.

For instance, if you don't want to feel even the slightest tingling sensation in your fingers, you need to avoid coming into contact with wires carrying DC voltages of 30 volts (that's 0.6 milliamps × 50,000 ohms).

Now, if you're not as careful and you wear a ring on your finger while tinkering around with electronics, or you step in a little puddle of water created by a dog or small child, you may accidentally lower your resistance to a dangerous level. If your resistance is 5000 ohms – and it may be even lower – you notice a sensation if you handle just 17.5 volts DC (because 5000 ohms × 0.0035 amps = 17.5 volts).

Household electrical systems in the UK operate at about 230 VAC. This very high voltage can and does kill. Don't even think about working on circuits that operate off mains electricity until you're properly trained in its use and dangers.

Circuits that run directly off mains electricity aren't for hobbyists and they're beyond the scope of this book. Besides, you can achieve so much with circuits that run off standard-size batteries. Unless you do something silly, like licking the terminal of a 9-volt battery (and, yes, that does deliver a shock!), you're fairly safe with these voltages and currents, but you must still be careful.

The main danger of household current is the effect it can have on your heart muscle. Just 65–100 milliamps can send your heart into fibrillation, which means that the muscles are contracting in an uncontrolled, uncoordinated fashion – and the heart isn't pumping blood.

At much lower levels (10–16 milliamps), AC current can cause severe muscle contractions, so what may start out as a loose grip on a high voltage wire (just to move it a little bit, or something like that) ends up as a powerful, unyielding grip. Trust us: you won't be able to let go. A stronger grip means a lower resistance (you're just making it easier for electrons to travel through your hand and into your body), and a lower resistance means a higher (often fatal) current. (Situations like this one really do happen. The body acts like a variable resistor, with its resistance decreasing sharply as the hands tighten around the wire.)

The potential dangers of DC currents aren't to be ignored either. Burns are the most common form of injury caused by high DC current. Remember that voltage doesn't have to come from a power plant to be dangerous. You need to respect even a 9-volt transistor battery: if you short its terminals, the battery may overheat and can even explode. Battery explosions often send tiny battery pieces flying out at high velocities, burning skin or injuring eyes.

Maximising your resistance – and your safety

When working with electronics, ensure that you maximise your resistance just in case you come into contact with an exposed wire. Make sure that any tools you pick up are insulated, and if possible, stand on a rubber mat.

This way, you add more resistance between you and any voltages you may encounter.

Take simple precautions to ensure that your work area starts out dry and stays dry. For example, don't place a glass of water or cup of coffee too close to your work area; if you accidentally knock it over, you may lower your own resistance or short-out circuit components.

Keeping a first-aid chart handy

Even if you're the safest person on earth, get one of those emergency first-aid charts that include information about what to do in case of electrical shock. You can find these charts on the Internet; try a search for 'first-aid wall chart'. You can also find them in school and industrial supply catalogues.

Helping someone who's been electrocuted may require using cardio-pulmonary resuscitation (CPR). Be sure that you're properly trained before you administer CPR on anyone. Check out `www.redcross.org` to get more information about CPR training.

Soldering safely

The soldering iron you use to join components in an electronics project operates at temperatures higher than 400° C. (You can read up on soldering in Chapter 11.) That's about the same temperature as an electric stove burner set at high heat. You can imagine how much that hurts if you touch it.

When using a soldering iron, keep the following safety tips in mind:

- **Always place your soldering iron in a stand designed for the job.** Never place the hot soldering iron directly on a table or workbench. You can easily start a fire or burn your hands.
- **Be sure that the electrical cord doesn't snag on the workbench or any other object.** Otherwise, the hot soldering iron can get yanked out of its stand and fall to the ground. Or worse, right into your lap!
- **Never solder a *live circuit* (a circuit to which you've applied voltage).** You may damage the circuit or the soldering iron – and you may receive a nasty shock.
- **Never grab a tumbling soldering iron.** Just let it fall and buy a new one if the iron is damaged.
- **Solder only in a well-ventilated area.** Soldering produces mildly caustic and toxic fumes that can irritate your eyes and throat.
- **Unplug your soldering iron when you're done.**

- **Use the appropriate soldering setting.** If your soldering iron has an adjustable temperature control, set the dial to the recommended setting for the kind of solder that you're using. Too much heat can spoil a good circuit.
- **Wear safety glasses when soldering.** Solder has been known to splutter.

Avoiding static like the plague

One type of everyday electricity that can be dangerous to both people and electronic components is static electricity. It's called static because it's a form of current that remains trapped in some insulating body, even after you remove the power source. Static electricity hangs around until it dissipates in some way. Most static dissipates slowly over time, but in some cases, it gets released all at once. Lightning is one of the most common forms of static electricity.

If you drag your feet across a carpeted floor, your body takes on a static charge. If you then touch a metal object, such as a doorknob or a metal sink, the static quickly discharges from your body, and you feel a slight shock. This is known as *electrostatic discharge* (ESD). The resulting current is small, because of the high resistance of the air that the charges arc through as they leave your fingertips, and doesn't last very long, so static shocks of the doorknob variety generally don't inflict bodily injury.

On the other hand, static shocks from certain electronic components can be harmful. The capacitor, an electronic component that stores energy in an electric field, is designed to hold a static charge. Most capacitors in electronic circuits store a very minute amount of charge, but some capacitors, such as those used in bulky power supplies, can store near-lethal doses.

Use caution when working around capacitors that can store a lot of charge, so that you don't get an unwanted shock.

Being sensitive to static discharge

The ESD that results from dragging your feet across the carpet or combing your hair on a dry day may be as high as a few thousand volts. Although you probably just experience an annoying tickle (and maybe a bad hair day), your electronic components may not be so lucky. Transistors and integrated circuits (ICs) that are made using metal oxide semiconductor (MOS) technology are particularly sensitive to ESD, regardless of the amount of current.

MOS devices contain a thin layer of insulating glass that can easily be zapped away by 50 volts of discharge or less. If you, your clothes and your tools aren't free of static discharge, that MOS field effect transistor (MOSFET) or complementary MOS (CMOS) IC you planned to use will be nothing more than a useless lump. Because bipolar transistors are constructed differently, they're less susceptible to ESD damage. Other components – resistors, capacitors, inductors, transformers and diodes – don't seem to be bothered by ESD.

We recommend that you develop static-safe work habits for all the components you handle, whether they're overly sensitive or not.

Minimising static electricity

You can bet that most of the electronic projects you want to build contain at least some components that are susceptible to damage from electrostatic discharge. You can take steps to prevent exposing your projects to the dangers of ESD:

- **Use an anti-static mat.** Available in both table-top and floor varieties, an anti-static mat looks like a sponge, but is really conductive foam. It can reduce or eliminate the build-up of static electricity on your table and your body.
- **Use an anti-static wrist strap.** Pictured in Figure 9-6, an anti-static wrist strap grounds you and prevents static build-up. The strap is one of the most effective means of eliminating ESD and costs only a fiver. To use one, roll up your shirt sleeves, remove all rings, watches, bracelets and other metals and wrap the strap around your wrist tightly. Then securely attach the clip from the wrist strap to a proper earth ground connection (which you can read about in the next sidebar 'Getting a good grounding').
- **Wear low-static clothing.** Whenever possible, wear natural fabrics, such as cotton or wool. Avoid polyester and acetate clothing because these fabrics have a tendency to develop a whole lot of static.

Getting a good grounding

If you have metal pipes, you can use the cold water pipe under a sink or connected to your water heater to connect your anti-static wrist strap to earth ground. This pipe comes into your house from under the ground and therefore is connected to the earth outside your house (that's what an earth ground is!). (Note that hot water pipes aren't grounded; they originate from your hot water heater.)

Attach a metal clamp to a cold water pipe and run a wire from the clamp to your work area. Make sure that you have a conductive connection all the way through (nothing interrupting the metal). Then secure the end of the wire on your workbench and connect the clip at the end of your anti-static wrist strap to the wire.

If you have plastic pipes or can't easily access a cold water pipe, jam a metal rod at least three feet into the ground outside your house. Attach a metal clamp to the rod and run a wire from the clamp to your workbench, as for the cold water pipe connection.

Usually, wearing cotton clothing and using an anti-static wrist strap is sufficient for preventing ESD damage.

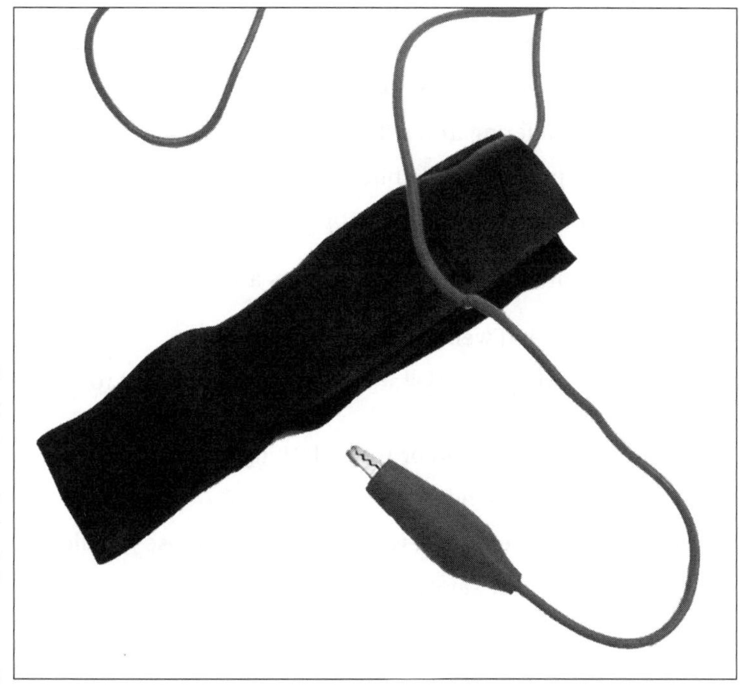

Figure 9-6: An anti-static wrist strap reduces or eliminates the risk of electrostatic discharge.

As long as you ground yourself by using an anti-static wrist strap, you generally don't need to ground your other metal tools, such as screwdrivers and wire cutters. Any static that these tools generate dissipates through your body and into the anti-static wrist strap.

Staying Safe: Safety Checklist

Before you get started on any electronics project, review the following simple checklist of *minimal* safety requirements. Better still, make a copy of this checklist, laminate it and post it at your workbench as a reminder of the simple steps that can ensure your safety – and the well-being of your electronics projects.

At your workspace, check that you have:

- Ample ventilation
- Dangerous tools and materials locked up
- Dry working surface, dry floor
- First-aid chart within view
- Grounded soldering iron with weighted stand
- No liquids, pets or small children within a 3-metre range
- Phone (and caring friend) nearby

Also make a point of checking these points about yourself:

- Are you alert and well-rested?
- Are you wearing an anti-static wrist band (attached to you and to earth ground)?
- Are you wearing cotton or wool clothing?
- Do you have dry hands (or gloves to use)?
- Have you removed rings, wristwatches and loose jewellery?
- Have you got your safety glasses handy?

Chapter 10

Sussing out Schematics

In This Chapter

- Understanding the role of schematics
- Getting to know the most common symbols
- Using (not abusing!) component polarity
- Diving into some specialised components

Imagine trying to drive around the country without a road map. The chances are that you get lost along the way and end up driving in circles. Road maps exist to help you find your way.

Electronic circuits have their own road maps. They're called *schematic diagrams* (or just *schematics*), and they show you how all the parts of the circuits are connected. Schematics display these connections with symbols that represent electronic parts and lines that show you how to attach the parts.

Although not all electronics circuits that you encounter are described in the form of a schematic, many are. If you're serious about electronics, sooner or later, you need to understand how to read a schematic. The good news is that the language of schematics isn't all that hard. Most schematic diagrams use only a small handful of symbols for components, such as resistors, capacitors and transistors.

In this chapter, we tell you everything that you really need to know so that you can read almost any schematic diagram you come across.

Approaching Schematics

A *schematic* is a circuit diagram that shows all the components of a circuit, including power supplies, and their connections. The most important things to focus on when reading a schematic are *the connections*, because the positioning of components in a schematic diagram doesn't necessarily correspond to the physical layout of components in a constructed circuit — just like the London Underground map. In fact, for a complex circuit, the physical circuit layout is highly unlikely to reflect the positioning shown in the schematic. Complex circuits often require separate *layout* diagrams, sometimes known as *artwork*.

Schematics use symbols to represent resistors, transistors and other circuit components, and lines to show connections between components. By reading the symbols and following the interconnections, you can build the circuit shown in the schematic. Schematics can also help you understand how a circuit operates, which comes in handy for testing and repairing the circuit.

Discovering how to read a schematic is a little like discovering a foreign language. On the whole, you find that most schematics follow fairly standard conventions. However, just as many languages have different dialects, the language of schematics is far from universal. Schematics can vary depending on the age of the diagram, its country of origin, the whim of the circuit designer and many other factors.

In this book, we use conventions commonly used in the UK and Europe. But electronics is one of the most global of industries, and so you may find yourself wanting to build circuits designed in other countries, especially the United States. To help you deal with some of the variations you may come across, we include some alternative symbols in our descriptions throughout this chapter and some American symbols in the later section 'Alternative Schematic Drawing Styles'.

Seeing the Big Picture

Imagine that you have a schematic in front of you. What do you see? Well, an unwritten rule in electronics tells you how to orient certain parts of a circuit schematic, especially when drawing diagrams of complex circuits. Batteries and other simple power supplies are almost always oriented vertically, with the positive terminal on top. In complex schematics, power supplies are split between two symbols (as we discuss in the later section 'Recognising Symbols of Power'), but the positive terminal is usually shown at the top of the schematic (sometimes extending across a horizontal line, or *rail*) and the negative terminal appears at the bottom (again, sometimes along a rail). Inputs are commonly shown on the left and outputs on the right.

Many electronic systems (for instance, the radio receiver we discuss in Chapter 2) are represented in schematics by several stages of circuitry (even though the system really consists of one mighty big circuit). The schematic for such a system shows the sub-circuits for each stage in a left-to-right progression (for instance, the tuner sub-circuit on the left, the detector in the middle and the amplifier on the right), with the output of the first stage feeding into the input of the second stage, and so forth. Organising schematics in this way helps to make complex circuits more understandable.

Following connections

In all schematics, whether simple or complex, components are arranged as neatly as possible, and connections within a circuit are always drawn as lines, with any bends shown as 90° angles. (No squiggles or arcs allowed!) Understanding what all the lines in a schematic mean is absolutely critical – and that meaning isn't always obvious.

The more complex the schematic, the more likely that some lines criss-cross each other (due to the two-dimensional nature of schematic drawings). You need to know when crossed lines represent an actual wire-it-together connection and when they don't. Ideally, a schematic clearly distinguishes connecting and non-connecting wires as follows:

- A break or a loop (think of it as a bridge) in one of the two lines at the intersection indicates wires that *shouldn't* be connected.
- A dot at the intersection of two lines indicates that the wires *should* be connected.

You can see some common variations in Figure 10-1.

This method of showing connections isn't universal, so you have to figure out which wires connect and which don't by checking the drawing style used in the schematic. If you see an intersection of two lines without a dot to positively identify a real connection, you simply can't be sure whether the wires should be connected or not.

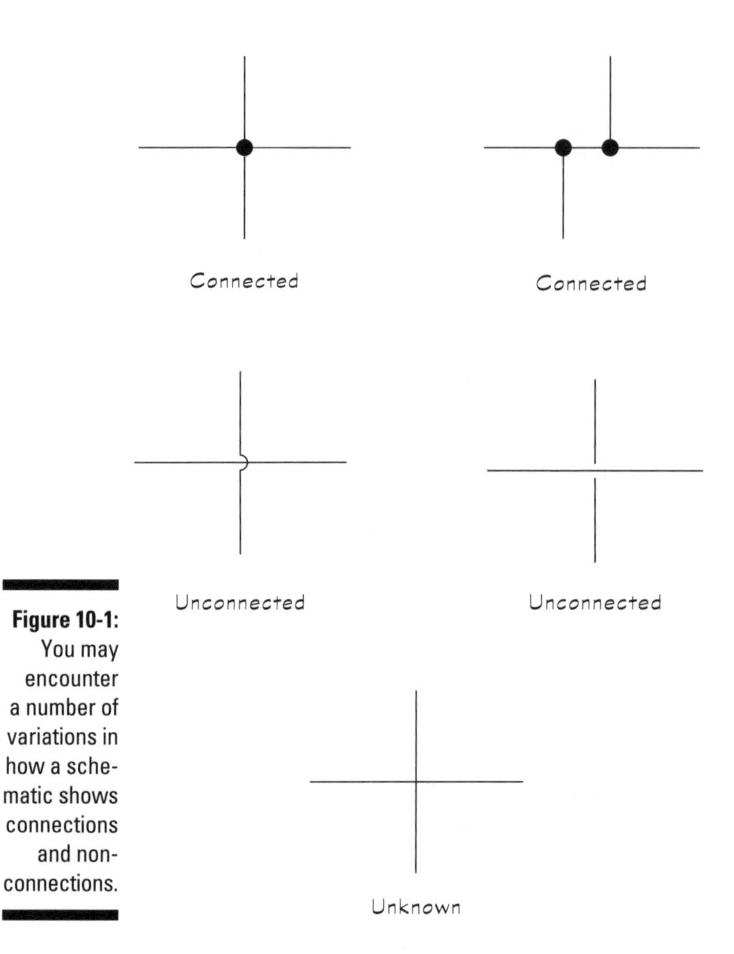

Figure 10-1: You may encounter a number of variations in how a schematic shows connections and non-connections.

To physically implement the connections shown in a schematic, you typically use insulated wires or thin traces of copper on a circuit board. Most schematics don't make a distinction about how you connect the components together; that connection is wholly dependent on how you choose to build the circuit. The schematic's representation of the wiring merely shows which wires go where.

Looking at a simple battery circuit

Figure 10-2 shows a simple DC circuit with a 5-volt battery connected to a resistor labelled R1. The positive side of the battery (+V) is connected to the lead on one side of the resistor, and the negative side of the battery is connected to the lead on the other side of the resistor. With these connections made, current flows from the positive terminal of the battery through the resistor and back to the negative terminal of the battery.

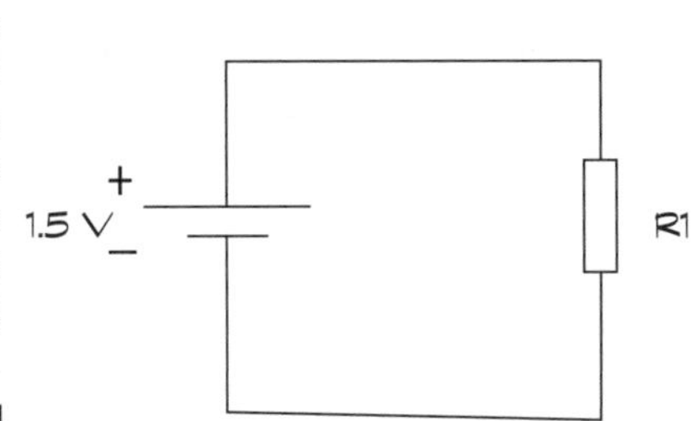

Figure 10-2: A simple schematic shows the connections between a battery and a resistor.

In schematics, 'current' is assumed to be conventional current, which is described as the flow of positive charges, travelling in a direction opposite to that of real electron flow. We discuss conventional current and electron flow in Chapter 2.

Recognising Symbols of Power

Power for a circuit can come from an alternating current (AC) source, such as the 230 VAC (volts AC) socket in your house or office (so-called *mains electricity*), or a direct current (DC) source, such as a battery or the low-voltage side of a wall transformer. DC supplies can be positive or negative with respect to the 0 volt reference (known as *common ground*, or simply *common*) in a circuit. Table 10-1 shows various symbols used to represent power and ground connections.

Figuring out the various power connections in a complex schematic is sometimes a task unto itself. This section aims to clear things up a bit. As you read through this section, refer to Table 10-1 to see the symbols we discuss.

Table 10-1 Symbols for Power and Ground

Name	*Symbol*
Battery (cell)	
Solar (photovoltaic) cell	
DC power supply	

continued

Table 10-1 (continued)

Name	Symbol
AC power supply	—o ~ o—
Earth ground	
Chassis ground	
Signal ground	

Showing where the power is

DC power supplies are shown in various ways around the world including:

- **Battery or solar cell symbol:** The symbols you're most likely to see. Each of these symbols represents a DC source with two leads. Technically, the battery symbol that includes two parallel lines represents a single electrochemical cell, whereas the symbol with multiple pairs of lines represents a battery (which consists of multiple cells); however, many schematics use the symbol for a cell to represent a battery. Each symbol includes a positive terminal (indicated by the larger horizontal line) and a negative terminal. The polarity symbols (+ and –) and nominal voltage are usually shown next to the symbol. The negative terminal is often assumed to be at 0 (zero) volts, unless clearly distinguished as different from the zero voltage reference (*common ground*, which we discuss later in the section 'Marking your ground'). Conventional current flows out of the positive terminal and into the negative terminal when the battery is connected to a complete circuit.
- **'Split' DC power and ground symbols:** To simplify schematics, a DC power supply is sometimes shown using two separate symbols: a small circle at the end of a line representing one side of the supply, with or without a specific voltage label, and the symbol for ground (vertical line with three horizontal lines at the bottom) representing the other side of the supply, with a value of 0 volts. In complex circuits with multiple connections to power, you may see the positive side of the supply represented by a rail labelled +V extending across the top of the schematic. These split power-symbol representations are used to eliminate a lot of wire connections in a schematic.

- **Two terminals labelled with polarities:** This symbol resembles the DC version of the AC symbol, comprising two little terminal circles with a waveform in between.

The circuit shown in Figure 10-2 can also be drawn using separate symbols for power and ground, as in Figure 10-3. Note that the circuit in Figure 10-3 is, in fact, a complete circuit.

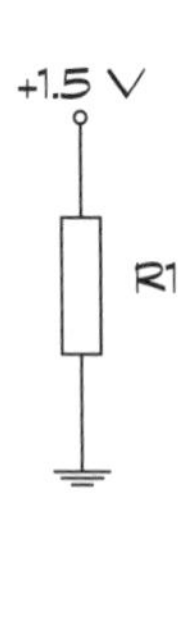

Figure 10-3: This circuit shows a DC power source symbol at the top and ground symbol at the bottom, which together represent connections to a battery.

Many DC circuits use multiple DC power supplies, such as +5 VDC, +12 VDC and even –5 VDC or –12 VDC, and so the voltage source symbols in the schematics are usually labelled with the nominal voltage. If a schematic doesn't specify a voltage, you're often (but not always!) dealing with 5 VDC. And, unless otherwise specifically noted, the voltage in a schematic is almost always DC, not AC.

Some circuits (for instance, op amp circuits, which we discuss in Chapter 7) require both positive and negative DC power supplies. You often see the positive supply represented by an open circle labelled +V and the negative supply represented by an open circle labelled –V. If the voltages aren't specified, they may be +5 VDC and –5 VDC. Figure 10-4 shows how these power supply connection points are implemented.

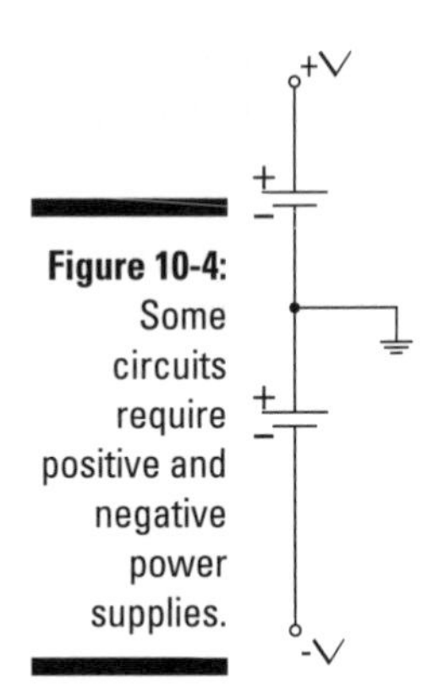

Figure 10-4: Some circuits require positive and negative power supplies.

An AC power supply is usually represented by a circle with two leads, with or without a waveform shape and polarity indicators:

- **Circle containing waveform:** A squiggly line or other shape inside an open circle represents one cycle of the alternating voltage produced by the power supply. Usually, the source is a sine wave, but it can be a square wave, a triangle wave or something else.
- **Circle with polarity:** Some schematics include one or both polarity indicators inside or outside the open circle. This symbol is just for reference purposes, so that you can relate the direction of current flow to the direction of voltage swings.

Power for a circuit can come from an AC source, such as the 230 VAC socket in your house or office (so-called *mains electricity*). You typically use an internal power supply to *step down* (or lower) the 230 VAC and convert it to DC. This lower-voltage DC power is then delivered to the components in your circuit. If you're looking at a schematic for a DVD player or some other gadget getting its power from a wall socket, that schematic probably shows both AC and DC power.

Marking your ground

Ready for some electronics schematic double-talk? As regards labelling ground connections in schematics, common practice is to use the symbol for *earth ground* (which is a real connection to the earth) to represent the *common ground* (the reference point for 0 volts) in a circuit. We discuss these two types of grounds in Chapter 2. More often than not, the ground points in low-voltage circuits aren't actually connected to earth ground, but are just tied to each other; hence the term *common ground* (or simply *common*). Any voltages labelled at specific points in a circuit are assumed to be relative to this common ground. (Remember, voltage is really a differential measurement between two points in a circuit.)

So what symbol should *really* be used for ground points that aren't truly connected to the earth? The answer is the symbol labelled *chassis ground* or *frame*. Common ground is sometimes called chassis ground, because in older equipment the metal chassis of the device (hi-fi, television or whatever) served as the common ground connection. Using a metal chassis for a ground connection isn't as common today.

You may also see the symbol for *signal ground* used to represent a 0-volt reference point for signals carried by two wires. One wire is connected to this reference point and the other wire carries a varying voltage representing the signal. Again, in many schematics, the symbol for earth ground is used instead.

In this book, we use only the schematic symbol for earth ground because most schematics you see these days use that symbol.

As you can see in Figure 10-5, a schematic may show the ground connections in a number of ways:

- **No ground symbol:** The schematic can show two power wires connected to the circuit. In a battery-powered circuit, common ground is assumed to be the negative terminal of the battery.
- **Single ground symbol:** The schematic shows all the ground connections connected to a single point. It doesn't often show the power source or sources (for instance, the battery), but you should assume that ground connects to the positive and/or negative DC power sources as shown in Figure 10-4.
- **Multiple ground symbols:** In more complex schematics, drawing the circuit with several ground points is usually easier. In the actual working circuit, all these ground points connect together.

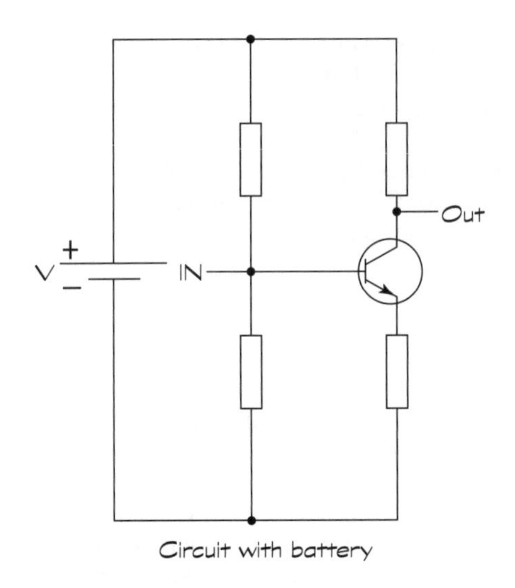

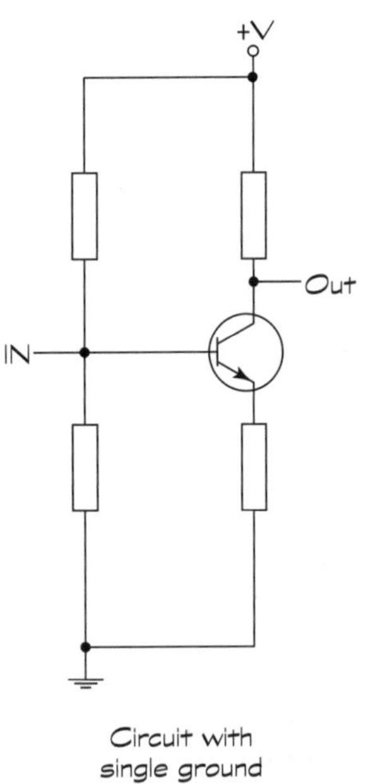

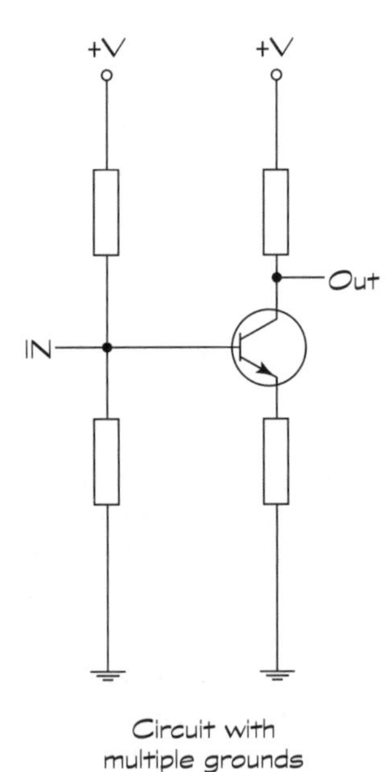

Figure 10-5: A circuit's common ground connections may be shown using multiple symbols, a single symbol or no symbol at all (when assumed to be the negative terminal of a battery).

Labelling Circuit Components

You can find literally hundreds of symbols for electronic components in schematics, for the simple reason that hundreds of them need to be depicted. Fortunately, you probably encounter only a small number of these symbols in schematics for hobby electronics projects.

Along with the circuit symbol for a particular electronic component, you may see additional information to help uniquely identify the part:

- **Reference ID:** An identifier, such as R1 or Q3. The convention is to use one or more letters to represent the type of component and a numerical suffix to distinguish one particular component from others of the same type. The most common type designators are R for resistor, C for capacitor, D for diode, L for inductor, T for transformer, Q for transistor and U or IC for integrated circuit.
- **Part number:** Used for standard components, such as a transistor or integrated circuit, or where you have a manufacturer's customer product part. For example, a part number may be something like 2N222 (that's a commonly used transistor) or 555 (a type of IC used in timing applications).
- **Value:** Component values are sometimes shown for passive parts, such as resistors and capacitors, which aren't designated by conventional part numbers. For example, when indicating a resistor, the value (in ohms) may be marked beside the resistor symbol and/or the reference ID. Most often, you see just the value without the units label (ohms, microfarads and so on). Normally, resistor values are assumed to be in ohms and capacitor values are assumed to be in microfarads.
- **Additional information:** A schematic may include additional specifics about one or more components, such as the wattage for a resistor when it isn't your typical ¼ or ⅛ watt value. If you see '10W' next to a resistor value, you know that you need a power resistor.

Many schematics show only the reference ID and the circuit symbol for each component, and then include a separate *parts list* to provide the details of part numbers, values and other information. The parts list maps the reference ID to the specific information about each component.

Reference ID primer

Components are often identified in a schematic using an alphabetic type designator, such as C for capacitor, followed by a numerical identifier (1, 2, 3 and so on) to distinguish multiple components of the same type. Together, these identifiers form a *reference ID* that uniquely identifies a specific capacitor or other component. The reference ID can be used in a parts list to indicate the precise value of the component to use, if that value isn't printed beside the component symbol. The following type designators are among those most commonly used:

C – Capacitor

D – Diode

IC (or U) – Integrated circuit

L – Inductor

LED – Light-emitting diode

Q – Transistor

R – Resistor

RLY – Relay

T – Transformer

XTAL – Crystal

Analogue electronic components

Analogue components control the flow of continuous (analogue) electrical signals. Table 10-2 shows the circuit symbols used for basic analogue electronic components. The third column of the table provides the chapter reference in this book where you can find detailed information about the functionality of each component.

Table 10-2 Symbols for Analogue Components

Component	Symbol	Chapter Reference
Resistor		Chapter 3
Variable resistor		Chapter 3
Light dependent resistor	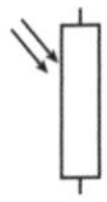	Chapter 8

Component	*Symbol*	*Chapter Reference*
Capacitor		Chapter 4
Polarised capacitor		Chapter 4
Variable capacitor		Chapter 4
Inductor		Chapter 5
Transformer		Chapter 5
NPN (bipolar) transistor		Chapter 6
PNP (bipolar) transistor		Chapter 6
Phototransistor (PNP)		Chapter 8
Diode		Chapter 6
Zener diode		Chapter 6
Light-emitting diode (LED)		Chapter 6
Photodiode		Chapter 8

The circuit symbol for an op amp is really a block diagram symbol, because it represents the interconnection of dozens of individual components in a nearly complete circuit (power is external to the op amp). Schematics always use a single symbol to represent the entire circuit, which is packaged as an integrated circuit (IC).

Digital logic and IC components

Digital electronic components, such as logic gates, manipulate digital signals that consist of just two possible voltage levels (high or low). Inside each digital component is a nearly complete circuit (power is external) consisting of individual transistors or other analogue components. Circuit symbols for digital components are really block diagram symbols used to represent the interconnection of individual components that make up the logic. You can build the logic from scratch or obtain it in the form of an IC. Logic ICs usually contain several gates (not necessarily all the same type) sharing a single power connection.

Table 10-3 shows the circuit symbols for individual digital logic gates and logic ICs. You can find detailed information about the functionality of each logic gate in Chapter 7.

Table 10-3 Symbols for Logic Gates and Digital ICs

Component	*Symbol*
AND	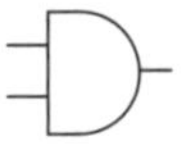
NAND	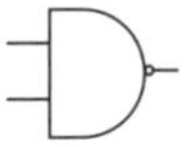
OR	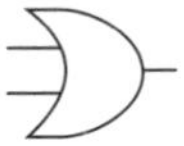
NOR	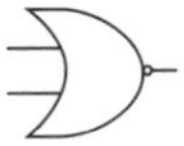
XOR	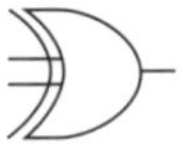
Inverter (NOT)	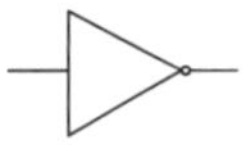

Some schematics show individual logic gates and others show connections to the full IC. You can see an example of each in Figure 10-6.

The 74HC00 IC shown in Figure 10-6 is a CMOS quad 2-input NAND gate. In the top circuit diagram, each NAND gate is labelled '¼ 74HC00' because it's one of four NAND gates in the IC. (This type of gate labelling is common in digital circuit schematics.) Note that the fourth NAND gate isn't used in this particular circuit (which is why Pins 11, 12 and 13 aren't used).

Whether the schematic uses individual gates or an entire IC package, it usually notes the external power connections. If it doesn't, you have to look up the pinout of the device on the IC datasheet to determine how to connect power. We discuss pinouts and datasheets in Chapter 7.

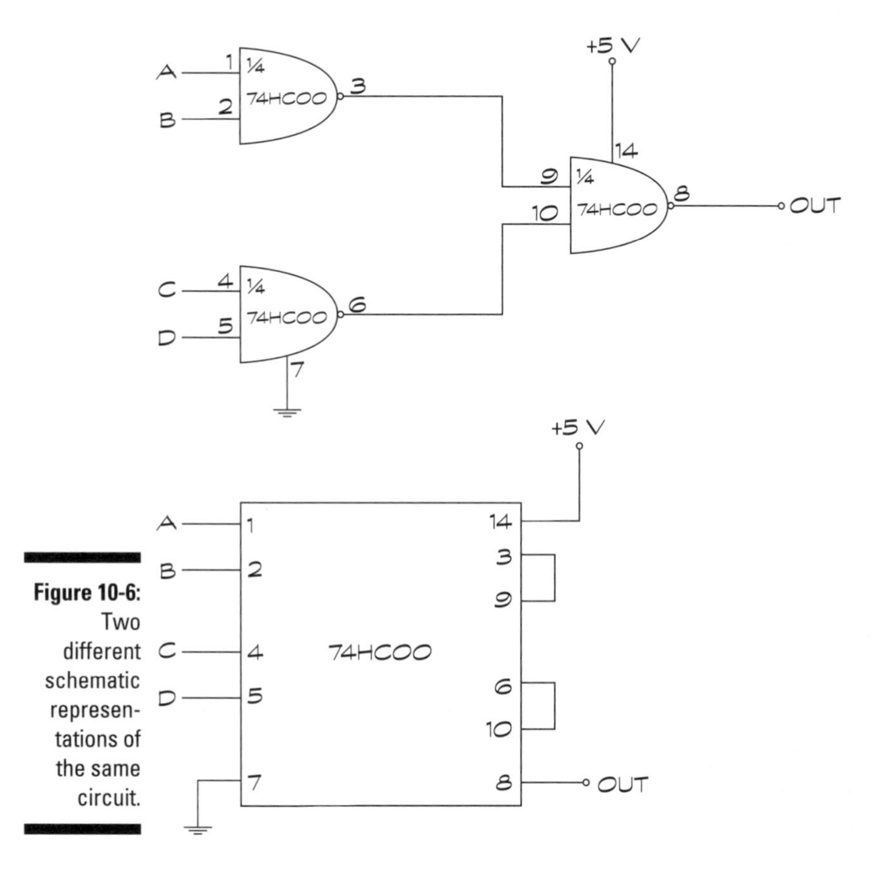

Figure 10-6: Two different schematic representations of the same circuit.

Many more digital ICs are available than just those containing logic gates. Most ICs (op amps, which are linear ICs, are an exception) are shown the same way in schematics: as a rectangle, labelled with a reference ID (such as IC1) or the part number (such as 74CH00), with numbered pin connections. The function of the IC is usually determined by looking up the part number, but the occasional schematic may include a functional label, such as 'one shot'.

Odds and ends

Table 10-4 lists the symbols for switches and relays. Refer to Chapter 8 for detailed information on each of these components.

Table 10-4 Symbols for Switches and Relays

Component	*Symbol*
SPST switch	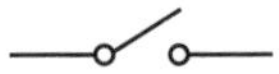
SPDT switch	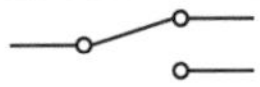
DPST switch	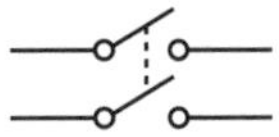
DPDT switch	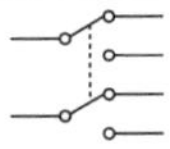
Switch spring return	
Relay	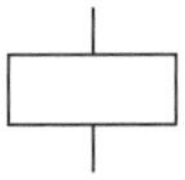

Table 10-5 lists the symbols for various input transducers (sensors) and output transducers. You can read about most of these components in Chapter 8, and you can read about LEDs in Chapter 6.

Table 10-5 Symbols for Input and Output Transducers

Component	*Symbol*
Microphone	
Thermistor	
Aerial	

Component	*Symbol*
Incandescent lamp	
Speaker	
Piezoelectric buzzer	

Some circuits accept inputs from and send outputs to other circuits or devices. Schematics often show what looks like a loose wire leading into or out of the circuit. Usually, it's labelled something like 'signal input', 'input from doodad #1' or 'output' so that you know you're supposed to connect something up to it. (You connect one wire of the signal to this input point and the other to signal ground.) Other schematics may show a symbol for a specific connector, such as a *plug* and *jack* pair, which connect an output signal from one device to the input of another device. (We discuss plugs and jacks in Chapter 8.)

Table 10-6 shows a few of the ways in which input and output connections to other circuits are shown in schematics. Symbols for input/output connections can vary greatly among schematics. The symbols that we use in this book are among the most commonly used. Although the exact style of the symbol may vary from one schematic to the next, the idea is the same: a connection is to be made to something external to the circuit.

Table 10-6 Symbols for Connections to Other Circuits

Name	*Symbol*
Unspecified input	
Unspecified output	

Exploring a Schematic

Now that you're familiar with the ABCs of schematics, the time is right to put all your knowledge together and walk through each part of a simple schematic. The schematic shown in Figure 10-7 shows the LED flasher circuit used in Chapter 14. This circuit controls the on/off blinking of an LED, with the blinking rate controlled by turning the knob of a potentiometer (variable resistor).

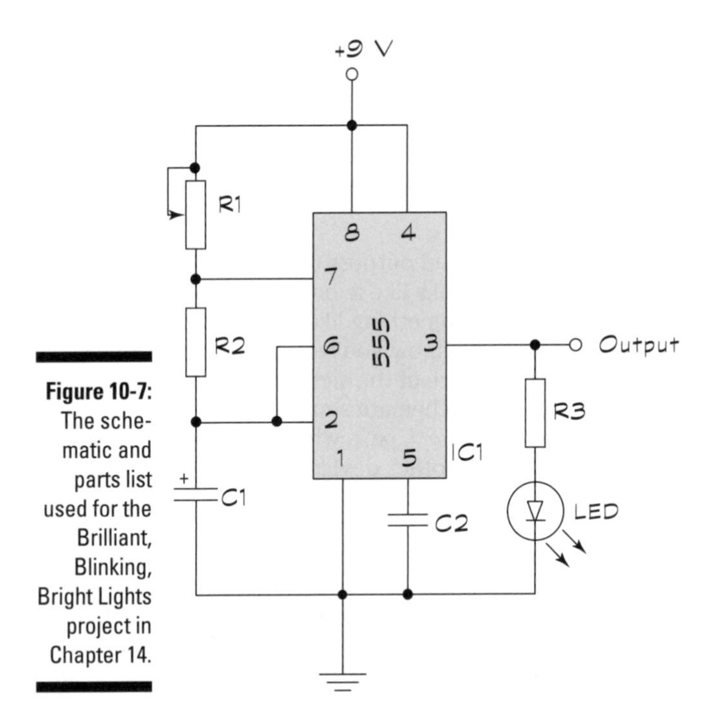

Figure 10-7: The schematic and parts list used for the Brilliant, Blinking, Bright Lights project in Chapter 14.

Here's what this schematic is saying:

- **IC1, an 8-pin 555 timer IC**, is at the heart of the schematic, with all eight pins connecting to parts of the circuit. Pins 2 and 6 are connected together.
- **A 9-volt power supply**, which can be a 9-volt battery, powers the circuit:
  - The positive terminal of the power supply is connected to Pins 4 and 8 of IC1, and to one fixed lead and the variable contact (wiper) lead of R1, which is a potentiometer (variable resistor).
  - The negative terminal of the power supply (shown as the common ground connection) is connected to Pin 1 of IC1, to the negative side of capacitor C1, to capacitor C2 and to the cathode (negative side) of the LED.

- **R1 is a potentiometer** with one fixed lead connected to Pin 7 of IC1 and to resistor R2, and both the other fixed lead and the wiper lead connected to the positive battery terminal (and to Pins 4 and 8 of IC1).
- **R2 is a fixed resistor** with one lead connected to Pin 7 of IC1 and to one fixed lead of R1, and the other lead connected to Pins 2 and 6 of IC1 and the positive side of capacitor C1.
- **C1 is a polarised capacitor** with its positive side connected to R2 and to Pins 2 and 6 of IC1, and its negative side connected to the negative battery terminal (as well as to Pin 1 of IC1, capacitor C2 and the cathode of the LED).
- **C2 is a non-polarised capacitor** connected on one side to Pin 5 of IC1 and on the other side to the negative battery terminal (as well as the negative side of capacitor C1, Pin 1 of IC1 and the cathode of the LED).
- **The anode of the LED** is connected to resistor R3 and the cathode of the LED is connected to the negative battery terminal (as well as to the negative side of capacitor C1, capacitor C2 and Pin 1 of IC1).
- **R3 is a fixed resistor** connected between Pin 3 of IC1 and the anode of the LED.
- **The output** shown at Pin 3 of IC1 can be used as a signal source (input) for another stage of circuitry.

Each item in the above walk-through list focuses on one circuit component and its connections. Although we mention the same connections multiple times in the list, check and double-check your circuit connections by making sure that *each lead or pin of each individual component* is connected correctly. (Ever hear the general rule 'measure twice, cut once'? Well, the same principle applies here.) You can't be too careful when it comes to connecting up electronic components.

Alternative Schematic Drawing Styles

The schematic symbols in this chapter are mostly those that electronics designers use in the UK and Europe, plus a few others you may come across tossed in so that you're not thrown when you see them. Different countries (or continents) use some slightly different schematic symbols. You sometimes may need to do a little bit of schematic translation work to understand all the components.

Figure 10-8 shows a few more schematic symbols that crop up in schematics drawn in America and elsewhere. Notice the obvious differences in the resistor symbols, both fixed and variable.

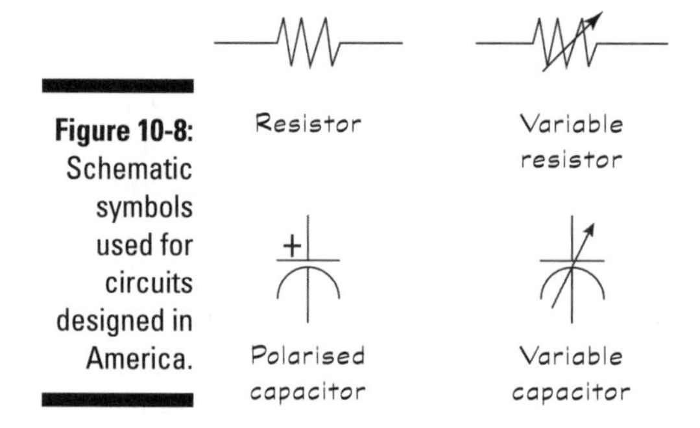

Figure 10-8: Schematic symbols used for circuits designed in America.

Americans and Europeans organise their symbols differently. In the United States, they express resistor values over 1,000 ohms in the form of 6.8K or 10.2K, with the K following the value. The European schematic style sometimes leaves out the decimal point. In the UK, you're quite likely to see resistor values expressed in the form 6K8 or 10K2. This style substitutes the K (which stands for kilohms, or thousands of ohms) for the decimal point.

You may encounter a few other variations in schematic drawing styles, but all are fairly self-explanatory, and the differences aren't substantial. After you discover how to use one style of drawing, the others come easily.

Chapter 11

Constructing Circuits

In This Chapter

- Creating no-fuss, no-mess circuits with solderless breadboards
- Soldering – safely – like the pros
- Owning up to and fixing soldering mistakes
- Making your circuits permanent

You've carefully set up your workbench, strategically positioned your shiny new toys – er, tools – to impress your friends, shopped around for the best deals on components and now you're ready to get down to work and build some loud, flashing circuits. So how do you transform a two-dimensional circuit diagram into a real live working (maybe even moving) electronic circuit?

In this chapter, we show you various ways to connect up electronic components into circuits that push electrons around at your command. We describe how to make quick and easy, temporary but flexible, circuits on solderless breadboards, which provide the ideal platform for testing and tweaking your designs.

We also give you the low-down on how to fuse components together safely using a molten metallic mix called solder. And we introduce you to several ways to make your circuits permanent using soldering and/or wire-wrapping techniques together with an assortment of today's most popular circuit boards.

So arm yourself with screwdrivers, needle-nose pliers and a soldering iron, and don your safety glasses and anti-static wrist strap: you're about to enter an electronics building site!

Taking a Look at Solderless Breadboards

Solderless breadboards, also called prototyping boards or plugblocks, allow you to build (and dismantle) temporary circuits easily. These reusable rectangular boards have several hundred square sockets, or *contact holes*, into which you plug your components (for instance, resistors, capacitors, diodes, transistors and integrated circuits). Flexible metal strips run underneath the surface to electrically connect together groups of these contact holes.

You poke a wire or lead into a hole in the breadboard, and it makes contact with the underlying metal. By plugging in components in just the right way and running wires from your breadboard to your power supply, you can build a working circuit without permanently fixing components together.

We highly recommend that you use a solderless breadboard (or two) when you first build a circuit. That way, you can test the circuit to make sure that it works properly and carry out any necessary adjustments.

Often, you can improve on the performance of a circuit just by tweaking a few component values. You can easily make such changes by simply removing one component and inserting another on the board – without having to unsolder and resolder. (We discuss soldering in the 'Soldering On' section later in this chapter.)

When you're sure that your circuit works the way you want it to, you can create a permanent circuit on other types of boards, as described in the later section 'Committing to a Permanent Circuit'.

Solderless breadboards are designed for low-voltage DC circuits. Never use a breadboard for 230 VAC household mains electricity because doing so is extremely dangerous, frying the board, you or both!

Exploring a solderless breadboard, inside and out

The photo in Figure 11-1 shows a basic solderless breadboard with white lines added to help you visualise the underlying connections between contact holes. In the centre of the board, the holes link vertically in blocks of five (for instance, A, B, C, D and E in column 30 are all connected together, and F, G, H, I and J in column 30 are all connected together). No connections exist across the centre gap between rows E and F. You can straddle an integrated circuit (IC) across that centre gap and instantly set up independent sets of connections for each of its pins.

Across the top and bottom of a breadboard, holes are linked horizontally, but you can't tell exactly how many holes are electrically connected just by looking at the board. For the 400-hole board in Figure 11-1, all 25 contacts in each of the four rows across the top and bottom are electrically connected.

In many larger breadboards, such as the 840-hole board pictured in Figure 11-2, a break exists in the connections halfway across each row. We placed small jumper wires between neighbouring contacts to create a 50-point connection in each row. In some boards, the two rows across the top are electrically connected, as are the two rows across the bottom.

You can use a multimeter to check whether or not two points within a row or between rows are electrically connected. Stick a jumper wire in each hole and then touch one multimeter probe to one wire and the other probe to the other wire. If you get a low ohm reading, you know that the two points are connected together. If you get an infinite ohms reading, you know that they're not connected. See Chapter 12 for more about testing things with your multimeter.

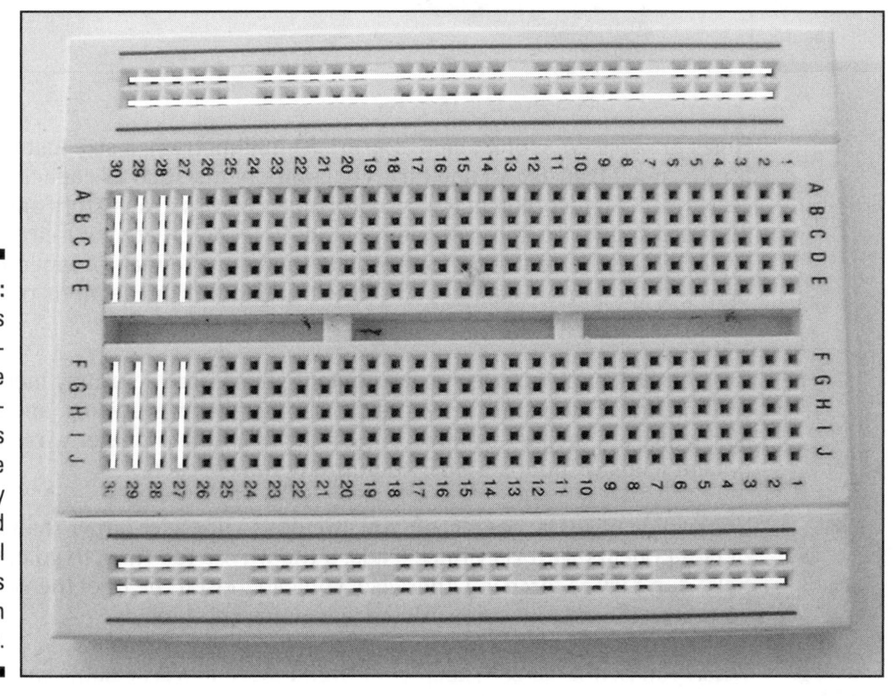

Figure 11-1: Solderless breadboards have lines of contact holes that are electrically connected in small groups underneath the surface.

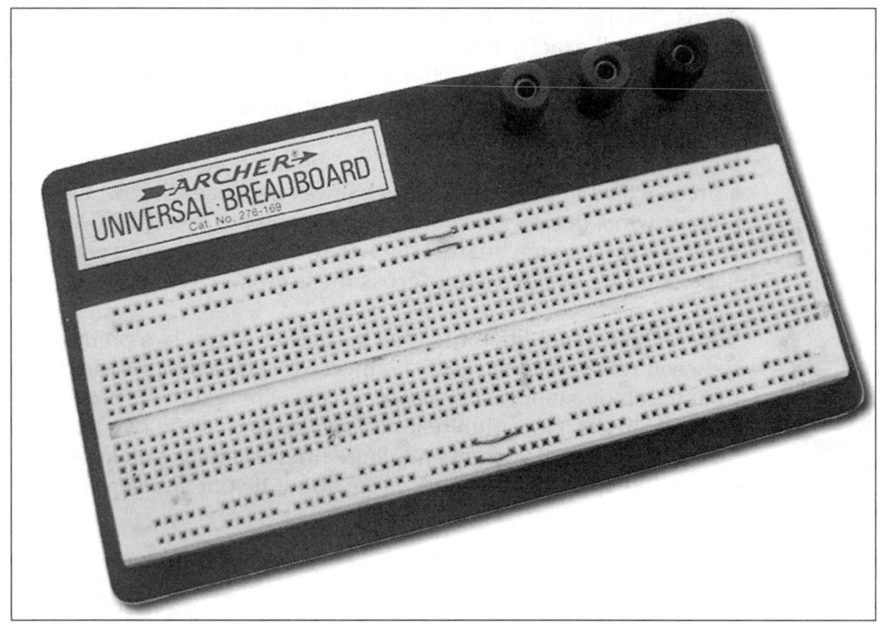

Figure 11-2: For larger circuits, you can use bigger solderless breadboards, such as this 840-hole board.

Holes are spaced 1/10 of an inch apart (2.54 millimetres), a size just right for ICs, most transistors and discrete components such as capacitors and resistors. You just plug in ICs, resistors, capacitors, transistors and 20- or 22-gauge solid wire in the proper contact holes to create your circuit. You typically use the centre two sections of the board to make connections between components, and you use the top and bottom sections of the board to connect power.

Breadboard manufacturers make contact strips from a springy metal coated with plating. The plating prevents the contacts from oxidising, and the springiness of the metal allows you to use different diameter wires and component leads without seriously deforming the contacts.

You can damage the contacts if you attempt to use wire larger than 20-gauge or components with very thick leads. If the wire is too thick to go into the hole, don't try to force it. Otherwise, you can loosen the fit of the contact, and your breadboard may not work the way you want it to.

When you're not using it, keep your breadboard in a resealable sandwich bag to keep out the dust. Dirty contacts make for poor electrical connections. Although you can use a spray-on electrical cleaner to remove dust and other contaminants, you make things easier on yourself by keeping the breadboard clean in the first place.

Sizing up the breadboards

Solderless breadboards come in many sizes. Smaller breadboards (with 400 to 550 holes) accommodate designs with up to three or four ICs plus a small handful of other discrete components. Larger boards, such as the 840-hole board shown in the earlier Figure 11-2, provide more flexibility and accommodate five or more ICs.

If you're into really elaborate design work, you can buy extra large breadboards with anywhere from 1,660 to more than 3,200 contact holes. These boards can handle one to three dozen ICs plus other discrete components.

Don't overdo it when buying a solderless breadboard. You don't need a breadboard the size of Buckinghamshire if you're only making small to medium-sized circuits, such as the ones we show you in Chapter 14.

If you get halfway through designing a circuit and find that you need a little more breadboard space, you can always make connections between two breadboards. Some solderless breadboards even have interlocking ridges so that you can put several together to make a larger breadboard.

Building Circuits on Solderless Breadboards

Essentially, breadboarding consists of sticking components into the board, connecting power to the board and making connections with wire. But you can do these things in a right way and a wrong way. This section gives you the low-down on what type of wire to use, efficient breadboarding techniques and how to give your board a neat, logical design.

Preparing your parts and tools

Before you start randomly sticking things into your breadboard, make sure that you have everything you need. Check the parts list – the list of electronic ingredients you need to build your circuit – and set aside the required components. Gather together essential tools, such as needle-nose pliers, wire cutters and a wire stripper. Ensure that all component leads are suitable for inserting into breadboard holes.

Clip long leads, if possible, so that the components lie flat and snug against the board. (Don't worry if you can't reuse them for another circuit – they're cheap enough.) You may need to solder single-core wires to the terminals of any components that don't already have leads (see the section 'Soldering On' later in this chapter for how to carry out this job).

Familiarise yourself with the polarity of parts; that is, which leads are what on transistors, potentiometers and ICs. And finally, get interconnect wires ready, as described in the very next section.

Saving time with pre-stripped wires

Many of the connections between components on your breadboard are made by the breadboard itself, underneath the surface, but when you can't make a direct connection via the board, you use interconnect wires (sometimes called *jumper wires*). You use solid 20- to 22-gauge (0.8 or 0.7 millimetres in diameter) insulated wire to connect components together on your breadboard. Thicker or thinner wire doesn't work well in breadboards: when it's too thick, the wire doesn't go into the holes; when it's too thin, the electrical contact is dodgy.

Don't use stranded wire in a breadboard. The individual strands can break off, lodging inside the metal contacts of the breadboard and causing short circuits. This would have you tearing your hair out!

While you're buying your breadboard, pick up a set of pre-stripped jumper wires, as we suggest in Chapter 9. (Believe us, the small cost is worthwhile.) These wires come in a variety of lengths and are already stripped (obviously) and bent, ready for you to use in breadboards. For a tenner, you can get an assortment of several hundred. Or you can buy a reel of wire, cut it into little bits and spend hours painstakingly stripping off about 8 millimetres of the insulation from each end of each little bit. The choice is yours!

Even if you purchase a large assortment of pre-stripped wires, the time comes when you have to make an interconnect wire or two of your own. You start with 20- or 22-gauge (0.8 or 0.7 millimetres in diameter) wire and cut it to the length you need.

If you have a wire stripper with a gauge selection dial, set the dial for the gauge of wire that you're using. Other wire strippers may have several cutting notches labelled for various gauges. Using one of these gauge-specific devices instead of a generic wire stripper prevents you from nicking the wire when stripping off the insulation.

Nicks weaken the wire, and a weak wire can get stuck inside a breadboard hole and ruin your whole day.

To make your own breadboard wire, follow these steps:

1. **Cut the wire to the length you need using a wire cutter.**
2. **Strip off about 0.5 to 1 centimetre of insulation from each end.**

   If you use a gauge-specific tool, insert one end of the wire into the stripping tool, hold the other end with a pair of needle-nosed pliers and draw the wire through the stripping tool. If you use a generic wire stripper, how much you squeeze the tool around the wire provides the gauge control: too much, and you nick the wire; too little, and you don't cut through all the insulation.
3. **Bend the exposed ends of wire at a right angle (90°).**

   You can use needle-nosed pliers to do this job.

Laying out your circuit

You've got your parts and tools ready, a schematic (the circuit diagram, as we discuss in Chapter 10) in your hand and now you want to build your circuit on your breadboard. But where should you start? What's the best way to connect everything?

Welcome to the world of circuit layout – figuring out where everything should go on the board so that it all fits together and is neat, tidy and works perfectly. Don't expect your circuit layout to look exactly like your schematic – that's not only difficult to do, but usually impractical. You can, however, orient key circuit elements so that your circuit is easier to understand and debug.

When building a circuit on a breadboard, concentrate on the connections between components rather than the position of components in your schematic.

Here are some guidelines for building your breadboard circuit:

1. **Use one of the top rails (long rows) for the positive power supply, and one of the bottom rails for ground (and the negative power supply, if one exists).**

   These rails give you plenty of interconnected sockets so that you can easily connect components to power and ground.

2. **Position any circuit inputs on the left side of the board and outputs on the right side.**

   Plan your component layout to minimise the number of jumper wires. The more wiring you have to insert, the more crowded and confusing the board becomes.

3. **Place ICs first, straddling the centre gap.**

   Allow at least three – preferably ten – columns of holes between each IC. You can use a chip inserter/extractor tool to implant and remove ICs to reduce the chances of damaging the IC while handling it. (If you're working with highly static sensitive CMOS chips, which we discuss in Chapter 9, you need to ground the tool to eliminate stray static electricity.)

4. **Work your way around each IC, starting from Pin 1, inserting the components that connect to each pin.**

   Then insert any additional components to complete the circuit. Use needle-nose pliers to bend leads and wires to a 90° angle and insert them into the sockets, keeping leads and wires as close to the board as possible to prevent them from getting knocked loose.

5. **If you're getting cramped, colonise empty areas of the board.**

   If your circuit calls for common connection points in addition to power and you don't have enough points in one column of holes, use longer pieces of wire to bring the connection out to another part of the board where you have more space. You can make the common connection point one or two columns between a couple of ICs, for instance.

Figure 11-3 shows a resistor, jumper wire and light-emitting diode (LED) inserted into a breadboard.

Don't worry about urban sprawl on your breadboards. You're better off placing components a little farther apart than jamming them too close together. Keeping a lot of distance between ICs and components also helps you to tweak and refine the circuit. You can more easily add parts without disturbing the existing ones.

Messy wiring makes debugging a circuit difficult, and a tangle of wires greatly increases the chance of mistakes. Wires pull out when you don't want them to, or the circuit can malfunction altogether. To avoid chaos, take the time to plan and construct your breadboard circuits carefully. The extra effort can save you lots of time and frustration later on.

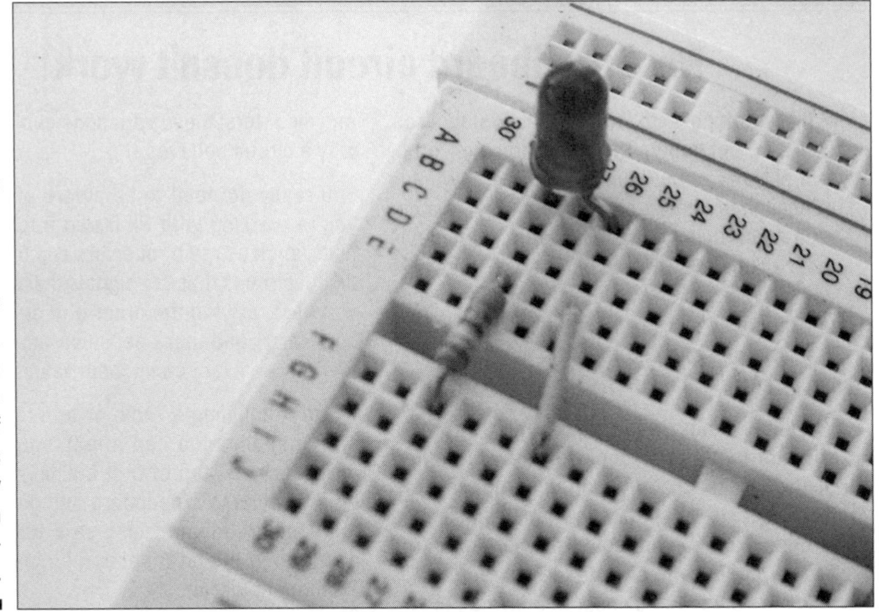

Figure 11-3: Strip and bend the ends of jump wire, and then clip the leads of components so that they fit neatly on the breadboard.

Avoiding damaged circuits

Here are a few other things you need to know in order to keep your breadboard and circuits in good working order:

- When you're using CMOS chips, build the rest of the circuit first. Be sure to provide connections for the positive and negative power supply and to connect all inputs (tie those inputs that you're not using to the positive or negative supply rail). When you're ready to test the circuit, put in the CMOS IC.
- Never expose a breadboard to heat, because you can permanently damage the plastic. ICs and other components that become very hot (because of a short circuit or excess current, for example) may melt the plastic underneath them. To check for any signs of overheating, power up your circuit for half a minute, turn it off again and lightly tap the top of the components with your fingertip.
- If a small piece of a lead or wire becomes lodged in a socket, use needle-nose pliers to pull it gently out of the hole – make sure that the power is switched off!

My breadboard circuit doesn't work!

While working with solderless breadboards, you may encounter the fairly common problem of *stray capacitance*, which is unwanted capacitance (stored electrical energy) in a circuit. All circuits have an inherent capacitance that can't be avoided, but when lots of wires are going every which way, the capacitance can unexpectedly increase. At a certain point (and it differs from one circuit to the next), this stray capacitance can cause the circuit to misbehave.

Because solderless breadboards contain strips of metal and require somewhat longer component leads, they tend to introduce a fair amount of stray capacitance into unsuspecting circuits. As a result, solderless breadboards have a tendency to change the characteristics of some components, most notably capacitors and inductors; these variations can change the way a circuit behaves.

You really do need to be aware of this fact if you're working with RF (radio frequency) circuits, such as radio receivers and transmitters, digital circuits that use signals that change at a very fast rate (on the order of a couple of million Hertz) and more sensitive timing circuits that rely on exact component values.

If you're building a radio or other circuit that stray capacitance can affect, you may have to forego the step of first building the circuit on a solderless breadboard and go straight to a solder breadboard or a strip board, as we describe in the later section 'Committing to a Permanent Circuit'.

- You won't always be able to finish and test a circuit in one sitting. If you have to put your breadboard circuit aside for a while, put it out of the reach of children, animals and the overly curious. (Chapter 9 describes how to set up a safe work area.)
- Never use a solderless breadboard to carry 230 VAC household mains electricity.

Soldering On

Soldering is the method you use to make conductive connections between components and/or wire. You use a device called a *soldering iron* to melt a soft metal called *solder* so that the solder flows around the two metal leads you're joining. When the soldering iron is removed, the solder cools and forms a conductive physical joint, known as a *solder joint*, between the wires or component leads.

You may be thinking: why should I care about soldering when I plan to use solder*less* breadboards for my circuit construction projects? The answer is because almost every electronics project involves a certain amount of soldering.

For instance, you can purchase components, such as potentiometers, switches and microphones, that don't come with leads, which means that you need to solder two or more wires to their terminals to create leads so that you can connect them to your breadboard.

Of course, you also use this technique extensively when you build permanent circuits on solder boards, strip boards or printed circuit boards, as we describe in the later section, 'Committing to a Permanent Circuit'.

Preparing to solder

To get your soldering jobs done, you need a soldering iron (25–30 watts), a spool of standard solder, a secure soldering stand and a small sponge. (Take a look at Chapter 9 for detailed information on how to choose soldering equipment for your electronics projects.)

Make sure that you secure your soldering iron in its stand and position it in a safe place on your workbench, where it's unlikely to be knocked over.

Gather up a few other items as follows:

- Anti-static wrist strap (which we describe in Chapter 9).
- Crocodile clip (which doubles as a heat sink to take heat away from temperature-sensitive components).
- Isopropyl alcohol.
- Piece of paper and a pencil.
- Safety glasses (to protect your eyes from spluttering solder).
- Some sticky tape.

Place each part that you need to solder on the paper, securing it with sticky tape. Write a label, such as R1, on the paper next to the part so that it matches the label on your schematic. Put on your safety glasses and anti-static wrist strap and make sure that your work area is properly ventilated.

Wet the sponge, squeezing out excess water. Turn on the soldering iron, wait a minute or so for it to heat up (to about 400° C), and then wet the tip of the soldering iron by briefly touching it to the sponge. If the tip is new, *tin* it before soldering to help prevent solder from sticking to the tip. You tin the tip by applying a small amount of molten solder to it. (Sticky solder can form an ugly globule, which can wreak havoc if it falls off into your circuit.) Then wipe off any excess solder on the sponge.

Periodically tin your solder tip to keep it clean. You can also purchase soldering tip cleaners if dirt becomes caked on and you just can't get it off during regular tip re-tinning.

Soldering for success

Successful soldering requires that you follow some simple steps and get a lot of practice. Remember that timing is the secret of good soldering (just like comedy). As you read through these steps, pay close attention to words such as 'immediately' and 'a few seconds' – and interpret them literally. Here are the steps for soldering a joint:

1. **Clean the metal surfaces to be soldered.**

   Wipe leads, wire ends or etched circuit board surfaces (which we describe in 'Committing to a Permanent Circuit', later in this chapter) with isopropyl alcohol so that the solder adheres better.

   Let surfaces dry thoroughly before soldering: you don't want them to catch on fire!

2. **Secure the items being joined.**

   You can use a third hand clamp (as we describe in Chapter 9) or a vice and a crocodile clip to hold a discrete component steady as you solder a wire to it, or use needle-nose pliers to hold a component in place over a circuit board.

3. **Position the soldering iron.**

   Hold the iron like a pen, with the tip at a 30 to 45° angle to the work surface (see Figure 11-4).

4. **Apply the tip to the joint you're working on (as shown in Figure 11-4).**

   Do not apply heat directly to the solder. Allow the metal a few seconds to heat up.

5. **Feed the cold solder to the heated metal area.**

   The solder melts and flows around the joint within a couple of seconds.

6. **Remove the solder immediately, and then the iron.**

   As you remove them, hold the component still until the solder cools and the joint solidifies.

7. **Place the soldering iron securely in its stand.**

Never place a hot soldering iron on your work surface.

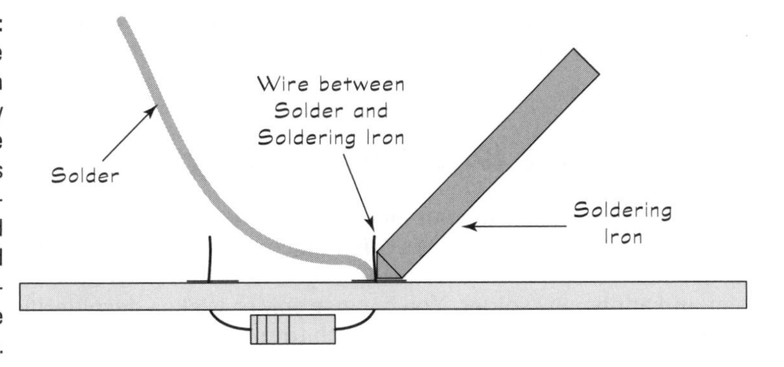

Figure 11-4: Holding the iron at an angle, apply heat to the metal parts you're soldering and then feed the cold solder into the joint.

Be careful to use just the right amount of solder (which means applying solder for just the right amount of time): use too little, and you form a weak connection; use too much, and the solder may form globs that can cause short circuits.

You can damage many electronic components if you expose them to prolonged or excessive heat, so take care to apply the soldering iron only long enough to heat a component lead for proper soldering – no more, no less.

To avoid damaging heat-sensitive components, such as transistors, attach a crocodile clip to the lead between the intended solder joint and the body of the component. This way, any excess heat is drawn away through the clip and doesn't damage the component.

Inspecting the joint

After soldering, you need to look at the joint and check that it's strong and conductive. The cooled solder joint should be shiny, not dull, and be able to withstand a gentle tug from one side. If you soldered a lead to a circuit board, you should be able to see a fillet (a volcano-shaped raised area of solder) at the joint. If you see dull solder or jagged peaks, you know that you've a *cold solder joint.* Cold joints are physically weaker than properly made joints, and they don't conduct electricity as efficiently.

Cold solder joints can form when you move the component while the solder is still cooling, if the joint is dirty or oily or when you fail to heat the solder properly. Resoldering without first desoldering often produces cold solder joints because the original solder isn't heated enough.

If you have a cold solder joint, the best solution is to remove the existing solder completely (as we describe in the next section, 'Desoldering when necessary'), clean the surfaces with isopropyl alcohol and re-apply fresh solder.

Desoldering when necessary

At some point or other in your electronics career, you're bound to run into a cold solder joint, a backwards-oriented component or some other soldering mishap. To correct these mistakes, you need to remove the solder at the joint and apply new solder. You can use solder wick, a desoldering pump (also known as a solder sucker) or both to remove solder from the joint.

Use a flat braid of copper known as *solder wick* (also called solder braid, desoldering braid and variations of those terms) to remove hard-to-reach solder. You place it over unwanted solder and apply heat. When the solder reaches its melting point, it adheres to the solder wick, which you then remove and dispose of.

Exercise care when using solder wick because if you touch the hot braid, you can get a serious burn.

A *desoldering pump* uses a vacuum to suck up excess solder that you melt with your soldering iron. Two types are available: a spring-loaded plunger and a bulb.

You use a spring-loaded pump as follows:

1. **Depress the plunger and position the nozzle over the joint that you want to remove.**
2. **Position the soldering iron tip carefully into the joint to heat the solder, avoiding contact with the end of the pump.**
3. **Release the plunger to suck up the solder as the solder begins to flow.**
4. **Expel the solder from the pump (into a waste receptacle) by depressing the plunger one more time.**

Repeat these steps as needed to remove as much of the old solder as possible.

Don't store a desolder pump with a cocked plunger because the rubber seal can become deformed, diminishing the vacuum to the extent that the pump may be unable to suck up any solder.

A bulb desolder pump works a lot like the spring-loaded variety, except that you squeeze the bulb to suck up the solder. You may find using this pump difficult unless you mount the bulb on the soldering iron.

You can also buy an all-in-one electric desolder pump that melts and removes the solder in one go.

Cooling down after soldering

Get into the habit of unplugging – not just shutting off – your soldering iron when you finish your soldering work. Brush the tip of the still warm iron against a damp sponge to clean off excess solder.

When the iron has cooled down, you can use tip cleaner paste to remove stubborn dirt. Make sure that the iron is completely cool before storing it. Place your solder reel in a plastic bag to keep it from getting dirty.

Ensuring safe soldering

Even if you plan to solder just one connection, you need to take the appropriate precautions to protect yourself – and those around you.

The iron reaches temperatures of over 400° C. You (or a nearby friend or pet) may be the unwittingly recipient of popping, spluttering solder if you meet the occasional air pocket or other impurity in your reel. Just one small drop of solder hitting you in the eye, or a tumbling soldering iron that finds rest on a foot, can ruin a day – and a friendship.

Set up your work area – and yourself – with soldering safety in mind (check out Chapter 9 for more on electronics safety).

Make sure that the room is well-ventilated, you have the iron placed snugly in its stand and you position the electrical cord to avoid snags. Wear shoes (no flip-flops!), safety glasses and an anti-static wrist strap when soldering.

Avoid bringing your face too close to hot solder, which gives off fumes and may splutter. Keep your face to one side and use a magnifying glass if necessary, to see tiny components that you're soldering.

Never solder a powered circuit; make sure that the battery or other power supply is disconnected before applying your soldering iron to components.

If your iron has an adjustable temperature control, dial up the recommended setting for the solder you're using.

If your soldering iron accidentally goes belly-up, stand back and let it fall. If you try to grab it, Murphy's Law says you'll grab the hot end.

Finally, always unplug your soldering iron when you're done and promptly wash your hands.

Committing to a Permanent Circuit

So you've perfected the world's greatest circuit, and you want to make it permanent. You have a few options for how to transfer your circuit from the solderless breadboard and make connections that last.

Solder breadboards and strip boards are two popular foundations for permanent circuits. You can solder your connections onto either surface, but if you want additional flexibility, you can use a technique called wire-wrapping to make connections on a strip board. This section covers the various methods of making a commitment to your circuit.

Moving your circuit to a solder breadboard

The solder breadboard (also called a solder board, an experimenter's PC board or a universal solder board) allows you to take any design that you create on a solderless breadboard and make it permanent. You can transfer your design to a solder board easily because the solder breadboard has the exact same layout as the solderless breadboard.

To transfer your design, you simply pick the parts off your solderless breadboard, insert them in the solder breadboard and solder them into place in the corresponding spots. Use wires as you did in the original solderless breadboard to connect components that aren't electrically connected by the metal strips of the circuit board.

If you design a really small circuit, you can use just one half of a solder breadboard. Before transferring the components, cut the solder breadboard with a hacksaw. Clean the portion of the board that you want to use and solder away.

Soldering sockets for ICs

When you build circuit boards that include integrated circuits, instead of soldering the IC directly onto the board, use an IC socket. You solder the socket onto the board, and then, when you're done soldering, you plug in the IC and hit the switch.

IC sockets come in different shapes and sizes so as to match the integrated circuits they're meant to work with. For example, if you have a 16-pin integrated circuit, choose a 16-pin socket.

Here are some good reasons for using IC sockets:

- Soldering a circuit board can generate static. By soldering to the socket rather than the actual IC, you can avoid ruining CMOS or other static-sensitive ICs.
- ICs are often one of the first things to go wrong when you're experimenting with electronics. The ability to pull out a chip you suspect is bad and replace it with a working one makes troubleshooting a whole lot easier.
- You can share an expensive IC, such as a microcontroller, among several circuits. Just pull the part out of one socket and plug it into another.

Sockets are available in all sizes to match the different pin arrangements of integrated circuits. They don't cost much – just a couple of pennies for each socket.

Leave space at the corners of the board so that you can drill mounting holes. You use these holes to secure the board inside whatever enclosure your project provides (such as the chassis of a robot). Alternatively, you can secure the board to a frame or within an enclosure by using double-sided foam tape. The tape cushions the board and prevents breakage, and the thickness of the foam prevents the underside of the board from touching the chassis.

Solder breadboards have one main disadvantage: they don't use space very efficiently. Unless you cram the components onto the board, the breadboard limits you to building circuits with only two or four integrated circuits and a handful of discrete components. In time, you can figure out how to conserve space and make good use of the area on a solder board.

Prototyping with strip boards

You can also use a pre-drilled strip board for your projects. These general-purpose circuit boards go by many names, such as perf, grid or matrix boards. Some are bare, but most contain copper pads and traces for wiring. All styles are designed for you to use with ICs and other modern-day electronics components. The holes are spaced 1⁄10 of an inch apart (2.54 millimetres). Lots of subtle variations are available, so look around and choose the one that suits your project (see Figure 11-5).

A buss runs throughout the circuit board so that you can easily attach components to it. Many strip boards have at least two busses, one for power and one for ground. The busses run up and down the board, as you can see in Figure 11-6. This layout works ideally for circuits that use many ICs. Alternating the busses for the power supply and ground also help to reduce undesirable inductive and capacitive effects.

To construct circuits using bare strip boards, which don't have coppers pads and traces for wire, you use the wire-wrapping method (which we discuss in the following section). Most strip boards contain pads and traces, so you can solder components directly onto them or use wire-wrapping.

You can use a strip board just as you use a solder breadboard (see the preceding section 'Moving your circuit to a solder breadboard'). After cleaning the board so that the copper pads and traces are bright and shiny, plug the components into the board and solder them into place. Use insulated wire to connect components that aren't adjacent to one another.

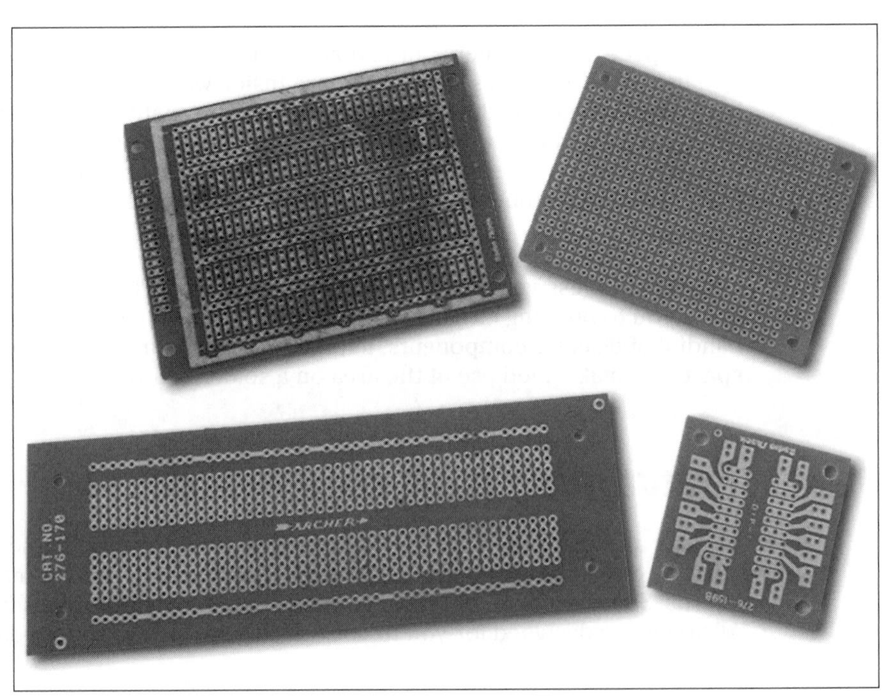

Figure 11-5: Pre-drilled boards come in many shapes, sizes and other variations, but all are ready for you to add electronic components after a quick clean if necessary.

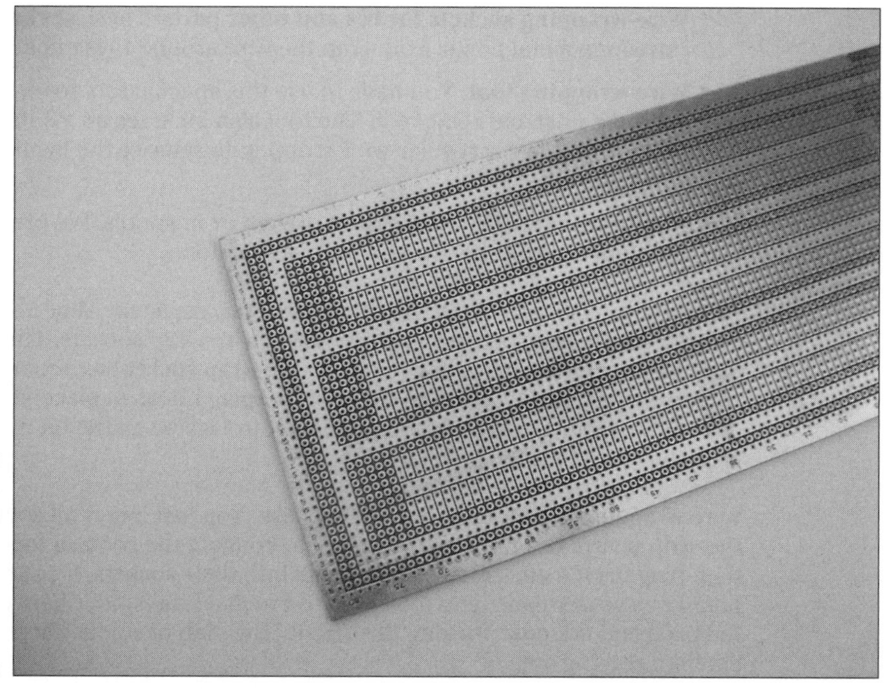

Figure 11-6: Several busses run up and down this strip board.

Wrapping with wires

Wire-wrapping is a point-to-point wiring system that uses a special tool and extra-fine 28- or 30-gauge wrapping wire. When you carry out the process properly, wire-wrapped circuits are as sturdy as soldered circuits. And you have the added benefit of being able to make modifications and corrections without the hassle of desoldering and resoldering.

You have to limit wire-wrapping to projects that use only low-voltage DC. The technique isn't for anything that requires a lot of current, because the wire you use isn't large enough to carry much current.

To wire-wrap, you need the following pieces of equipment:

- **Strip board:** You attach the components to this board. You can use a bare (no copper) board or one that has component pads for soldering. We personally prefer the padded board.
- **Tie posts:** These posts serve as common connection points for attaching components together.

- **Wire-wrapping sockets for ICs and other parts:** These sockets have extra-long metal posts. You wrap the wire around these posts.
- **Wire-wrapping tool:** You have to use this specific tool to wrap wire around a post and remove it. The tool also includes an insulation stripper; use this, not a regular wire stripper, to remove the insulation off wire-wrap wire.
- **Wrapping wire:** The wire comes pre-cut or in spools. We prefer pre-cut wire, but try both before you form an opinion.

Although you can wire-wrap directly to resistor, capacitor, diode and other component leads, most people prefer using wire-wrap sockets. The reason? Most components have round leads. A wire-wrap socket has square posts. The square shape helps to bite into the wire, keeping things in place. If you wrap directly to component leads, you may want to tack on a little bit of solder to keep the wire in place.

Wire-wrapping is a straightforward process. You just insert all sockets into the strip board, use the wire-wrap tool to connect the sockets together and then plug the ICs and other components into their sockets. If your strip board has solder pads, you'd be wise to touch a little solder between one of the pads and the post sticking through it. This dab of solder keeps the sockets from coming out.

A big advantage of wire-wrapping over soldering is that you can make changes relatively easily. Simply unwrap the wire and re-route it to another post. If the wire gets cruddy, just replace it with a new one.

Wire-wrapping involves more than we can cover here. If it sounds like a method you think may be useful to you, do an Internet search on 'wire wrapping electronics' (or similar terms) to find many websites that can help you become an expert wire-wrapper-upper.

Making a custom circuit board

When you become experienced in designing and building electronics projects, you may want to graduate to the big time and create your own custom circuit board. You can make – yes, make – your very own printed circuit board (PCB), just like electronics manufacturers do, which is geared for a particular circuit design.

A *printed circuit board* is a surface that supports your circuit and includes interconnections along the surface of the board. PCBs are reliable, rugged, allow for higher density circuits and enable you to include non-standard-size components that may not fit in other types of circuit boards.

To make the board, you glue or laminate a thin sheet of copper, known as *cladding*, onto the surface of a plastic, epoxy or phenolic base. Then you form the circuit layout by etching away specific portions of the copper, leaving behind just the printed circuit design that consists of pads (contact points for components) and traces (interconnects).

Next, you drill holes into the centre of each pad so that you can mount components on the top of the board with leads poking through the holes. Finally, you solder each component lead to the board's pads.

To discover more about the ins and outs of making your own PCB, you can search the Internet for 'make printed circuit board' and find tutorials, illustrations and even videos that explain the process in great detail.

Chapter 12

Measuring and Analysing Circuits

In This Chapter

- Meeting and measuring with multimeters
- Making sure that electronic components are working properly
- Probing around your circuits
- Identifying the sources of circuit problems

Your excitement builds as you put the finishing touches to your circuit. With close friends standing beside you, eager to witness the first of your ingenious electronics exploits, you hold your breath as you flip the power switch, and . . . nothing happens! At least, nothing at first. Then disbelief and disappointment as your confidence – and your friends – begin to back away.

'What's gone wrong?' you ask yourself. Then you notice the smoke emanating from what used to be a resistor. And you realise that you used a 10-Ω resistor instead of a 10-KΩ resistor, trusting your old eyes and your worn-out mind to read and interpret resistor stripes properly. D'oh!

In this chapter, you discover how to use an incredibly versatile tool – the multimeter – to perform important face-saving checks on electronic circuits and components. You come to see that your multimeter is as important to you as an oxygen tank to a scuba diver: you can both get along all right on your own for a time, but sooner or later, you're bound to suffer unless you get some help. We also introduce you to two other testing tools – the logic probe and the oscilloscope – which you may find useful.

The tests that all these instruments can do help you to determine if everything is okay before you go showing off your circuitry to friends and family.

Multitasking with a Multimeter

You can't follow the electrons in your circuits with your bare eyes, and you certainly don't want to follow them with your bare hands, so you need help from a versatile electronics test instrument known as a multimeter.

A *multimeter* is an inexpensive handheld testing device that can function as a voltmeter to measure voltage, an ammeter to measure current and an ohmmeter to measure resistance; some multimeters can also test diodes, capacitors and transistors.

With this one handy tool, you can verify proper voltages, test whether you have a short circuit, determine whether a break exists in a wire or connection, and much more.

Get to know your multimeter: it can help you make sure that your circuits work properly, and it's an invaluable tool for detecting circuit problems.

Figure 12-1 shows a typical digital multimeter. You turn a dial to select the type of measurement you want to make and a range of values for that measurement. You then apply the metal tips of the two test leads (one red, one black) to a component or some part of your circuit, and the resulting measurement is displayed.

Multimeter test leads have conical tips that you hold in contact with the component you're testing. You can purchase special test clips that slip over the tips, which allow you to attach the test leads easily to the component leads or other wires.

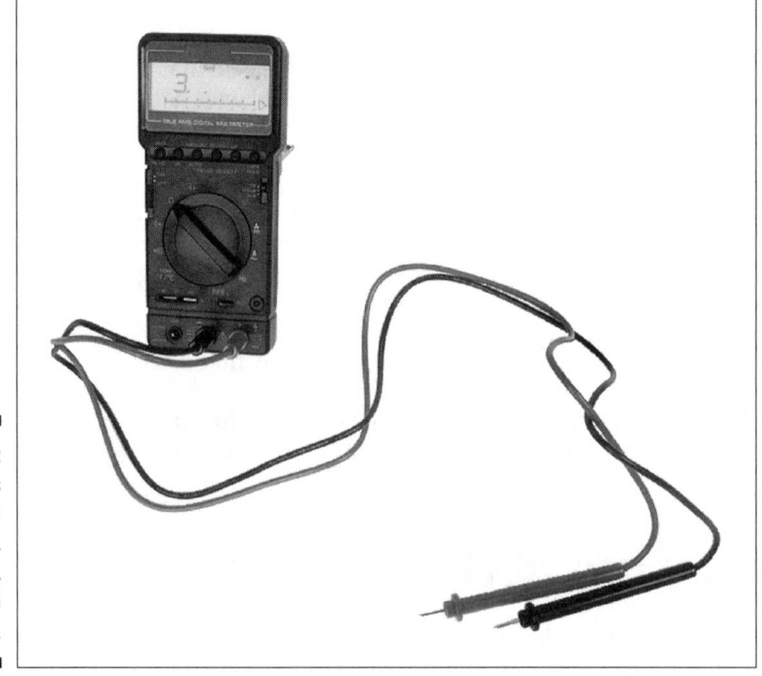

Figure 12-1: Multimeters measure voltage, resistance, current and continuity.

It's a voltmeter!

Multimeters can measure both DC and AC voltages. They provide a variety of voltage measurement ranges, from 0 volts to a maximum voltage. A typical set of DC voltage ranges is 0–0.25 volts, 0–2.5 volts, 0–10 volts, 0–50 volts and 0–250 volts.

You can use the multimeter as a *voltmeter* to measure the voltage of a battery outside of a circuit or *under load* (meaning when it's providing power within a circuit). You can also use your multimeter with your circuit powered up to test voltages dropped across circuit elements and voltages at various points within a circuit with respect to ground.

Voltmeters are so important in electronics that they even have their very own circuit symbol:

You may see this symbol with leads touching points in a circuit you read about on a website or in an electronics book. The symbol is telling you to take a voltage measurement across those two points.

Your multimeter can help you pinpoint the location of a problem in your circuit. It can verify if the proper voltage reaches a component, such as a light-emitting diode (LED) or a switch. You use multimeter tests to narrow down the field of suspects until you find the culprit causing all your headaches.

Now, it's an ammeter!

Your multimeter also functions as an *ammeter*, which is the term for a device that measures current going through a circuit. You may want to check that the current flowing at various points is what you expected from the circuit you've built, but the most likely reason for checking is to determine if a circuit or component is drawing too much current.

If your circuit has more current going through it than it's designed to handle, the components may overheat, which can permanently damage your circuit. In this instance, switch off the circuit to stop it overheating any further.

Ohm my! It's an ohmmeter, too!

You can measure the resistance (measured in ohms, as we discuss in Chapter 3) of an individual component or an entire circuit with your multimeter functioning as an *ohmmeter*. With your ohmmeter, you can check up on wires, resistors, motors and many other components. You almost always test resistance with the circuit *unpowered*.

When you measure the resistance of an individual component, take it out of the circuit before you test it. If you test a resistor that's wired in to a circuit, you get the equivalent resistance between two points, which isn't necessarily the resistance of just your resistor. See Chapter 3 for more on equivalent resistance.

Resistance, or lack of resistance, can reveal short circuits and open circuits, and therefore you use your ohmmeter to suss out problems such as breaks in wires and hidden shorts between components. A short circuit generates an ohmmeter reading of zero (or virtually zero) resistance, whereas an open circuit generates an infinite resistance readout. If you test the resistance from one end of a wire to the other and you get an infinite readout, you know that a break must exist somewhere along the length of the wire. Such tests are known as *continuity tests*.

By measuring resistance, you can tell whether or not the following circuit elements and connections are working properly:

- **Circuit board traces:** A bad copper trace (line) on a printed circuit board acts like a broken wire and generates an infinite resistance reading.
- **Fuses:** A blown fuse generates an infinite resistance reading, indicating an open circuit.
- **Solder joints:** A bad joint may generate an infinite resistance reading.
- **Switches:** An 'on' switch should generate a zero (or low) resistance reading, whereas an 'off' switch should generate an infinite reading.

Many multimeters include an audible continuity testing feature. By turning the meter's selector to Continuity or Tone, you can hear a beep whenever the meter detects continuity in a wire or connection. If the wire or connection doesn't have continuity, the meter stays silent. The audible tone gives you a convenient way to check an entire circuit without having to keep your eye on the meter.

Meeting Your First Multimeter

Multimeters range from bare-bones handheld models that cost around a tenner to sophisticated industrial worktop models that cost thousands of pounds. Even a basic multimeter can really help you understand what's going on in low-voltage circuits, but if you can afford to splash out a little more on a multimeter with more features, that's going to come in handy in the future. Multimeters can be digital or analogue.

Digital multimeters, like the one in Figure 12-1 earlier, display each measurement result as a precise number, taking the guesswork out of the reading process. Most handheld digital multimeters are accurate to within 0.8 per cent for DC voltages, whereas the pricey benchtop varieties are over 50 times more accurate.

Many digital multimeters also include an auto-ranging feature, which means that the meter automatically adjusts itself to display the most accurate result possible, and some have special testing features for checking diodes, capacitors and transistors.

You may see older-style *analogue multimeters* like the one in Figure 12-2, which have a needle on a scale. Analogue multimeters outperform digital multimeters as regards detecting changing readings, but if you don't have much of a need for that function, we recommend that you get a digital multimeter because of its ease of use and more accurate readings.

Figure 12-2: An analogue multimeter uses a needle to indicate voltage, current and other values, but we suggest you start with a digital model.

Unpacking your digital multimeter

All digital multimeters perform the basic voltage, resistance and current measurements. Where they differ is in the range of values they can measure, the additional measurements they can perform, the resolution and sensitivity of their measurements and the extra bells and whistles they come with.

Be sure to at least browse through the manual for the multimeter you purchase, to check the description of its features and specifications, as well as important safety precautions.

Here's what you can expect to find when you explore a digital multimeter:

- **Power switch/battery/fuse:** The on/off switch connects and disconnects the battery that powers the multimeter. Many multimeters use standard-size batteries, such as a 9-volt or AA cell, but pocket-size meters use a coin-type battery. (Avoid using rechargeable batteries, which may produce erroneous results for some models.)

  Most multimeters use an internal fuse to protect themselves against excessive current or voltage, and some come with a spare fuse (if yours doesn't, we suggest that you buy one).

- **Function selector:** Dial this knob to the test you want to perform (voltage, current, resistance or something else) and, on some models, the range setting you want to use. Some multimeters also include one or more of the following categories: AC amperes, capacitance, transistor gain (h_{FE}) and diode test.

  Many models further divide some measurement categories into 3–6 different ranges; the smaller the range, the greater the sensitivity of the reading. Figure 12-3 shows a close up of a function selector dial.

- **Test leads and receptacles:** Inexpensive multimeters come with basic test leads, but you can purchase higher-quality coiled leads that stretch out to a metre or so, and yet recoil to a manageable length when not in use. You may also want to purchase spring-loaded clip leads that stay in place and are insulated to prevent the metal tip from coming in contact with other parts of your circuit.

  Some multimeters with removable test leads provide more than two receptacles for the leads. You insert the black test lead into the receptacle labelled GROUND or COM, but the red lead may be inserted into a different receptacle depending on what function and range you've dialled up. Most meters provide additional input sockets for testing capacitors and transistors, as shown in the top right corner of Figure 12-3. Refer to your multimeter manual for details.

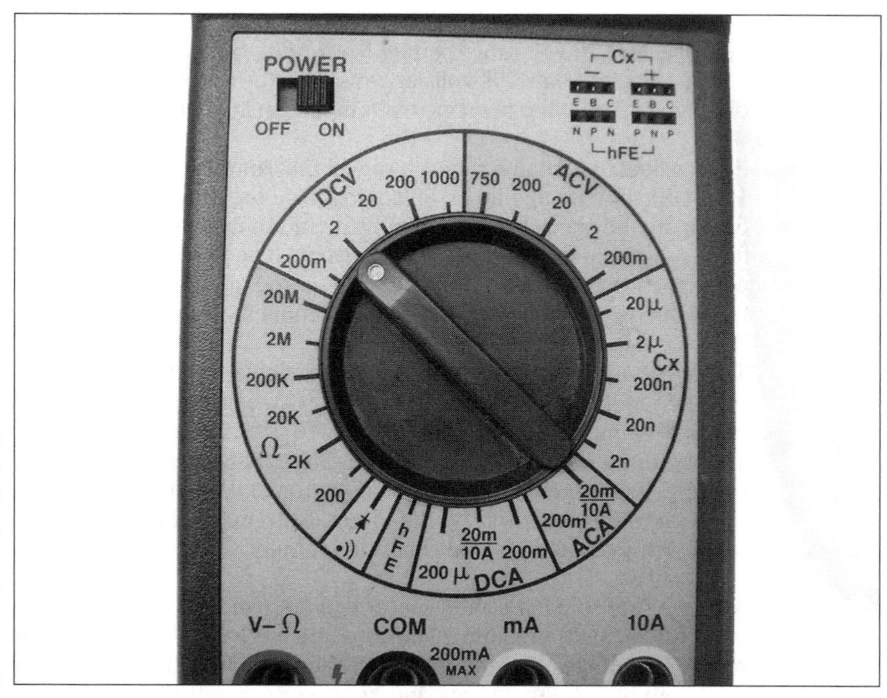

Figure 12-3: Digital multimeters provide a wide variety of measure-mentoptions.

- **Digital display:** The readout is given in units specified by the range you dial up. For instance, a reading of 15.2 means 15.2 volts if you dial a 20-volt range, or 15.2 millivolts if you dial a 200-millivolt range.

  Most digital multimeters designed for hobbyists have what's called a *3½ digit display*: this means that the readout contains three or four digits, where each of the three rightmost digits can be any digit from 0 to 9, but the optional fourth digit (the leftmost, or most significant, digit) is limited to 0 or 1. For instance, if set to a 200-volt range, such a multimeter can give readouts ranging from 00.0 to 199.9 volts.

Homing in on the range

Many digital multimeters require that you select the range before the meter can make an accurate measurement. For example, if you're measuring the voltage of a 9-volt transistor battery, you set the range to the setting closest to, but above, 9 volts. For most meters, this requires that you select the 20- or 50-volt range.

If you select too large a range, the reading you get isn't as accurate. (For instance, on a 20-volt range setting, your 9-volt battery may produce a reading of 8.27 volts, but on a 200-volt range setting, the same battery produces a reading of 8.3 volts. You often need as much precision in your readings as possible.)

If you select too small a range, a digital multimeter typically displays a flashing 1 (or OL), and the needle on an analogue meter shoots off the scale, possibly damaging the precision needle movement (so make sure that you start with a large range and dial it down, if necessary). If you see an over range indicator when testing continuity, the resistance is so high that the meter can't register it; in which case you can fairly safely assume that's an open circuit.

The auto-ranging feature found on many digital multimeters makes getting a precise reading even easier. For instance, when you want to measure voltage, you set the meter function to volts (to DC or AC) and take the measurement. The meter automatically selects the range that produces the most precise reading. If you see an over range indicator (a flashing 1 or OL), the value is too high to be measured by the meter. Auto-ranging meters don't require range settings, so their dials are a lot simpler.

A limit exists to what a multimeter can test, and you call that limit its maximum range. Most consumer multimeters have roughly the same maximum range for voltage, current and resistance. For your hobby electronics, any meter with the following maximum ranges (or better) works just fine:

- AC volts: 500 volts
- DC volts: 1000 volts
- DC current: 200 milliamps
- Resistance: 2 MΩ (two megohms, or 2 million ohms)

What if you need to test higher currents?

Most digital multimeters limit current measurements to less than one amp. The typical digital multimeter has a maximum range of 200 milliamps. Attempting to measure substantially higher currents may cause the fuse in the meter to blow. Many analogue meters, especially older ones, support current readings of 5 or 10 amps, maximum.

You may find analogue meters with a high ampere input handy if you're testing motors and circuits that draw a lot of current. If you have only a digital meter with a limited milliamp input, you can still measure higher currents indirectly by using a low-resistance, high-wattage resistor. To do this, you place a 1-Ω 10-watt resistor in series with your circuit so that the current you want to measure passes through this test resistor. Then you use your multimeter as a voltmeter, measuring the voltage dropped across the 1-Ω resistor. Finally, you apply Ohm's Law to calculate the current flowing through the test resistor as follows:

$$\text{Current} = \frac{V}{R} = \frac{V}{1\Omega}$$

Because the nominal value of the resistor is 1 Ω, the current (in amps) through the resistor has roughly the same value as the voltage (in volts) you measure across the resistor. (Note that the resistor value is not going to be exactly 1 Ω in practice, so your reading may be off by as much as 5–10 per cent, depending on the tolerance of your resistor and the accuracy of your meter.) You can read about Ohm's Law in Chapter 3.

Setting Up Your Multimeter

Before testing your circuits, you must make sure that your meter is working properly. Any malfunction gives you incorrect testing results – and you may not even realise it. To test your multimeter, follow these steps:

1. **Make sure that the test probes at the end of the test leads are clean.**

   Dirty or corroded test probes can cause inaccurate results. Use an electronic contact cleaner to clean both ends of the test probes and, if necessary, the connectors on the meter.

2. **Turn on the meter and dial it to the Ohms (Ω) setting.**

   If the meter isn't auto-ranging, set it to low ohms.

3. **Plug both test probes into the proper connectors of the meter and then touch the ends of the two probes together (see Figure 12-4).**

   Avoid touching the ends of the metal test probes with your fingers while you're performing the meter test. The natural resistance of your body can throw off the accuracy of the meter.

4. **The meter should read 0 (zero) ohms or very close to it.**

   If your meter doesn't have an auto-zero feature, press the Adjust (or Zero Adjust) button. On analogue meters, rotate the Zero Adjust knob until the needle reads 0 (zero). Keep the test probes in contact and wait a second or two for the meter to set itself to zero.

5. **If you don't get any response at all from the meter when you touch the test probes together, recheck the dial setting of the meter.**

Nothing happens if you have the meter set to register voltage or current. If you make sure that the meter has the right settings and it still doesn't respond, you may have faulty test leads. If necessary, repair or replace any bad test leads with a new set.

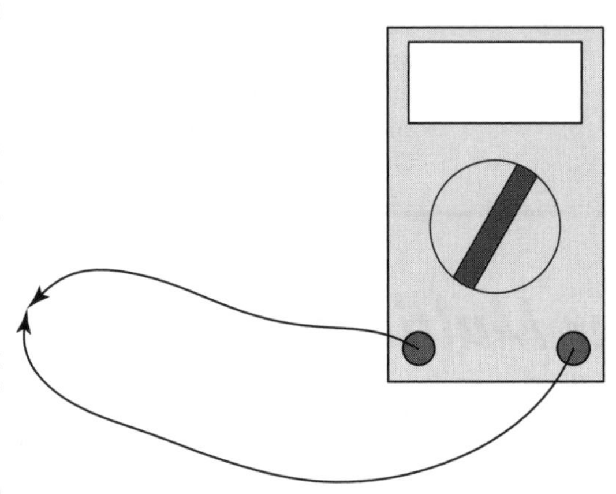

Figure 12-4: Touch the test probes of the meter together and verify a 0 (zero) ohms reading to make sure that the meter is working properly.

You can consider the meter calibrated when it reads zero ohms with the test probes shorted together (held together so that they're touching each other). Carry out this test each time you use your meter, especially if you turn off the meter between tests.

If your meter has a Continuity setting, don't use it to zero-adjust (calibrate) the meter. The tone may sound when the meter reads a few ohms, so it doesn't give you the accuracy that you need. Recalibrate the multimeter using the Ohms setting, and not the Continuity setting, to ensure proper operation.

Operating Your Multimeter

When you use your multimeter to test and analyse circuits, you must consider what settings to dial up, whether you're testing components individually or as part of a circuit, whether the circuit should be powered up or not and where you place the test leads (in series or in parallel with whatever you're testing).

Think of your multimeter as an electronic component in your circuit (because in a way, it is). If you want to measure voltage, your meter must be placed in parallel with the section of the circuit you're measuring, because voltages across parallel branches of a circuit are the same. If you want to measure current, your meter must be placed in series with the section of the circuit you're measuring, because components in a series circuit carry the same current.

In the following sections, we explain exactly how to measure voltage, current and resistance using a multimeter.

Measuring voltage

To examine voltage levels throughout your circuit using a multimeter, your circuit must be powered up. You can test the voltage at almost any point in a circuit, not just the battery connections.

After setting up your meter as described in the earlier section 'Setting Up Your Multimeter', dial up the type of voltage (AC or DC) and the range. Choose the range that gives you the most sensitivity. Attach the black lead of the meter to the ground connection of the circuit and the red lead of the meter to the point in the circuit that you want to measure. (This process places your multimeter in parallel with the voltage drop you want to measure.)

Figure 12-5 shows an example of using a multimeter to measure the voltage at a couple of different points in a simple circuit containing a 555 timer integrated circuit (IC). The multimeter is set to DC volts with a range of 0–20 volts.

In the top image the meter is measuring the voltage that powers the entire circuit, while in the bottom image the meter is measuring the voltage at the output of the 555 timer IC. Because the output of the 555 timer IC is a low or high voltage, the reading on the multimeter alternates between 0 (zero) and 9 volts (the positive power supply voltage).

Depending on the resistor and capacitor values in the 555 timer circuits shown in Figure 12-5, the output may change so rapidly that your multimeter can't keep up with the voltage swings. To test fast-changing signals, you need a logic probe (for digital signals only) or an oscilloscope. We discuss these two test instruments in later sections 'Probing the depths of logic' and 'Scoping out signals with an oscilloscope'.

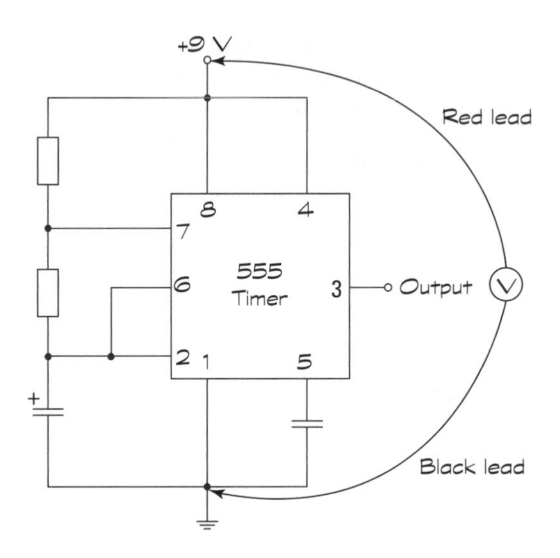

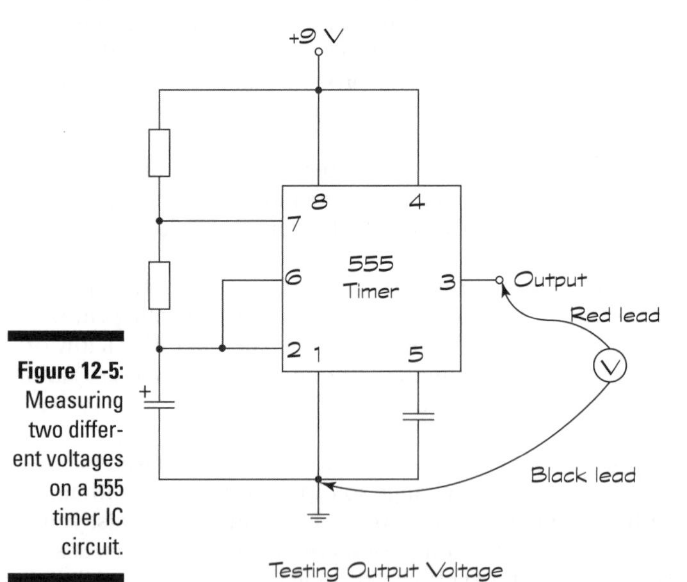

Figure 12-5: Measuring two different voltages on a 555 timer IC circuit.

Measuring current

When you use the multimeter as an ammeter, you connect the meter *in series with* the component you want to measure current through, so that the exact same current you want to measure passes through the meter. This set up, as shown in Figure 12-6, is very different from the voltmeter configuration.

After setting up your meter as described in the earlier section 'Setting Up Your Multimeter', dial up the type of current (AC or DC) and the range. Choose the range that gives you the most sensitivity. Then interrupt the circuit at the point where you want to measure current.

You attach the black lead of the meter to the more negative side of the circuit and the red lead to the more positive side, which places your multimeter in series with the component you want to measure current through. Make sure that the circuit is powered up. If you don't get a reading, try reversing the connections of the leads to the multimeter.

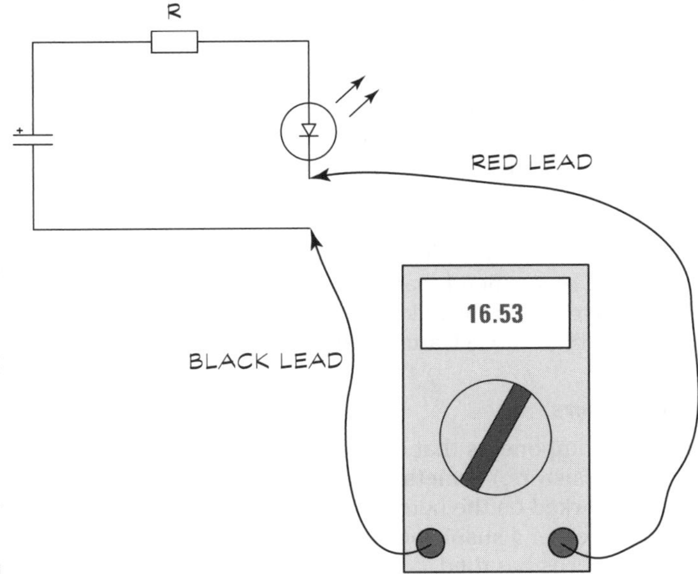

Figure 12-6: Measuring current involves connecting the meter in series with the circuit or component.

To measure how much overall current an entire circuit draws, insert your meter in series with the positive power supply. But remember that many digital meters are limited to testing currents of 200 milliamps or less. Be careful: don't test higher current if your meter isn't equipped to do so.

Never leave your multimeter in an ammeter position after measuring current. You can damage the meter. Get into the habit of turning the meter off immediately after running a current test.

Don't blow your fuse!

Many meters provide a separate input (test lead receptacle) for testing current, usually marked as A (for amps) or mA (for milliamps). Some multimeters provide an additional input for testing higher currents, typically up to 10 amps. The multimeter shown in the earlier Figure 12-3 has two inputs for testing current, labelled mA and 10A.

Be sure to select the appropriate input before making any current measurement. Forgetting to do this step may blow a fuse (if you're lucky) or damage your meter (if you're unlucky).

Measuring resistance

You can run lots of different tests using your multimeter as an ohmmeter to measure resistance. Obviously, you can test resistors to check their values or see if they've been damaged, but you can also examine capacitors, transistors, diodes, switches, wires and more using your ohmmeter. Before you measure resistance, make sure that you calibrate your ohmmeter as described in the earlier section 'Setting Up Your Multimeter'.

If your multimeter has specific features for testing capacitors, diodes and/or transistors, we recommend that you use those features rather than the methods that we give you in the following sections. But if you have a bare-bones multimeter without those features, the following methods can really help you.

Testing resistors

Resistors are components that limit current through a circuit. (See Chapter 3 for all about resistors.) Sometimes you need to verify that the nominal resistance value marked on the body of a resistor is accurate, or you may want to investigate whether a suspicious-looking resistor with a bulging centre and third-degree burns has died.

Here's how to test a resistor:

1. **Disconnect it from your circuit.** Make sure that you turn the power off before you touch it!
2. **Set your multimeter to read ohms.** If you don't have an auto-ranging meter, start at a high range and dial down the range as needed.
3. **Position the test leads on either side of the resistor.** Take care not to let your fingers touch the metal tips of the test leads or the leads of the resistor. (If you touch them, you add the resistance of your body into the reading, producing an inaccurate result.)

The resistance reading should fall within the tolerance range of the nominal value marked on the resistor. For instance, if you test a resistor with a nominal value of 1 K ohms and a tolerance of 10 per cent, your test reading should fall in the range of 900 to 1,100 ohms. A bad resistor can be completely open inside, in which case you may get a reading of infinite ohms, or it can be shorted out, in which case you get a reading of zero ohms.

Testing potentiometers

Like a resistor, you can test a *potentiometer* (also called a *pot*), which is a variable resistor, using the Ohms setting on your multimeter. (We discuss pots in Chapter 3.)

Remove the potentiometer from your circuit. Figure 12-7 shows multimeter test leads applied to one fixed end (point 1) and the *wiper*, or variable lead (point 2), of a pot. Turning the dial shaft in one direction increases the resistance; turning the dial shaft in the other direction decreases the resistance. With the meter applied to the wiper (point 2) and the other fixed end (point 3), the opposite resistance variation happens. If you connect the meter to both fixed ends (points 1 and 3), the reading that you get should be the maximum resistance of the pot, no matter how you turn the dial shaft.

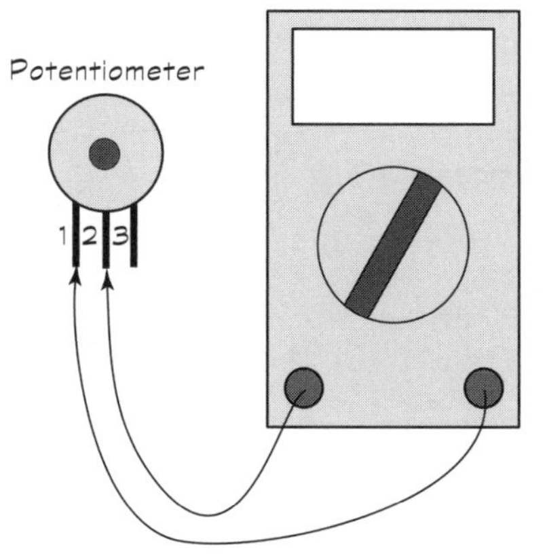

Figure 12-7: Connect the test leads to the first and centre, centre and third, and first and third terminals of the pot.

As you turn the shaft of the potentiometer, take note of any sudden changes in resistance, which may indicate a fault inside the pot. Should you find such a fault, replace the pot with a new one.

Testing capacitors

You use a *capacitor* to store electrical energy for a short period of time. (We cover capacitors in Chapter 4.) If your multimeter doesn't have a capacitor-testing feature, you can still use it in the ohmmeter setting to help you decide whether or not to replace a capacitor.

Before you test a capacitor, you must discharge it to clear all the electrical energy out of it. That's because large capacitors can retain a charge for long periods of time – even after you remove power.

To discharge a capacitor, you short out its terminals through an insulated *bleeder jumper* (as shown in Figure 12-8), which is simply a wire with a large (1 or 2 megohm) resistor attached. The resistor prevents the capacitor from being shorted out, which would make it unusable.

After discharging the capacitor and removing the bleeder jumper, dial your multimeter to ohms and touch the test leads to the capacitor leads. Unpolarised capacitors don't care which way you connect the leads, but when you're testing a polarised capacitor, connect the black lead to the negative terminal of the capacitor and the red lead to the positive terminal.

Wait a second or two and then note the reading. A good capacitor shows a reading of infinity when you perform this step. A zero reading may mean that the capacitor is shorted out. A reading of between zero and infinity can be indicative of a leaky capacitor, one that's losing its ability to hold charge.

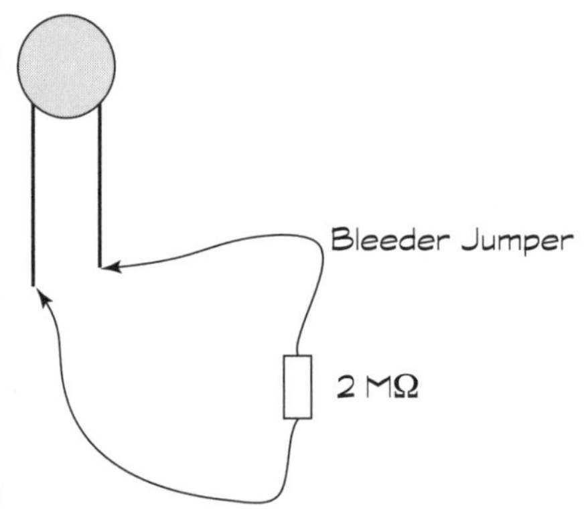

Figure 12-8: Purchase or make a bleeder jumper, used for draining excess charge from a capacitor.

This test doesn't tell you when the capacitor is open, which can happen if the component becomes structurally damaged inside or if its dielectric (insulating material) dries out. An open capacitor reads infinite ohms just like a good capacitor. For a conclusive test, use a multimeter with a capacitor-testing function.

Testing diodes

A *diode* is a semiconductor component that acts like a one-way value for current. (We discuss diodes in gory detail in Chapter 6.) If your multimeter doesn't have a diode-check setting, you can use the Ohms setting to test most types of diodes.

Set your meter to a low-value resistance range. Connect the black lead to the cathode (negative side, with a stripe) and the red lead to the anode (positive side). The multimeter should display a low resistance. Then reverse the leads, and you should get an infinite resistance reading.

If you're not sure which end is up in a diode you've got on hand, you can use your multimeter to identify the anode and the cathode. Run resistance tests with leads connected one way, and then the other. For the lower of the two resistance readings, the red lead is connected to the anode and the black lead is connected to the cathode.

Testing transistors

A *bipolar transistor* is essentially two diodes in one package, as illustrated in Figure 12-9. (For a PNP transistor, both diodes are reversed.) If your multimeter has neither a transistor-checking feature nor a diode-checking feature, you can use the Ohms setting to test most bipolar transistors, in a similar fashion to how you test diodes (see the preceding section). You simply set your meter to a low-value resistance range and test each diode within your transistor in turn.

For instance, if you're testing an NPN transistor, such as the one shown in Figure 12-9, connect the black lead to the collector of the transistor and the red lead to the base. If everything is working correctly, the multimeter displays a low resistance. Then reverse the leads, and you get an infinite resistance reading. Next, connect the black lead to the emitter and the red lead to the base and the meter displays a low resistance. Reverse the leads and the meter displays infinite resistance.

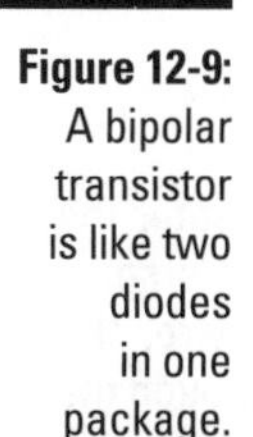
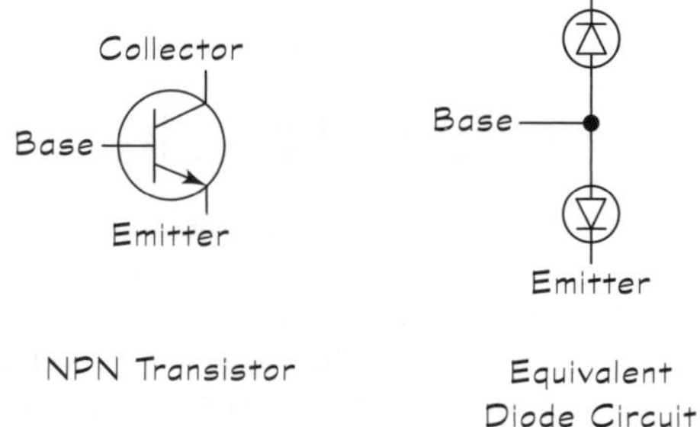

Figure 12-9: A bipolar transistor is like two diodes in one package.

For a PNP transistor, the readings should be the opposite of what they are for an NPN transistor.

Testing with a multimeter can permanently damage some types of transistors, especially field effect transistors (FETs). Use this test with bipolar transistors only. If you're not sure what type of transistor you have, look it up in a datasheet before testing. You can often find the datasheet by searching the Internet for the component identification number (for instance, search on '2n3906 datasheet').

Testing wires and cables

You can use your multimeter as an ohmmeter to run continuity tests on wires and cables. You may want to do this so that you can detect breaks inside wires and *short circuits*, or unintended continuity, between two wires in a cable.

To test for continuity in a single wire, you connect the multimeter test leads to either end of the wire and dial up a low-range Ohms setting. You should get a reading of zero ohms, or a very low number of ohms. A reading of more than just a few ohms indicates a possible break in the wire, causing an *open circuit*.

To test for a short between different wires that shouldn't be electrically connected, you set the meter to measure ohms, and then connect one of the test leads to an exposed end of one wire and the other test lead to an exposed end of the other wire. If you get a reading of zero or a low number of ohms, you may have a short circuit between the wires. A higher reading usually means your wires aren't shorted together.

Note that you may get a reading other than infinite ohms if the wires are still connected to your circuit when you make the measurement. Rest assured that your wires aren't shorted unless you get a very low or zero reading.

Even wire resists the flow of electrons

Why don't you always get zero ohms when testing wire, especially a long wire? Well, all electrical circuits exhibit some resistance to the flow of current, and the ohms measurement tests this resistance. Even short lengths of wire have resistance, although usually well below 1 ohm, so these aren't an important test subject for continuity or shorts.

However, the longer the wire, the greater the resistance, especially if the wire has a small diameter. Usually, the thicker the wire, the lower its resistance per metre. Even though the multimeter doesn't read exactly zero ohms, you can assume proper continuity in this instance if you get a low ohms reading.

Testing switches

Mechanical switches can get dirty and worn, or sometimes even break, making them unreliable or completely unable to pass electrical current. In Chapter 8, we describe four common types of switches: single-pole single-throw (SPST), single-pole double-throw (SPDT), double-pole single-throw (DPST) and double-pole double-throw (DPST). Depending on the switch, you may see zero, one or two 'off' positions, and one or two 'on' positions.

You can use your multimeter, set to ohms, to test any of these switches. You need to familiarise yourself with the on/off position(s) and the terminal connections of the particular switch you're testing, and run tests for each possibility. With your test leads connected across the terminals of any input/output combination placed in the 'off' position, the meter should read infinite ohms; the 'on' position should give you a reading of zero ohms.

You can most easily test switches by taking them out of a circuit. If you have the switch wires in a circuit, the meter may not show infinite ohms when you place the switch in the 'off' position. If, instead, you get a reading of some value other than zero ohms, you can assume the switch is operating properly as an open circuit when in the 'off' position.

Testing fuses

Fuses are designed to protect electronic circuitry from damage caused by excessive current flow and, more importantly, to prevent a fire if a circuit overheats. A blown fuse is an open circuit that's no longer providing protection, so it must be replaced.

To test a fuse, set your multimeter to the Ohms setting and touch one test lead to either end of the fuse. If the meter reads infinite ohms, you have a burned-out fuse.

Running other multimeter tests

Many digital multimeters include extra functions that test specific components, such as capacitors, diodes and transistors. These tests provide more definitive results than the resistance measurements we discuss in the previous few sections.

If your multimeter has a capacitor-testing feature, it displays the value of the capacitor, which can come in handy because not all capacitors follow the industry-standard identification scheme. Refer to your multimeter manual for the exact procedure because the specifics vary from model to model. Be sure to observe the proper polarity when connecting the capacitor to the test points on the meter.

If your multimeter has a diode-check feature, you can test a diode by attaching the red test lead to the anode (positive terminal) of the diode, and the black lead to the cathode (negative terminal). You should get a fairly low, but not zero, reading (for instance, 0.5). Then reverse the leads and you should get an over range reading. If you get two zero readings or two over range readings, chances are your diode is bad (short or open).

You can use the diode-check feature to test bipolar junction transistors, treating them as two individual diodes, as shown in the earlier Figure 12-9.

If your multimeter has a transistor-checking feature, follow the procedure outlined in the manual, which varies from one model to another.

Using a Multimeter to Check Your Circuits

One of the top benefits of a multimeter is that it can help you analyse the rights and wrongs of your circuits. By using the various test settings, you can verify the viability of individual components and confirm that voltages and currents are what they should be. Say that you hook up a circuit that doesn't work right away: your multimeter can help you sniff out the problem when you can't resolve the problem by physically checking all your connections.

To troubleshoot your circuit, mark up your circuit diagram with component values, estimated voltage levels at various points in the circuit and expected current levels in each branch of the circuit. (Often, the process of marking up the diagram uncovers a maths error or two.) Then, use your multimeter to probe around.

Here are some of the things to check as you troubleshoot your circuits:

- Check power supply voltages.
- Check individual component functionality and actual values (out of the circuit).
- Check continuity of wiring.

- Check voltage levels at various points in the circuit.
- Check current levels through part of the circuit (without exceeding the current capabilities of your multimeter).

Using a step-by-step procedure, you can test various components and parts of your circuit and narrow down the list of suspects until you uncover the cause of your circuit problem or wave the white flag of surrender (never!).

Introducing Logic Probes and Oscilloscopes

Your multimeter is the most important tool on your workbench, but it's not the only instrument you can use to test your circuits. If you're really, really serious about electronics, you can buy a logic probe or an oscilloscope. Neither is a must-have tool when you're starting out, but as you work your way into intermediate and advanced electronics, you should consider adding them to your electronics workbench.

The detail of how these tools work is beyond this book and here we just introduce you to them briefly so that you can decide when and if to invest in them.

Probing the depths of logic

A *logic probe* (a fairly inexpensive tool costing under £15), like the one shown in Figure 12-10, is specifically designed to test digital circuits, which handle just two voltage levels: *low* (zero volts or thereabouts), indicating logical 0 (zero); or *high* (12 volts or less, but most commonly 5 volts), indicating logical 1 (one). (You can read all about digital logic in Chapter 7.)

The logic probe simply checks for high or low signals and turns on a different LED indicator depending on the signal it detects. A third LED indicator glows when the logic probe detects *pulsing*, which is when a signal alternates between high and low very quickly. Most logic probes also include a tone feature, so you can hear highs and lows while keeping your eyes on your circuit.

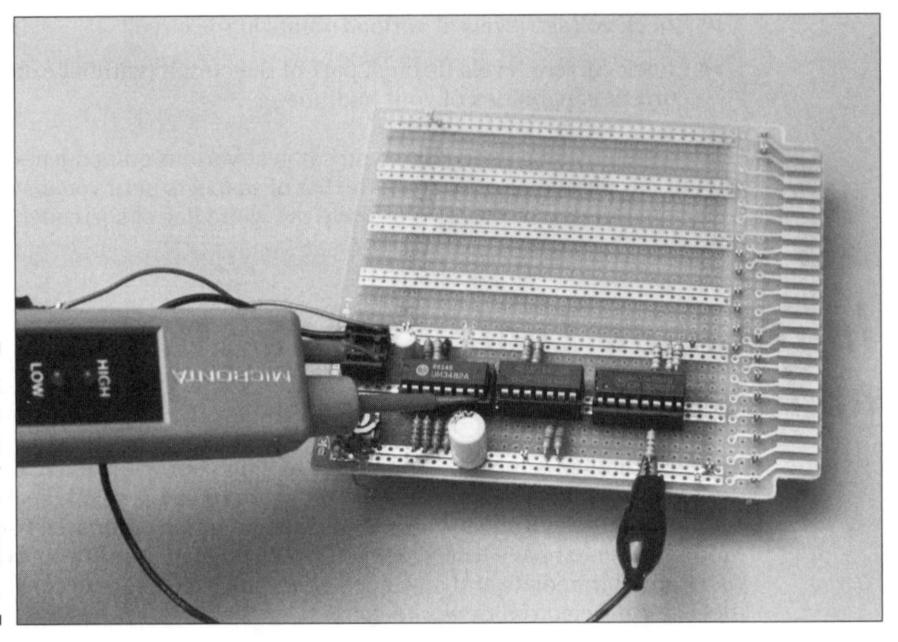

Figure 12-10: The logic probe is useful for trouble-shooting digital circuits.

Logic probes receive power from the circuit under test. Most probes work with a minimum supply voltage of about 3 volts and a maximum of no more than 15 volts (sometimes more, sometimes less). Check whether the supply voltage of your circuit falls within the acceptable range for a logic probe.

Scoping out signals with an oscilloscope

For the average amateur electronics hobbyist working at home or at school, the oscilloscope is a nice tool to have around but isn't absolutely necessary. An *oscilloscope* is a piece of test equipment that displays how a voltage varies with time as a *trace* across a cathode-ray tube (CRT) or other display.

This ability makes an oscilloscope useful for checking whether digital signals have the right timing, or for testing pulsating signals that change too quickly for a logic probe to detect (faster than about 5 MHz).

With a *dual trace oscilloscope*, which is a scope with two input channels, you can test the relationship between two signals. You may need to do this test when you work with some digital circuits, for example. Often, one signal triggers the circuit to generate another signal. Being able to see both signals together helps you determine whether the circuit is working as required.

Serious electronics engineers spend small fortunes on serious bench oscilloscopes like the one shown in Figure 12-11, but you can purchase a battery-operated handheld scope with a liquid crystal display for under £100. You don't get the advanced features of the better bench scopes, but you're still able to view signal variations over time – something that your multimeter simply can't do.

Another option is to spend £100 or so on a PC-based oscilloscope that uses your personal computer to store and display the electrical signals that you measure. Most PC-based scopes are self-contained in a small external module that connects to the desktop PC or laptop through a parallel, serial, or USB port.

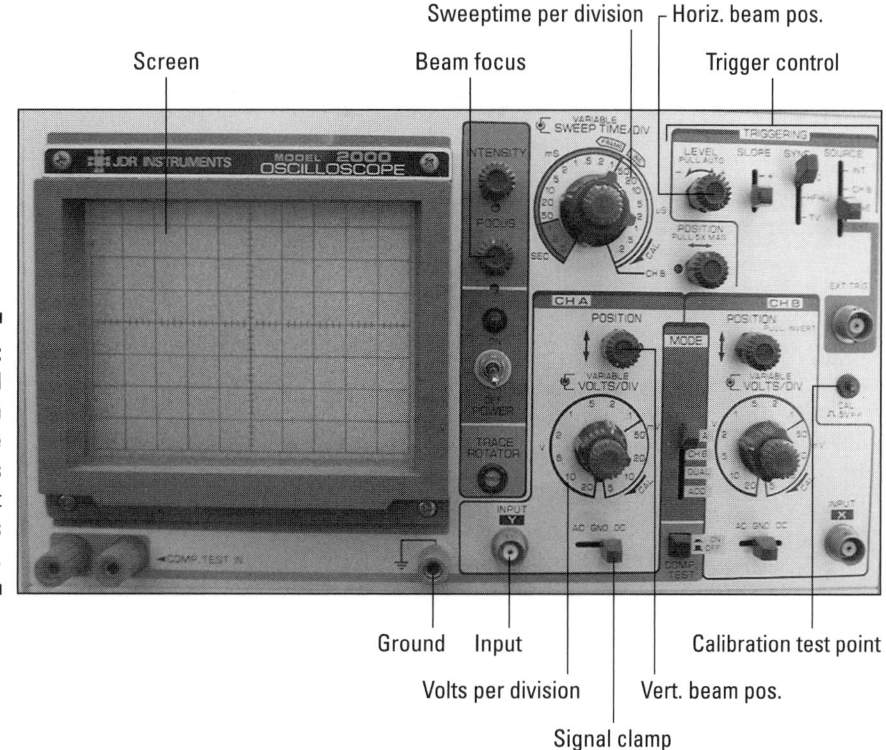

Figure 12-11: A typical bench oscilloscope with its important controls identified.

Part III
Putting Theory into Practice

'Are you sure this invention you want me to try is going to be the electronics breakthrough the hairdressing industry is waiting for, Sidney?'

In this part . . .

Finding out how electronics components control electric current in theory is great, but seeing the process in action lets you discover what electronics can really do.

In this part, you find out how to set up simple circuits to demonstrate the principles of electronics. We show you how to use your multimeter – and a little bit of maths – to compile enough evidence to prove to anyone that electronic components really can do what they're supposed to. You may even make a few LEDs glow to show where and when electric current is flowing through your test circuits.

We also share with you the plans for several fun electronics projects you can build in 30 minutes or less – without breaking open your piggy bank. Using sensors, lights and buzzers, you can put electronics to work to create a variety of practical – and sometimes entertaining – circuits. You may find that these projects spark some creative ideas for other circuits that you can build to impress your friends and family.

Chapter 13

Exploring Some Simple Circuits

In This Chapter

- Putting Ohm's Law through its paces
- Following the ups and downs of capacitor voltage
- Zeroing in on Zener diodes
- Testing out current with transistors

You may be thinking that electronics theory (which we cover in Chapters 1–6) is well and good, but does it really work in your own home-built circuits? Are your resistors really going to resist and are logic gates always strictly logical? Do your semiconductors actually do what you want them to? When you charge up a capacitor, does it faithfully hang on to its electrical energy until you give it the go-ahead to discharge?

In this chapter, we put theory to the test. We show you some simple circuits that answer all your electronic concerns by demonstrating how various components and the laws that govern them work. You can build any of these circuits in less than ten minutes, and see for yourself how one or more components shape electrical current. Using your multimeter as your guide – with some visual cues from a few strategically placed LEDs – you can get a real sense of what's going on inside each circuit.

Getting Equipped

To prepare to build the circuits in this chapter, make sure that you have a few supplies at the ready. At a minimum, you need the following items to build and explore the circuits we discuss in this chapter:

- **Batteries:** All the circuits in this chapter run off one or two 9-volt batteries. Make sure that you have a few extra ones around in case you leave a circuit powered up for a long time. To make circuit building easier, buy two 9-volt battery clips and, if they don't already have them, solder 22-gauge solid wires onto the clip leads so that you can easily connect and disconnect your batteries from any circuit (avoiding the need for a switch). You can find information on how to solder leads in Chapter 11.

- **Calculator:** If you don't trust your brain to perform mathematical calculations accurately, bring your calculator to your workbench. Remember to check your units carefully and convert values to ohms, amps and volts before you calculate anything.
- **Multimeter:** You need a multimeter to measure voltage, resistance and current as you watch what's happening to the electrons your circuits are manipulating. Review Chapter 12 if you need a refresher on how to wield a multimeter.
- **Solderless breadboard:** You need at least one solderless breadboard to enable you to build, update, tear down and re-build the circuits we discuss. If necessary, review the breadboarding techniques discussed in Chapter 11, so that you know which holes are connected to which before you start to plug components in at random.

Just to make your shopping trip a little easier, we provide you with a list of all the electronic components that we use in the simple circuits in this chapter:

- **Capacitor:** One 470μF electrolytic capacitor rated at a minimum of 25 volts.
- **Diodes:** One 4.3V 1W Zener diode or similar ratings.
- **IC:** One quad 2-input NAND gate IC such as the 4011.
- **LEDs:** Two standard 5-millimetre LEDs of any colour (same as each other, all different or a mix of colours).
- **Potentiometers (pots):** One each of 10 kΩ, 100 kΩ and 1 MΩ. If they don't already come with connectors suitable for a solderless breadboard, solder wires around 7 or 8 centimetres long to each of their three leads. Remember that the middle lead is connected to the wiper arm inside the device, and the leads on either end are fixed.
- **Resistors:** With one exception, which is noted, resistors rated at ¼ watt or ⅛ watt with 10 or 20 per cent tolerance are fine. You need the following fixed resistors: one 220 Ω, one 330 Ω, two 470 Ω ½ watt, two 1 kΩ, two 10 kΩ and one 100 kΩ.
- **Transistors:** Two general-purpose NPN bipolar transistors, such as a 2N3904 or BC548.

In the next section, we walk you through the process of building the first circuit on a solderless breadboard. For circuits after that, we just give you the schematic diagram and leave the construction up to you. As you build each circuit, use a row along the top of your solderless breadboard for connections to the positive side of the power supply, and a row along the bottom for ground connections (the negative side of the power supply), as we explain and illustrate for the first simple circuit.

Seeing Is Believing: Ohm's Law Really Works!

In Chapter 3, we discuss one of the most important principles in electronics: Ohm's Law. This law governs all resistive electronic components and tells us that the voltage, V, dropped across a component with a fixed resistance, R, is equal to the current, I, flowing through the component multiplied by the value of the resistance (R). You can sum it up with this equation:

$$V = I \times R$$

You can also rearrange the terms of that equation to get two other equations that really say the same thing in two different ways:

$$I = \frac{V}{R} \text{ and } R = \frac{V}{I}$$

You use Ohm's Law to help you analyse circuits, from simple series circuits to more complex series-parallel circuits. In this section, you can test Ohm's Law to make sure that it works (it does!) and to take your first steps in circuit analysis.

As you explore these circuits, keep in mind another important rule that applies to every circuit: the voltage rises and drops around a circuit always add up to zero. In DC-powered circuits, such as the ones we explore in this section, another way to say the same thing is that the voltage drops across all the components in a circuit add up to the source voltage. If you'd like to review these concepts, refer back to Chapter 2.

Analysing a series circuit

Figure 13-1 shows a simple series circuit containing a 9-volt battery, a 470-Ω resistor (R1) and a 10-kΩ potentiometer, or variable resistor (R2). We also label the voltage drops across R1 and R2. For this circuit, you use the middle lead and one outer lead of the potentiometer, leaving the other outer lead unconnected. This way, you can vary the resistance of R2 between 0 (zero) ohms and 10 kΩ.

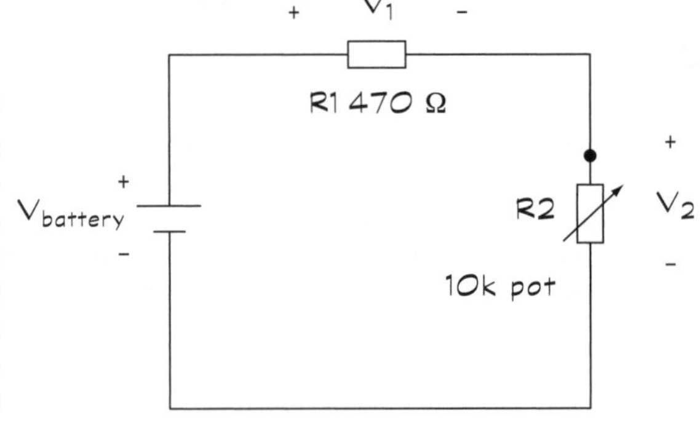

Figure 13-1: With a simple series circuit, you can witness Ohm's Law in action.

To avoid confusion, bend the wire extending from the unused lead out of the way, so you're clear about which outer lead you're using.

Before using a pot in your circuit, use your multimeter set on ohms to measure the resistance from one lead of the pot to the other. Turn the pot all the way down so that your multimeter reads 0 (zero) ohms.

Building the simple series circuit

Refer to the photo in Figure 13-2 for making connections on your solderless breadboard. Build the circuit one step at a time, connecting the battery last, as follows:

1. **Insert the fixed resistor (R1).** Plug one lead (either one) into a hole along the top row of the breadboard and the other lead into a hole in the centre section of the breadboard.
2. **Insert the potentiometer (R2).** Plug one lead (either one) into a hole in the centre section, in the same column as the lead for resistor R1, and plug the other lead into a hole along the bottom row.
3. **Insert the battery.** Plug the lead for the negative terminal into a hole along the bottom row and then plug the lead for the positive terminal into a hole along the top row to complete the circuit.

Figure 13-2: Setting up the simple series circuit on a solderless breadboard is a snap.

So, now the time has come to probe around your live circuit using your multimeter.

Adding up voltages around the circuit

We start by exploring some voltages. Set your multimeter to DC volts with a range of 10 volts.

With a simple test, you can determine the exact voltage of your battery. Touch the black voltmeter probe to the negative terminal of the battery and the red voltmeter probe to the positive terminal of the battery. You do this by gently inserting the tip of each probe into a hole in the breadboard that's connected to the point you're measuring in the circuit. You should see a reading of roughly 9 volts (for a fresh battery). Note the reading.

Next, test the voltage across the pot (R2). Place the black voltmeter probe at the point where the R2 is connected to the negative terminal of the battery, and place the red probe at the point where R1 and R2 are connected. What reading do you get? What reading would you expect to get, and why? (Hint: the pot is turned all the way down to zero ohms.)

Test the voltage across R1. Take care to orient the voltmeter probes so that you get a positive voltage reading, placing the black probe at the point where R1 meets R2, and the red probe at the point where R1 is connected to the battery. What reading do you get, and why? When you add the voltages you've measured together, you should get the full battery voltage, as follows:

$$V_1 + V_2 = V_{battery}$$

Try adjusting the pot to a few different settings and measuring V_1 and V_2 again. Do they always add up to $V_{battery}$?

Zeroing in on Ohm's Law

To test Ohm's Law, zero the pot again and think about what current you expect to see in this series circuit with the pot turned all the way down to zero ohms. With the probes disconnected from the circuit, change the multimeter setting to DC amps and set the range to 200 milliamps. (You'll probably turn this down to a 20-milliamp range, but starting with a larger range and turning it down is preferable to doing things the other way around.)

To measure current, you need to insert your meter in series with whatever you're measuring current through. But before you break the circuit, disconnect the battery by simply removing one of its leads from the breadboard. Get into the habit of removing power before fiddling with circuit components.

With the power disconnected, simply move the R2 (pot) lead that's connected to R1 to another column in your breadboard. Now you've broken the circuit. Then connect the red lead of your meter to the open side of R1 and the black lead of your meter to the open side of R2. Then connect the battery to the circuit, and you should get a current reading. If the reading is less than 20 milliamps, you can turn down your meter setting to the 20-milliamp range to get a more precise reading.

Now do a quick little bit of maths and use Ohm's Law, the values of your components and the readings you've just taken to check that everything adds up.

With your meter still connected between R1 and R2, slowly turn up the potentiometer and observe the change in current. Is the reading heading in the right direction? Turn the pot all the way to the end and check that your readings and component values all add up under the equation for Ohm's Law.

The current through a series circuit is limited by all the resistances in the circuit, and resistances in series add up. (Keep in mind, too, that the meter adds a certain amount of resistance, albeit tiny, to the circuit.)

Now, turn the pot down until the current reading is about 10 milliamps. Calculate the value of the resistance R2 that you expect. Next, disconnect the battery and then remove the pot from the circuit. Switch your multimeter setting to ohms with a range of 2 kΩ. Measure the resistance across the pot leads. Again, check that these figures add up under Ohm's Law.

You can experiment as much as you want, varying the pot and measuring currents and voltages to verify that Ohm's Law really does work. You'll also see that the voltage drops across R1 and R2 vary as the pot resistance changes, and that those voltages always add up to the battery voltage.

Try filling in the chart below as you run your tests, adding rows for other pot settings.

R2	*Current*	V_2	V_1
0 Ω			
	10 mA		
		5 volts	
10 kΩ			

Dividing up voltage

Using the same series circuit (as shown in Figure 13-1), you can test the concept of the voltage divider that we discuss in Chapter 3. Say, for instance, you need to supply a circuit with 5 volts, but you've got a 9-volt power supply. You can divide the battery voltage between two or more resistive components, adjusting the resistances so that one of the voltages is 5 volts.

Using Ohm's Law, you know that the voltage across R2 is equal to the current, I, through the circuit times the resistance, R2. You also know that the current, I, is equal to the battery voltage divided by the total resistance of the circuit, which is R1 + R2. From these two equations, you can write an expression for the voltage across R2:

$$
\begin{aligned}
V_2 &= I \times R2 \\
&= \frac{V_{battery}}{(R1 + R2)} \times R2 \\
&= \frac{R2}{(R1 + R2)} \times V_{battery}
\end{aligned}
$$

Notice that the expression on the right side of the above equation is equal to a ratio of resistances multiplied by the supply voltage. With the pot out of the circuit, turn its dial to some point and measure its resistance. Given this resistance, what would you expect the voltage across the pot to be? Insert the pot into the circuit, power it up and measure the voltage across the pot. Was your estimate close?

Use the chart below to test your voltage divider for a range of R2 values.

R2	*Estimated* $V_2 = R2/(R1 + R2) \times V_{battery}$	*Measured* V_2
50 Ω		
100 Ω		
470 Ω		
1 kΩ		
5 kΩ		
8 kΩ		
10 kΩ		

Now, imagine that you want to design a voltage divider circuit to produce 5 volts across R2. You calculate the value of R2 that results in 5 volts across R2, which means plugging in 5 volts for V_{R2} in the voltage divider equation shown above to determine the value of R2. (Note that you don't need to know the current passing through the circuit in order to design the voltage divider.)

After a bit of mathematical manipulation (which we don't go into here), and assuming your battery measures 9 volts, you end up with the following equation for R2:

$$R2 = \frac{5}{4} \times R1 = 1.25 \times R1$$

R1 equals 470 ohms (nominally, anyway), so you expect to produce 5 volts across R2 when the pot is set to about 588 ohms (that's 1.25 × 470 ohms).

To test this calculation, set your multimeter to DC volts with a range of 10 volts and measure the voltage across the pot. Next, turn the pot until the voltage reading is about 5 volts. Then remove the pot from the circuit and measure the resistance, which should be in the vicinity of 588 ohms.

You can test the voltage divider concept on your own. Pick a value of R2 and adjust the pot to that value. Calculate the expected voltage across R2 and then measure the actual voltage with the pot in the circuit. You get the idea.

Parallel parking resistors

Want to see current split right before your very eyes? Set up the circuit in Figure 13-3 and measure each current, following these steps:

1. **Set up your multimeter.** Set it to DC amps, with a range of 20 milliamps.
2. **Measure the supply current, I_1.** Break the circuit between the battery and R1 and then insert the multimeter in series with the battery and R1, with the red probe connecting to the positive battery terminal and the black probe connecting to the open side of the resistor. Note the current reading.
3. **Remove the multimeter.** Reconnect the battery and the resistor.
4. **Measure the branch current, I_2.** Remove one of the leads of resistor R2 and then insert the multimeter in series with R2, using the proper lead orientation. Note the current reading.
5. **Remove the multimeter.** Reconnect the R2 lead into the circuit.
6. **Measure the other branch current, I_3.** Remove one of the leads of R3 and then insert the multimeter in series with R3, using the proper lead orientation. Note the current reading.
7. **Remove the multimeter and turn it off.** Reconnect the R3 lead into the circuit.

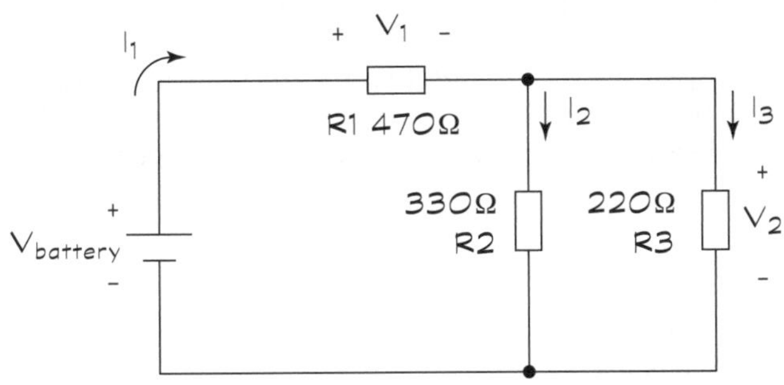

Figure 13-3: The supply current splits between the two branches of this circuit.

Unless the forces of nature have changed (or your multimeter has gone berserk), the branch currents you've measured add up to the supply current you measured: $I_2 + I_3 = I_1$.

You can use maths to calculate these currents by applying Ohm's Law and the rules for series and parallel resistances. To calculate the total supply current, I_1, you determine the total resistance of the circuit, R_{total}, and apply Ohm's Law using the battery voltage.

In Chapter 3, we discuss how to calculate the equivalent resistance, R_{total}, of a combined series and parallel circuit like the one in Figure 13-3, as follows:

$$R_{total} = R1 + (R2 \parallel R3) = R1 + \frac{R2 \times R3}{R2 + R3}$$

If you run the numbers for the resistor values we've chosen here, you should find that $R_{total} = 602\ \Omega$.

Then, you calculate the total supply current as follows:

$$I_1 = \frac{V_{battery}}{602 \text{ ohms}}$$

Assuming $V_{battery} = 9$ volts, you get:

$$I_1 = \frac{9 \text{ volts}}{602 \text{ ohms}} \approx 0.015 \text{ amps} = 15 \text{ mA}$$

Now that you know I_1, you can calculate the voltage V_1 across resistor R1, as follows:

$$V_1 = I_1 \times R1 = 0.015 \text{ amps} \times 470 \text{ ohms} \approx 7 \text{ volts}$$

If 7 volts is dropped across R1, and then V2, the voltage across the parallel resistors must be 2 volts (9 volts – 7 volts). Using V2 and applying Ohm's Law to each parallel resistor, you can calculate the branch currents:

$$I_2 = \frac{V_2}{R2} = \frac{2 \text{ volts}}{330 \text{ ohms}} \approx 0.006 \text{ amps} = 6 \text{ milliamps}$$

$$I_3 = \frac{V_2}{R3} = \frac{2 \text{ volts}}{220 \text{ ohms}} \approx 0.009 \text{ amps} = 9 \text{ milliamps}$$

Amazingly, the two branch currents add up to the supply current.

Use your multimeter as a voltmeter to verify your voltage calculations. Then take out the fixed resistor R3 and substitute in a 10 kΩ potentiometer. Vary the resistance and watch what happens to the currents and voltages. Having completed all these tests, you're an Ohm's Law expert!

Charging and Discharging a Capacitor

In this experiment, you can observe a capacitor charging up, holding its charge and discharging. You can also vary the time a capacitor takes to charge and discharge. Take a peek at Chapter 4 to review how capacitors work and how you can control their operation.

Watching your charges go up and down

The circuit in Figure 13-4 is really two circuits in one. The changeover switch alternates between positions labelled 'charge' and 'discharge', creating two circuit options:

- **Charging circuit:** When the switch is in the charge position, the circuit consists of the battery, resistor R1 and the capacitor, C. Resistor R2 is not connected to the circuit.
- **Discharge circuit:** When the switch is in the discharge position, the capacitor is connected to resistor R2 in a complete circuit. The battery and R1 are disconnected from the circuit (they're open).

Use a jumper wire for the changeover switch. Poke one end of the jumper wire into your breadboard so it's electrically connected to the positive side of the capacitor. Then, you can use the other end to alternately connect the capacitor to R1 or R2. You can also leave the other end of the jumper wire unconnected, which we suggest you do later in this section. You'll see why.

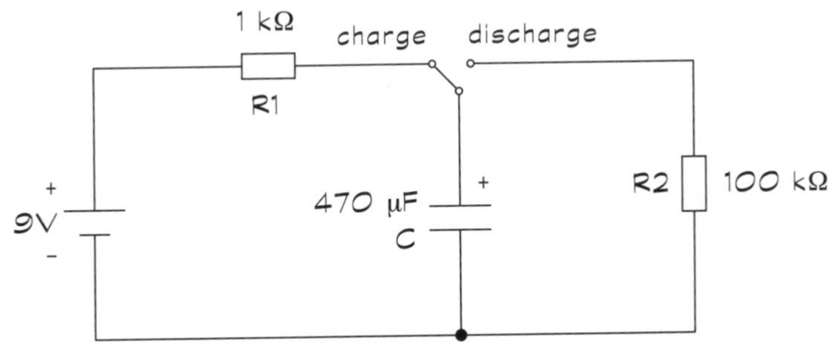

Figure 13-4: Move the switch in this circuit alternately to charge and discharge the capacitor.

Set up the circuit using a 9-volt battery, but don't connect the free end of the jumper wire switch to anything just yet.

Make sure that you insert the electrolytic capacitor the right way around by connecting the negative side of the cap to the negative terminal of the battery. Get it wrong and it may explode!

Charging the capacitor

Set your multimeter to DC volts with a range of 10 volts and connect it across the capacitor (red lead connected to the positive side of the cap, black lead connected to the negative side of the cap). Now, connect the changeover switch to the 'charge' position (at R1) and observe the voltage reading on your meter. You should see the reading rise to approximately 9 volts – but not instantaneously – as the capacitor charges up through resistor R1. This charging takes a couple of seconds.

Holding the charge

Remove the end of the jumper wire and just let it hang. Observe the reading on your voltmeter, which should continue to read 9 volts or thereabouts. (You may see the reading decrease a tiny bit, because capacitors can and do leak charge.) The capacitor is holding its charge (really, holding electrical energy), even without the battery connected.

Letting the charge go

Now, connect the free end of the jumper wire to the 'discharge' position (at R2) and observe the voltage reading on your meter. You should see it decrease fairly slowly as the capacitor discharges through resistor R2 to zero volts. This discharging takes several minutes.

Calculating the charge and discharge times

In Chapter 4, we explain that a capacitor in a simple RC circuit reaches nearly its full charge at approximately five times the RC time constant, T. T is simply the value of the resistance (in ohms) times the value of the capacitance (in farads). So, you can calculate the time taken to charge and discharge the capacitor in your circuit as follows:

$$\begin{aligned}\text{Charge time} &= 5 \times R1 \times C \\ &= 5 \times 1{,}000 \text{ ohms} \times 0.000470 \text{ farads} \\ &= 2.35 \text{ seconds} \\ \text{Discharge time} &= 5 \times R2 \times C \\ &= 5 \times 100{,}000 \text{ ohms} \times 0.000470 \text{ farads} \\ &= 235 \text{ seconds} \approx 3.9 \text{ minutes}\end{aligned}$$

Did you observe this result? Repeat the charging and discharging experiment and see if your calculations seem about right.

Varying the RC time constant

Inserting a couple of potentiometers in your charging and discharging circuits allows you to observe different RC time constants and watch your capacitor charge and discharge at different time intervals.

Set up the circuit in Figure 13-5, leaving the switch open (one end of the jumper wire disconnected from the board). Notice that a 1-kΩ resistor is in series with the capacitor; that's to protect the capacitor by limiting current flow regardless of the potentiometer setting.

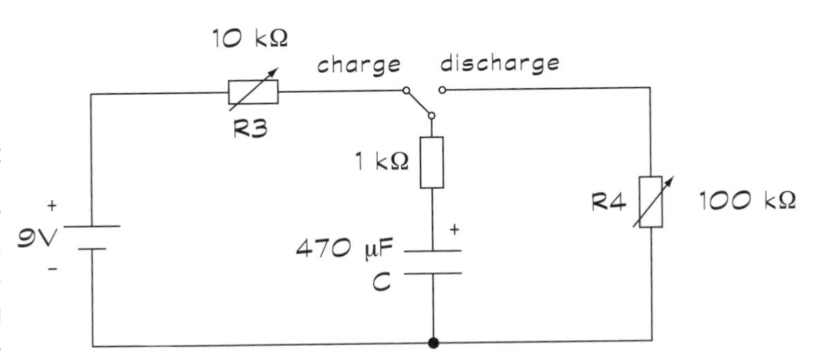

Figure 13-5: Adjust the potentiometers in this circuit to vary the capacitor charging and discharging times.

The total resistance in the charging circuit times the capacitance determines the RC time constant of your charging circuit. This total resistance is the sum of the fixed resistance (1 kΩ) and the variable resistance dialled up on potentiometer R1. So the capacitor charge time is given by:

$$\text{Charge time} = 5 \times (R3 + 1000) \times C$$

If you set R3 to zero ohms, your total resistance is simply 1000 ohms, and the capacitor charges up in approximately 2.35 seconds, just as it did in the earlier section ('Watching your charges go up and down'). Move the jumper wire to the charge position and observe the reading on the voltmeter.

The total resistance of the discharge circuit times the capacitance determines the RC time constant of your discharge circuit. This total resistance is the sum of the fixed resistance (1 kΩ) and the variable resistance of potentiometer R4. So, the capacitor discharge time is given by:

$$\text{Discharge time} = 5 \times (R4 + 1000) \times C$$

If you set R4 to its maximum resistance, 100 kΩ, expect a discharge time of approximately 237 seconds or 4 minutes (nearly the same as in the earlier section 'Watching your charges go up and down'). Switch the jumper wire to the discharge position and observe the voltmeter reading.

When the capacitor has discharged, remove the jumper wire. Adjust R3 to its maximum value, 10 kΩ. Move the switch to the charge position and observe the meter reading. The capacitor will take longer to charge, starting off quickly but then slowing down.

In Chapter 4, you can see the voltage versus time waveform of a charging/discharging capacitor. This waveform shows that, at first, a capacitor charges quickly and then charges more slowly. The same holds true when a capacitor discharges: it initially discharges quickly and then more slowly.

Now remove the jumper wire. Adjust R4 to zero ohms. Move the switch to the discharge position and observe the meter reading. Did the capacitor discharge rather quickly? You would expect it to discharge in about 2½ seconds.

Try adjusting each potentiometer to different values and alternately charging and discharging the capacitor. You can even get out your kitchen timer and try to time the charge/discharge cycles. Then remove each potentiometer, measure each resistance and calculate 5RC. Do your calculations roughly match your observations?

Dropping Voltages across Diodes

Diodes, which we discuss in Chapter 6, are like one-way valves for electrical current. By applying a small voltage from anode to cathode, current flows in the forward direction and the voltage drop across the diode remains fairly constant, even as the current increases.

In this section, you see some of the ways diodes are used in electronic circuits. You can also use your multimeter to explore voltage drops and current in circuits with diodes.

Turning on an LED

For red, yellow and green LEDs, a forward voltage of about 2 volts turns on the valve, so to speak, allowing current to flow and the LED to light up. LEDs can carry currents of up to about 20 milliamps. (Check the ratings of the particular LED you choose.)

The circuit in Figure 13-6 is designed to demonstrate the on/off operation of an LED, and how increasing current strengthens the light emitted by the diode. Turn a 10-kΩ pot to its maximum resistance. Then set up the circuit shown in Figure 13-6, using a standard red, yellow or green LED. Make sure that you orient the LED properly, with the cathode (negative side) connected to the negative battery terminal. On many LEDs, the shorter of the two leads is the cathode.

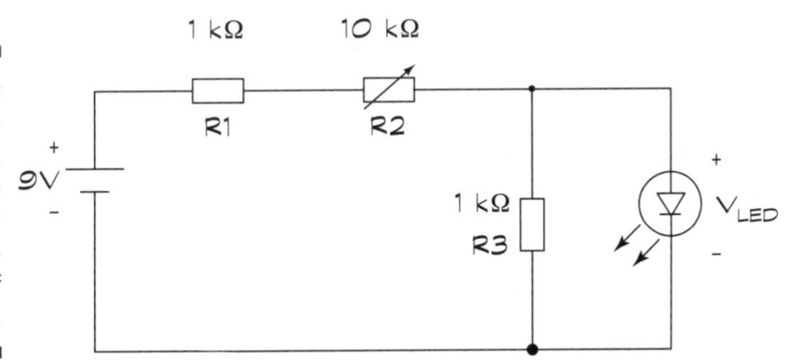

Figure 13-6: Use this circuit to turn an LED on and off, and to vary the intensity of the light.

Set your multimeter to DC volts with a range of 10 volts and place the leads across the LED. Is the LED lit up? What voltage reading do you get? The LED voltage should be less than 1 volt, which isn't enough to turn the diode on.

Now, turn the pot down slowly, keeping your eye on the LED. When the LED turns on, stop turning the pot. Observe the reading on the multimeter. The LED voltage should be close to 2 volts. Now, continue to turn the pot down as you watch the LED. Observe what's happening to the light. Turn the pot all the way down to zero ohms and watch the brightness of the LED. Note the voltage reading on your multimeter. Did the LED voltage change much as the light got brighter?

To understand why the LED was off when the pot was set to 10 kΩ, and then turned on as you lowered the pot resistance, consider the circuit as if you'd removed the LED. The circuit is a voltage divider, and the voltage across resistor R3 (which is the same as the LED voltage) is given by a ratio of resistance times the supply voltage:

$$\begin{aligned} V_{LED} &= \frac{R3}{(R1 + R2 + R3)} \times 9 \text{ volts} \\ &= \frac{1000}{(1000 + R2 + 1000)} \times 9 \text{ volts} \\ &= \frac{1000}{(2000 + R2)} \times 9 \text{ volts} \end{aligned}$$

If the resistance of the potentiometer is high (say, 10 kΩ), the voltage across the LED is rather low (about 0.75 volts). When the resistance of the pot is low enough, the voltage across the LED rises enough to turn the LED on. V_{LED} climbs to about 2 volts when R2 falls down to about 2.5 kΩ. (Plug in 2,500 for R2 in the equation above and see for yourself!)

Of course, your particular LED may turn on at a slightly different voltage, say in the range of 1.7 volts to 2.2 volts. If you measure the resistance of your pot at the point at which your LED turns on, you may see a somewhat lower or higher resistance value than 2.5 kΩ.

You can also observe the current flowing through the LED. Break the circuit between the cathode (negative side) of the LED and the negative battery terminal. Insert your multimeter in series where you broke the circuit and set it to measure DC amps.

Start with the pot turned all the way up to 10 kΩ. As you turn the pot down, observe the current reading. Note the reading when the LED first turns on. Then continue to turn the pot down and observe the current readings. You should see the current increase to over 5 milliamps as the light intensifies.

Overall, you expect to notice that the LED turns on when its voltage approaches 2 volts, with just a tiny current passing through it at this point. As you increase the current through the LED, the light brightens but the voltage across it remains fairly steady.

Clipping voltages

When a large enough reverse-bias voltage is applied to a Zener diode, it maintains a steady voltage drop, even as the current through it increases within a certain range. (Check out Chapter 6 for more detail on Zener diodes.)

Set up the simple voltage divider circuit in Figure 13-7, connecting two 9-volt batteries in series to create the 18-volt DC source. Be sure to use resistors rated at ½ watt for this circuit.

Measure the total supply voltage and then measure the voltage across R2, which should be roughly half the supply voltage, or about 9 volts. Measure the voltage across R1; you'd expect it to be about 9 volts as well. The resistors are dividing the supply voltage equally.

Now modify the circuit, as shown in Figure 13-8, by placing a 4.3V 1W Zener diode across R2, with the anode (positive side) of the diode connected to the negative battery terminal.

Figure 13-7: This simple voltage divider splits the supply voltage evenly between the two resistors.

If you can't find a Zener diode with exactly this rating, use one with a rating close to that but remember to adjust your calculations accordingly.

Measure the voltage across R2, which is the same as the voltage across the Zener diode. The voltage has changed from 9 volts to roughly 4.3 volts. Zener diode tolerance can be ±10 per cent, so the voltage can vary from 3.9–4.7 volts.

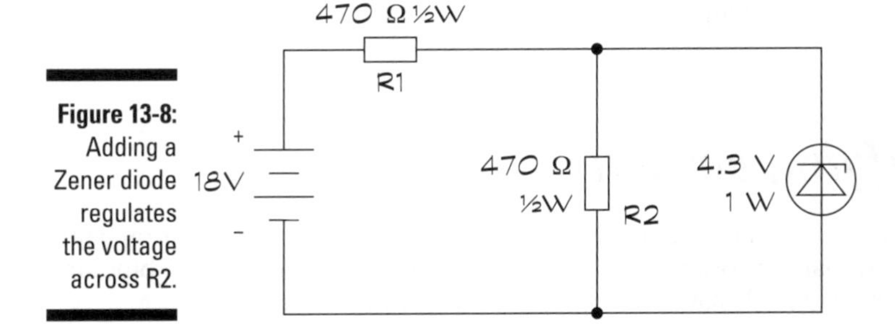

Figure 13-8: Adding a Zener diode regulates the voltage across R2.

The Zener diode is regulating the voltage across R2. So where did the rest of the supply voltage go? Measure the voltage across resistor R1. Did the result increase after you added the Zener diode? (It should measure roughly 13.7 volts.)

Now remove the power and replace resistor R2 with a 10-kΩ potentiometer (see Figure 13-9). Power up the circuit and measure the voltage across R2. Vary the potentiometer from zero ohms to 10 kΩ and observe the voltage reading.

You should see the voltage rise steadily from zero volts until it reaches the Zener voltage, and then remain at that voltage as you continue to increase resistance R2. The Zener diode holds the output voltage steady, even when the load on the circuit (represented by the potentiometer) varies.

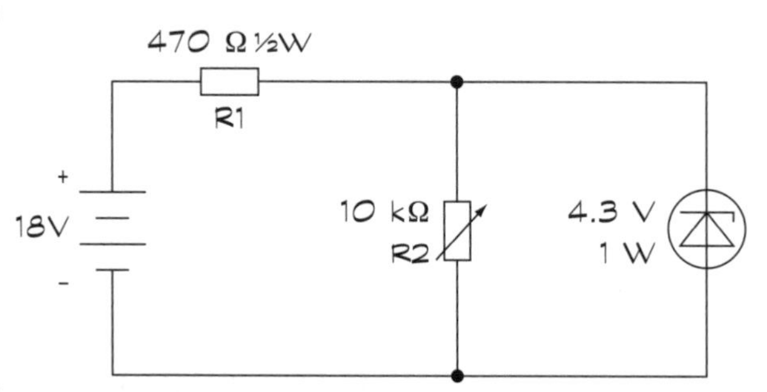

Figure 13-9: Even when the load resistance varies, the Zener diode holds the output voltage steady.

Gaining Experience with Transistors

In this section, you see how tiny transistors are used to control the current in one circuit (at the output of the transistor) using electronic components in another circuit (at the input of the transistor). That's what transistor action is all about!

Amplifying current

You can use the circuit in Figure 13-10 to demonstrate the amplification capabilities of a transistor. Build the circuit using a general-purpose NPN bipolar transistor, such as a 2N3904 or one of the many others available.

Be careful to connect the base, collector and emitter leads properly.

Turn the potentiometer all the way up so that the resistance is 1 MΩ. You probably see a tiny glow from LED2, but may not see any light coming from LED1 although a tiny current is passing through LED1.

Now slowly turn the pot down and watch the LEDs. You should see LED2 getting steadily brighter as you turn the pot down. At some point, you start to see light from LED1 as well. As you continue to turn the pot down, both LEDs glow brighter, but LED2 is clearly much brighter than LED1.

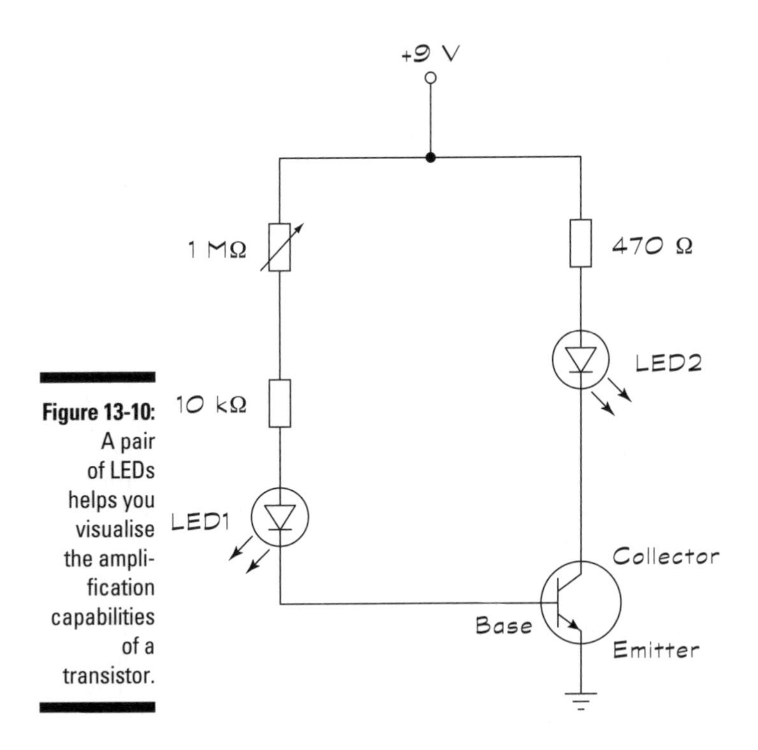

Figure 13-10: A pair of LEDs helps you visualise the amplification capabilities of a transistor.

You're witnessing transistor action. The tiny base current passing through LED1 is amplified by the transistor, which allows a much larger current to flow through LED2. You see a dim glow from LED1 due to the tiny base current, and a bright glow from LED2 due to the stronger collector current.

You can measure each current if you like. (See the following section 'Measuring tiny currents' for how to measure the small base current.)

With the pot set to 1 MΩ, we measured a base current of 6.1μAmp (that's 0.0000061 amps) and a collector current of 0.8 milliamps. Dividing the collector current by the base current, we find that the current gain of this transistor circuit is 131. With the pot set to 0 Ω, we measured a base current of 0.61 milliamps and a collector current of 14 milliamps, for a current gain of 23.

Measuring tiny currents

The base current of the bipolar transistor in Figure 13-10, which passes through LED1, is very small, especially when the pot is set to its maximum resistance. If you want to measure this tiny current, you can do so in two different ways.

You can make the measurement directly, by breaking the circuit on one side of LED1, inserting your multimeter and setting it to DC amps. The current is so small that it may not register on your meter.

Or, you can also measure the current indirectly, using Ohm's Law to help you. The same current that passes through LED1 and into the base of the transistor also passes through two resistors: the 10-kΩ resistor and the potentiometer. You can measure the voltage drop across either resistor and divide the voltage reading by the resistance. (Remember, Ohm's Law tells you that the current passing through a resistor is equal to the voltage across the resistor divided by the resistance.)

If you really want an exact measurement, power the circuit down, pull the resistor out of the circuit and measure its exact resistance with your multimeter. Then perform the current calculation. Using this method, we measured a base current of 6.1μAmp (that's 0.0000061 amps).

Creating light at your fingertip

The circuit in Figure 13-11 is a touch switch. It uses a pair of NPN transistors to amplify a really tiny base current enough to light the LED. This piggyback configuration of two bipolar transistors, with their collectors connected and the emitter of one feeding into the base of the other, is known as a *Darlington pair.*

Set up the circuit, using any general-purpose NPN transistors (such as 2N3904). Close the circuit by placing your finger across the open circuit shown in the figure (don't worry, it doesn't hurt).

The LED should turn on. When you close the circuit, your skin conducts a teeny tiny current (a few microamps), which is amplified by the pair of transistors, lighting the LED. That's the light touch!

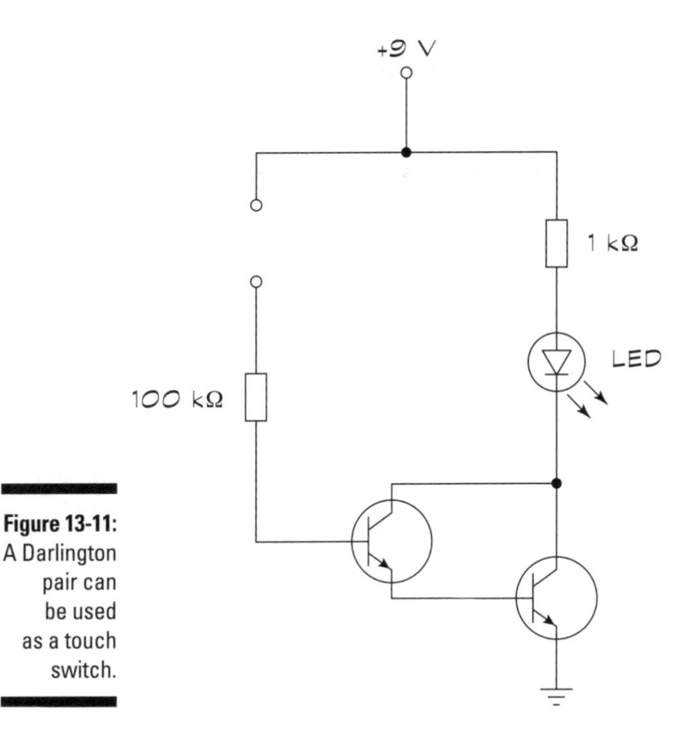

Figure 13-11: A Darlington pair can be used as a touch switch.

Using Your Logic

Tiny digital circuits called logic gates accept one or more bits (binary digits) as inputs and produce an output bit that depends on the function of the particular gate. (You can read up on logic gates in Chapter 7.) Logic gates form the basic building blocks of advanced digital systems, such as the microprocessor that does all the thinking in your computer.

Inside each logic gate is a bunch of electronic components strung together in just the right way to perform the designated logic function. You find logic gates housed in integrated circuits (ICs) complete with several pins that allow you to access the inputs, outputs and power connections of the digital circuit inside.

In this section, you discover how to make the right connections to a NAND logic gate IC, and you watch the output change as you fiddle around with various combinations of inputs. Then you find out how to create another type of logic gate, an OR gate, by combining NAND gates in just the right way.

Seeing the light with a logic gate

The circuit in Figure 13-12 uses an LED to indicate the high or low state of the output of a two-input NAND gate. Set up the circuit, using one of the four NAND gates on the quad 2-input NAND 4011 IC.

The NAND 4011 IC is a CMOS chip that's sensitive to static, so be sure to review the precautions outlined in Chapter 9 to avoid damaging it.

For the switches, you can simply use jumper wires, inserting one end into your solderless breadboard and moving the other end to close or open the switch.

The output of a NAND (NOT AND) gate is high whenever either one or both inputs are low, and the output of a NAND gate is low only when both inputs are high. (You can review the functionality of basic logic gates in Chapter 7.) *High* in this case is defined by the positive power supply (9 volts) and *low* is 0 volts.

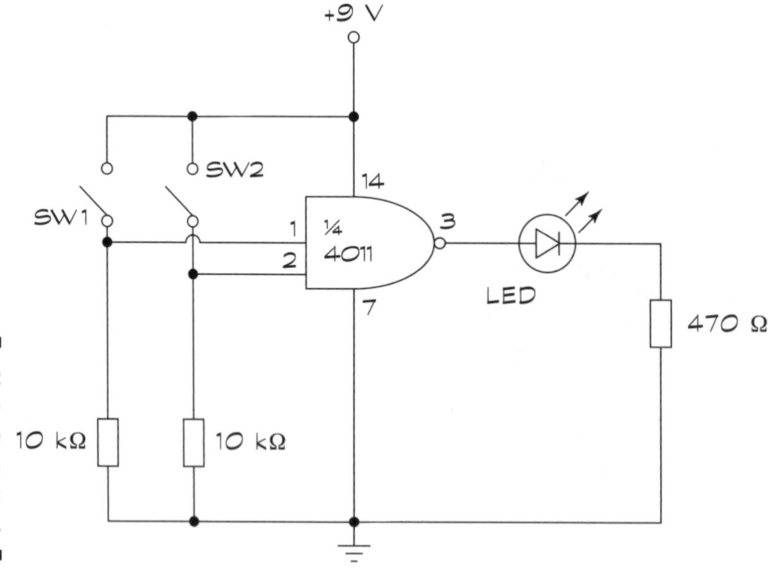

Figure 13-12: Use an LED to show the output of a NAND logic gate.

When you close one of the switches, you make that input high because you connect the positive power supply voltage to the input. When you open one of the switches, you make that input low because the connection is through a resistor to ground (0 volts).

Test the functionality of the NAND gate by trying all four combinations of open and closed switches, filling in the chart below (which is essentially a truth table).

Input 1	*Input 2*	*Output (High = LED on; Low = LED off)*
Low (SW1open)	Low (SW2 open)	
Low (SW1 open)	High (SW2 closed)	
High (SW1 closed)	Low (SW2 open)	
High (SW1 closed)	High (SW2 closed)	

The LED lights up when one or both switches are open and turns off when both are closed.

Turning three NAND gates into an OR gate

You can combine several NAND gates to create any other logical function. In the circuit in Figure 13-13, three NAND gates are combined to create an OR gate. Switches SW3 and SW4 control the inputs to the OR gate and the on/off state of the LED indicates the output of the OR gate.

Each of the two NAND gates on the left functions as a NOT gate (or inverter); it ties the inputs together so that a low input produces a high output and a high input produces a low output. The NAND gate on the right produces a high output when one or both of its inputs are low, which happens when one or both switches (SW3 and SW4) are closed.

The bottom line is that when one or both switches are closed, the output of the circuit is high. That's an OR gate!

Set up the circuit, being careful to avoid static.

You can use the remaining three NAND gates on the quad 2-input NAND IC that you used to build the circuit in Figure 13-12.

Open and close the switches to test that everything works. If so, the LED turns on when either one or both switches are closed.

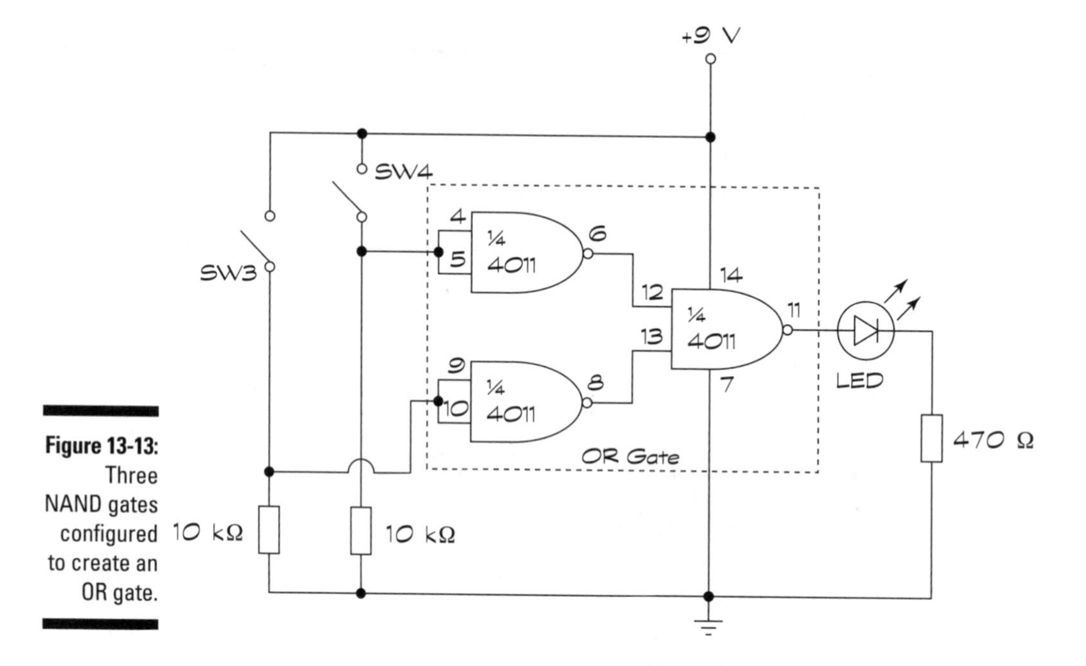

Figure 13-13: Three NAND gates configured to create an OR gate.

Chapter 14

Great Projects You Can Build in 30 Minutes or Less

In This Chapter

- Stocking up on project supplies
- Playing around with light and sound
- Exploring the extraordinary world of piezoelectrics
- Wiring up some new projects

Getting up to speed on electronics really begins to pay off when you get to the point of actually building a project or two. In this chapter, you get to play with some fun, surprising and educational electronics gadgets, which are simple enough to build within half an hour. We keep the number of required parts to a minimum, and the most expensive project costs under £20 or so to build.

We give you some detailed steps for the first project, so we suggest that you work through that first. Then, you should be able to follow the circuit schematics and build the rest of the projects on your own. Check out Chapter 10 if you need a little help reading or understanding the schematics.

What to Get and Where to Get It

You can build all the projects in this chapter, except for the electronic compass in the later section 'Lighting the Way with an Electronic Compass', on a solderless breadboard (see Chapter 11 for all about breadboards). Of course, feel free to build any of the projects on a regular soldered circuit board if you want to make them permanent.

We cover all the parts you need for these projects, such as capacitors, transistors, integrated circuits and even wire, in Chapters 4, 6, 7 and 8 respectively, and you can find more detail about breadboarding and building circuits in Chapters 11 and 12. If you get stuck on any of these projects, hop to one of those chapters to help you through.

With one exception (that tricky but worthwhile electronic compass project, again), you can find the parts that you need to construct the projects in this chapter at an electronics hobby shop such as Maplin. Check out both Chapter 16 and the Appendix for the details of electronic components suppliers online, too.

Unless we tell you otherwise:

- All resistors have ¼ or ⅛ watt, 5 per cent or 10 per cent tolerance.
- All capacitors are rated at a minimum of 25 volts. We note the type of capacitor that you need (ceramic, electrolytic or tantalum) in the parts list for each project.

Building Brilliant, Blinking, Bright Lights

The first project that one of us ever built was a light that blinked on and off. That's all it did, but that was enough. The project involved soldering together all the transistors, resistors and diodes. Start to finish, the whole project took two days. Today, thanks to one specific integrated circuit, making your first blinking light project doesn't take you more than a few minutes and costs only a few pounds in parts.

The special ingredient that makes the blinking light circuit easy to build is the 555 timer IC. This particular chip is the cornerstone of many projects that you build, including several in this chapter. You can use the 555 in a variety of ways, but the most important use is to provide pulses at regular intervals, like a kind of electronic metronome. Throughout this chapter we show you several ways to use this feature to produce a number of cool effects.

You can easily find the 555 timer IC in Maplin shops or order online. The price for one may be anywhere from 25p to 60p, but when you purchase in quantity, the price goes down to nearly half that price per circuit. Online retail stores that sell the 555 by the tube tend to offer the best price. These tubes contain as many as a couple of dozen chips. Don't worry about overstocking; you use them up soon enough.

Following the 555 fast flasher

You can see the schematic of the blinking light project in Figure 14-1, which shows you how to connect a 555 timer IC to a light-emitting diode (LED). By turning variable resistor (potentiometer) R1, you change the rate of blinks from a slow march to a machine-gun strobe.

If you need a quick course on reading schematics, head to Chapter 10.

This circuit provides a useful demonstration of how you can use the 555 as an *astable multivibrator*. This name is just a fancy term for a timer that goes off (not turns off, but goes off like an alarm clock) over and over again, forever (or until it runs out of juice to power it).

This circuit also makes a handy piece of test equipment. Connect the output of the 555 (Pin 3 on the chip) to some other project and use this circuit as a signal source. You see how this works in several of the other projects in this chapter, which are built around the 555 chip.

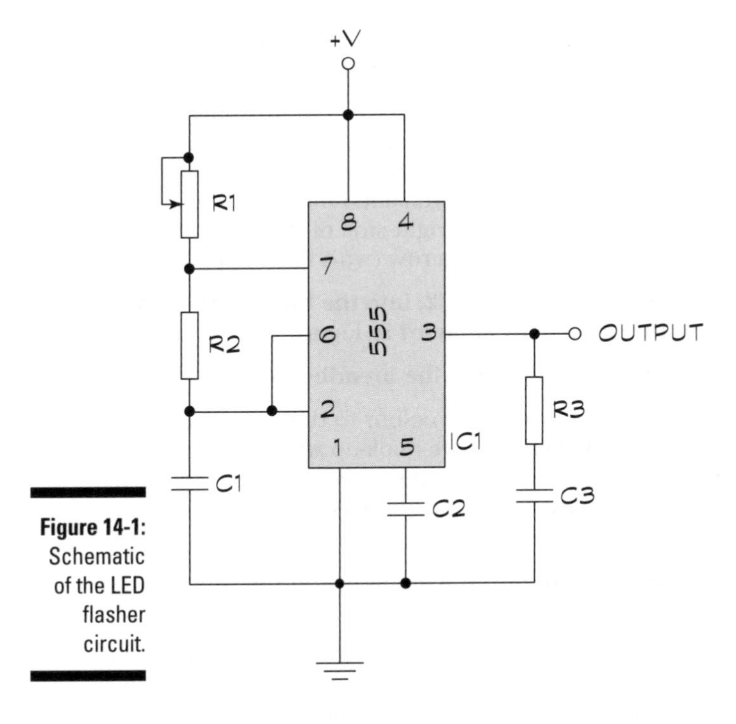

Figure 14-1: Schematic of the LED flasher circuit.

Building the LED flasher circuit is easy. Use the schematic you see in Figure 14-1 and the picture in Figure 14-2 to guide you. Note that we add a bit more space between components to help you see where all the parts go. You should usually build in a little bit of space, rather than squeezing things together, so that you can see what you're doing.

Follow these steps to build the circuit:

1. **Collect all the components you need for the project ahead of time. See the parts list below for a rundown of what you need.**

   Nothing is worse than starting a project, only to have to stop halfway through because you don't have everything to hand!

2. **Carefully insert the 555 timer chip into the middle of the board.**

   The IC should straddle the empty middle row of the breadboard. The clocking notch of the chip (that little indentation or dimple on one end) should face the left of the board. Though this positioning isn't mandatory, it's considered good practice in electronics.

3. **Insert the two fixed resistors, R2 and R3, into the board, following the schematic and the sample breadboard in Figure 14-2.**

   As noted in Chapter 7, the pins on integrated circuit chips are numbered counter-clockwise, starting at the clocking notch. So, if you're facing the breadboard with the 555 on it, and the clocking notch is on your left, Pin 1 is to the left of the clocking notch and Pins 2, 3 and 4 run in a row down the left side of the IC. On the right side of the IC, Pin 5 is opposite Pin 4, and Pins 6, 7 and 8 run up in a row (with Pin 8 opposite Pin 1).

4. **Insert the two capacitors, C1 and C2, into the board, following the schematic and the sample breadboard in Figure 14-2.**

5. **Connect the potentiometer (R1) to the breadboard.**

   If necessary, first solder wire of any colour to the potentiometer's three connections. Use 22-gauge solid core hook-up wire. One connection (the first or the third – it doesn't matter) goes to Pin 7 of the 555; the other two connections are joined (or *bridged*) and attach to the V+ of the power supply.

6. **Connect the LED as shown in the schematic and the sample breadboard.**

   You must insert this component the right way round. Connect the cathode of the LED to ground. Check the packaging of your LED to make sure that you get it right. (If you don't, and you insert the LED backwards, nothing bad happens, but the LED doesn't light up. Simply remove the LED, and reinsert it the other way around.)

7. **Finish making the connections.**

   People commonly refer to the wires used to make the circuit connections as jump leads or jumpers. Most of the circuits you build have at least one or two. You can make your own from 22-gauge single strand wire, but we recommend the packets of ready-made leads designed for breadboards. Use the schematic in Figure 14-1 and the sample breadboard in Figure 14-2 as a guide to making these connections.

8. ***Before* you switch on the power, double-check your work. Verify all the proper connections by cross-checking your wiring against the schematic.**

9. **Attach a 9-volt battery to the V+ and ground rows of the breadboard.**

   The V+ row is at the top, and the ground row is at the bottom. Using a 9-volt battery clip is easier, and you can get one at any electronics shop. A good idea is to solder 22-gauge solid hook-up wire to the ends of the leads from the clip, which makes inserting the wires into the solderless breadboard easier. Remember: the red lead from the battery clip is V+; the black lead is ground.

When you apply power to the circuit, the LED should flash. Rotate the R1 knob to change the speed of the flashing. If your circuit doesn't work, disconnect the 9-volt battery and check the connections again.

Here are some common mistakes to look out for:

- You inserted the 555 IC backwards. This error can damage the chip, and you may want to try another 555.
- You inserted the LED backwards. Pull it out and reverse the leads.

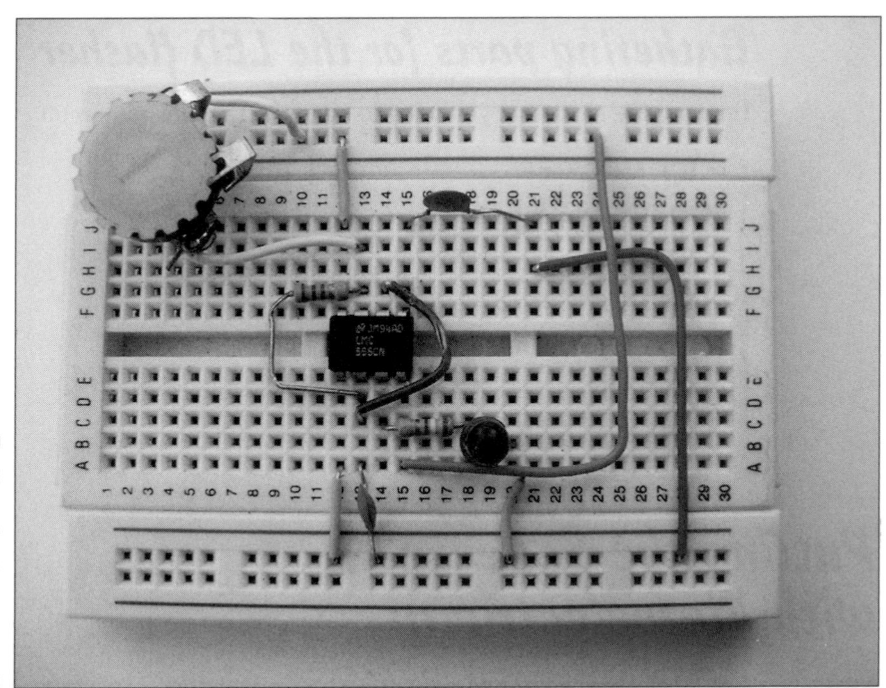

Figure 14-2: An LED flasher with parts mounted on a solderless board.

- You didn't press the connection wires and component leads into the breadboard sockets firmly enough. Be sure that each wire fits snugly into the breadboard, so that no loose connections remain.
- The component values are wrong. Double-check, just in case!
- The battery is dead. Try a new one.
- Change the LED for one with a lower operating voltage – 1 volt should be fine.
- If the LED lights up but doesn't flash, check the polarity of both the power supply and the LED. You may need to reverse both.
- You wired the circuit wrong. Have a friend take a look. Fresh eyes can catch mistakes that you may not notice.

Good electronics practice requires that you build a circuit that's new to you on a breadboard first, because you often need to tweak a circuit to get it working just the way you want it to. When you have it working to your satisfaction on a breadboard, you can make the circuit permanent if you like. Just take your time and remember to double- and even triple-check your work. Don't worry – you'll be a pro in no time, and soon be building fairly complex circuits on your solderless breadboard.

Gathering parts for the LED flasher

Here are the parts that you need to build the LED flasher circuit:

- **IC1:** 555 timer IC
- **R1:** 1-MΩ potentiometer
- **R2:** 47-kΩ resistor
- **R3:** 330-Ω resistor
- **C1:** 1-μF tantalum (polarised) capacitor
- **C2:** 0.1-μF ceramic (non-polarised) capacitor
- **LED:** Light-emitting diode (any colour)

Putting the Squeeze on with Piezoelectricity

Not all electronic circuits require batteries, resistors, capacitors, transistors or any of the other usual components that you find in an electronic circuit. This project generates its own electricity, and you end up with a light drum consisting of a neon light that glows when you tap on a piezo disc. This project serves as a great demonstration of something called piezoelectrics.

Experimenting with piezoelectricity

A simple and fun way to experiment with piezoelectricity is to get a piezo transducer. You can get these discs in various sizes, all for under a pound each.

Get a disc with the two wires already soldered onto it. Some discs have only one wire and these discs work just fine, too. You can clip a wire to the edge of the disc's metal for the ground connection.

Figure 14-3 shows a demonstrator circuit with one disc and one neon bulb. Neon bulbs are special in that they don't light up unless you feed them at least 90 volts. That's a lot of juice! But the piezo disc easily generates this much voltage.

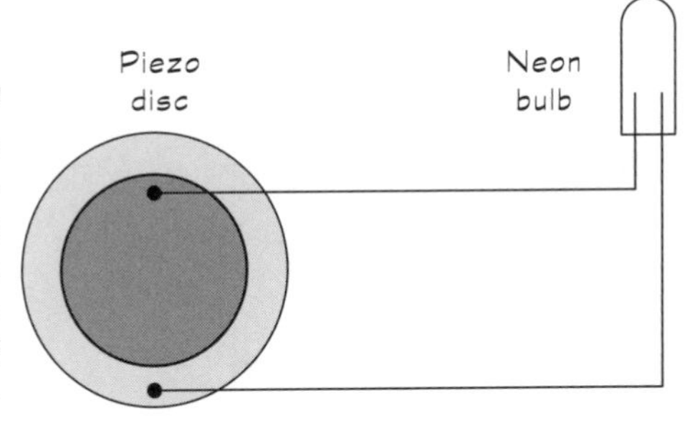

Figure 14-3: Try this simple circuit to demonstrate the properties of piezoelectricity.

Piezo – what?

The term *piezo* comes from a Greek word meaning to press or squeeze. A century ago some smart people discovered that you can generate electricity when you press certain kinds of crystals really hard. Lo and behold, these same crystals change shape – though only slightly – when you apply electricity to them. At first these people didn't know what to do with their discovery, but it turned out to be really important because we now use piezoelectricity in loads of everyday gadgets, such as quartz watches, alarm buzzers, barbecue grill starters and much more.

To build the circuit in Figure 14-3, follow these steps:

1. **Place the disc on an insulated surface.**

   A wooden or plastic table surface works fine, but don't use a surface made of metal.

2. **Connect the disc and neon bulb together by using two crocodile clips, as shown in Figure 14-4.**

   Place one test lead from the red wire of the disc to one connection of the neon lamp (it doesn't matter which connection). The other test lead goes from the black wire of the disc to the other connection of the neon lamp.

3. **Place the disc flat on the table.**

4. **Rap very hard on the disc with the plastic end of a screwdriver.**

   Each time you rap the disc, the neon bulb flickers.

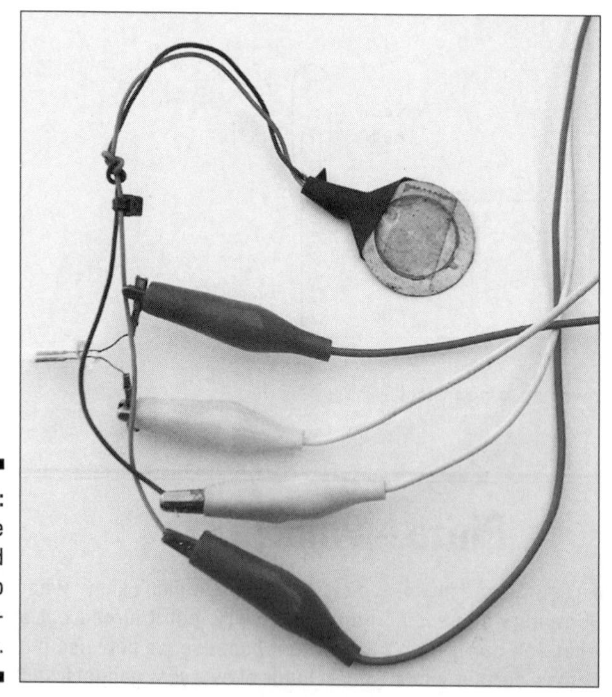

Figure 14-4: Connect the disc and neon lamp using crocodile clips.

Avoid touching the two wires that come from the disc. Although the shock you get isn't dangerous, it definitely doesn't feel very good!

What on earth can you do with this strange but surprising contraption? Now you're ready to build your very own light drum to dazzle friends and family. Just follow these steps:

1. **String up a whole bunch of discs and bulbs in a row.**
2. **Tape or glue these disc-bulb sets to a plastic base.**
3. **Get a pair of drumsticks, turn down the lights and tap on the discs in time with your favourite music.**

You can save money by building the circuit with LEDs rather than neon bulbs. Just substitute each bulb for an LED and 330-Ω resistor.

Getting the parts for the piezoelectric circuit

For the circuit that demonstrates piezoelectricity, you just need:

- A bare piezo transducer disc (around 50 millimetres in diameter, preferably with two wires soldered on)
- A neon bulb
- Two crocodile clips
- Something to whack the disc with, such as a screwdriver or drumsticks (not a cricket bat)

Assembling the Amazing See-in-the-Dark Infrared Detector

Did you ever want to see in the dark like a cat? Now you can, by building this simple infrared detector. The circuit uses just three parts (plus a battery). You can make the circuit a little fancier by adding an SPST (single-pole single-throw) toggle switch between the + (positive) side of the battery and the phototransistor to turn the detector on and off; or you can just unplug the battery when you aren't using the detector.

Figure 14-5 shows the schematic for the infrared detector. Be sure to use a phototransistor, and not a photodiode, in this circuit. Also, make sure that you get the proper orientation for both the phototransistor and the LED. If you hook up either of them back to front, the circuit fails.

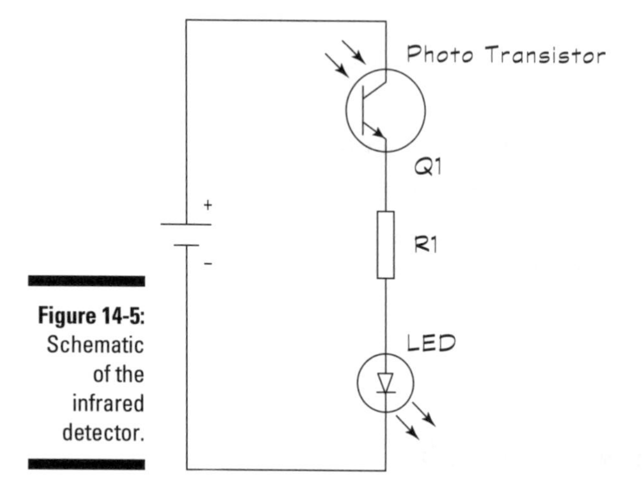

Figure 14-5: Schematic of the infrared detector.

Searching for infrared light

Using the infrared detector, you can test for infrared light from a number of sources. Here are just two ideas to try:

- **Getting to the bottom of a remote control dilemma:** Because remote controls use invisible infrared light, you have a hard time figuring out what's wrong when they stop working. Does the remote have a problem, or should you blame your TV or other appliance? To test the remote control, place it up against the infrared phototransistor. Press any button on the remote; if the LED on your project flashes, you know that you have a working remote.
- **Carrying out counter-surveillance:** Check to see if somebody's hidden camera is in your room. These days, covert cameras (such as the one in Figure 14-6) can see in the dark by using a built-in bright infrared light source. You can use the infrared detector circuit to find these sources, even if you can't see them yourself. Turn off the lights and wave it around slowly to scan the room. If you see the LED brighten, even though you can't see a light source, you may have just found the infrared light coming from a hidden camera!

Although the infrared phototransistor is most sensitive to infrared light, it also responds to visible light. For best results, use the infrared detector in a dimly lit room. Sunlight, and direct light from desk lamps and other sources, can influence the readings.

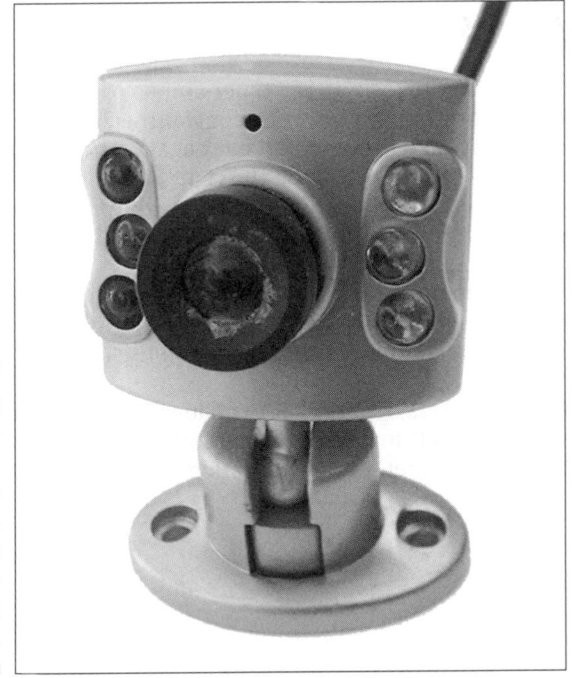

Figure 14-6: This miniature camera can see in the dark, thanks to its six infrared-emitting diodes.

Tracking down parts for the infrared detector

Here's what you need to build this project:

- **Q1:** Infrared phototransistor (almost any phototransistor should work fine)
- **R1:** 330-Ω resistor
- **LED:** Light-emitting diode (any colour)

Keeping People Away with a Siren!

Okay, so you don't really get home any quicker, but the warbling siren that you build in this project encourages other members of your family to steer clear of you. You'll certainly have fun playing around with it, but other people in the house may not like the sound quite as much! The warbler sounds cool, and you can build it into an alarm to alert you if somebody's trying to snatch your football cards, *Star Wars* DVDs or whatever.

Making your siren sound

This circuit (see Figure 14-7) uses two 555 timer chips. You rig both chips to act as astable multivibrators; that is, they constantly change their output from low to high to low to high . . . over and over again. The two timers run at different frequencies. The timer chip on the right in the figure produces an audible tone. If you connect a speaker directly to the output of this timer, you hear a steady medium-pitch sound.

The output of the 555 chip on the left, which produces a slower rising and falling tone, connects into Pin 5 of the 555 chip on the right in the figure. You connect the speaker to the output of the 555 chip on the right.

Adjust the two potentiometers, R2 and R4, to change the pitch and speed of the siren. You can produce all sorts of siren and other weird sound effects by adjusting these two potentiometers and operating this circuit at any voltage between 5 and about 15 volts.

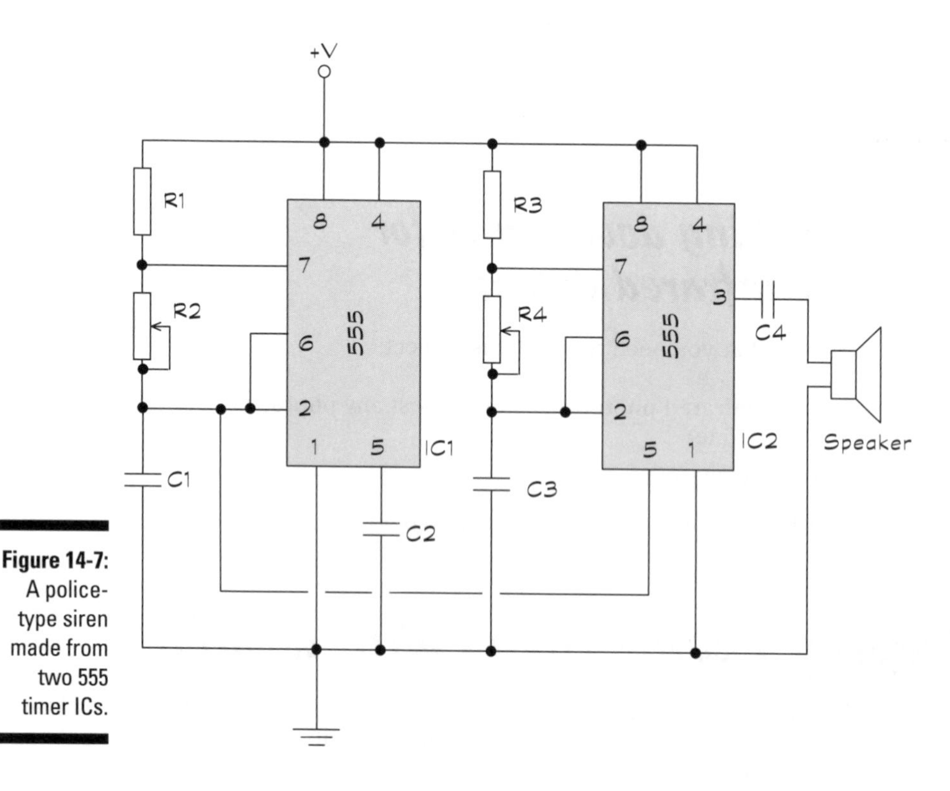

Figure 14-7: A police-type siren made from two 555 timer ICs.

Sorting the siren parts list

To start alarming your friends, gather these parts together to build the circuit:

- **IC1, IC2:** 555 timer IC
- **R1, R3:** 2.2-kΩ resistor
- **R2:** 50-kΩ potentiometer
- **R4:** 100-kΩ potentiometer
- **C1:** 47-μF electrolytic (polarised) capacitor
- **C2:** 0.01-μF ceramic (non-polarised) capacitor
- **C3:** 0.1-μF ceramic (non-polarised) capacitor
- **C4:** 1-μF electrolytic or tantalum (polarised) capacitor
- **Speaker:** 8-Ω, 1-watt speaker

Lighting the Way with an Electronic Compass

Discover where in the world you are with this very cool electronic compass! This magnetic compass uses the same technology that manufacturers build into many cars to show you your direction electronically. Four LEDs light up to show you the four points on the map: N (north), S (south), E (east) and W (west). The circuit illuminates adjacent LEDs to show the in-between directions, SW, SE, NW and NE.

Checking under the compass bonnet

At the heart of this project is a special compass module, the Dinsmore 1490. This module isn't a common, everyday part. You have to special-order it, but you can have a lot of fun with the project, making it worth the £10 to £15 that you pay for the compass module. Check out the manufacturer's website `www.dinsmoresensors.com` for more details. The module can be bought online from a UK-based supplier that for some strange reason is called Chocolate Labrador (`www.choclab.eu`), and don't forget to try other possible sources by doing a Google or Yahoo! Search on 'dinsmore compass'.

The 1490 compass module is about the size of a small thimble. The bottom of the sensor has a series of 12 tiny pins, as you can see in the pinout drawing in Figure 14-8. The pins are arranged in four groups of three and each group consists of the following connection types:

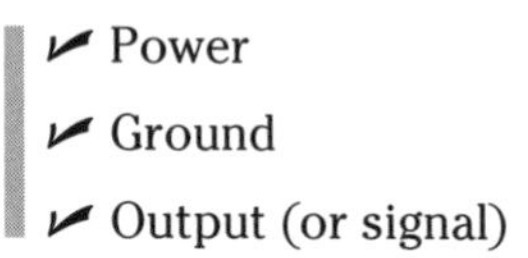

- Power
- Ground
- Output (or signal)

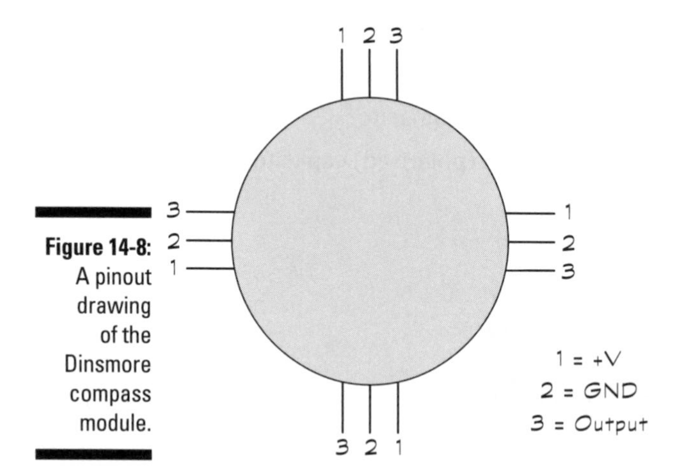

Figure 14-8: A pinout drawing of the Dinsmore compass module.

You can see the schematic for the electronic compass in Figure 14-9. By doing some careful soldering, you can build a nice portable electronic compass that you can take anywhere. Put it in a small enclosure, with the LEDs arranged in typical clockwise N, E, S, W circular orientation. You can buy enclosures at good electronics shops. They come in many different shapes, sizes and finishes. Select an enclosure large enough to contain the circuit board and batteries.

You can power the compass with a 9-volt battery. Add a switch from the + (positive) terminal of the battery to turn the unit on or off, or simply remove the battery from its clip to cut the juice and turn off your compass.

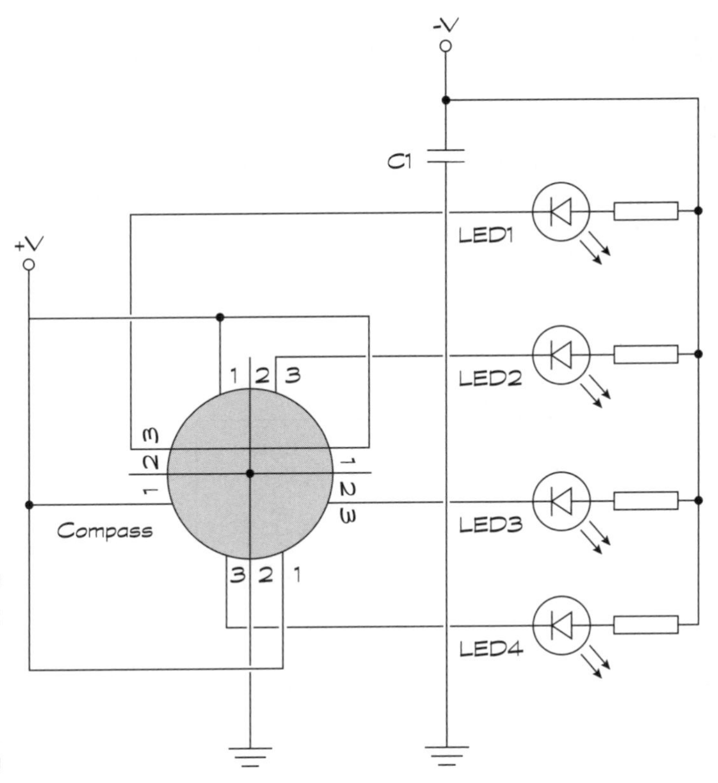

Figure 14-9: Schematic of the handheld compass.

Locating your electronic compass parts

To point you in the right direction, here are the parts that you need to gather to build your compass:

- **COMPASS:** Dinsmore 1490 magnetic electronic compass (see the preceding section 'Checking under the compass bonnet')
- **R1–R4:** 1-kΩ resistors
- **C1:** 10-μF electrolytic (polarised) capacitor
- **LED1–LED4:** Light-emitting diodes (any colour)
- **MISC:** Project box, switch, battery clip (all optional)

Alarming Way to Sense the Light

Figure 14-10 shows you a schematic of a light alarm. The idea of this project is simple: if a light comes on, the alarm goes off. You build the alarm around a 555 timer chip, which acts as a tone generator. When light hits the photoresistor, the change in resistance triggers transistor Q1. This response turns the 555 on, and it squeals its little heart out. You can adjust the sensitivity of the alarm by turning R1, which is a variable resistor (potentiometer).

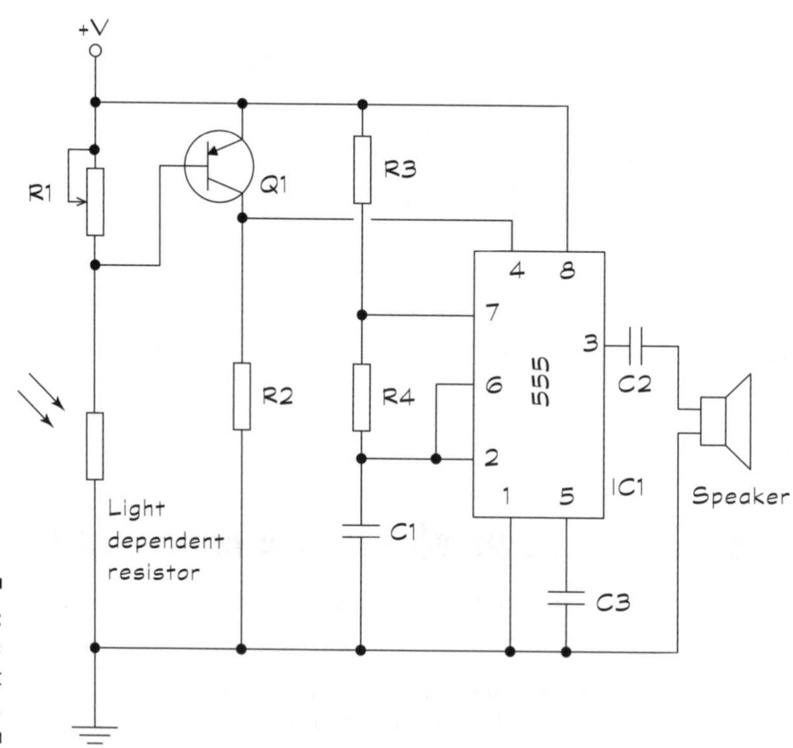

Figure 14-10: Schematic of a light alarm.

Making your alarm work for you

What's the use of an alarm that goes off whenever it senses light? Well, this clever little circuit has several practical uses:

- Put the light alarm inside a pantry so that it goes off whenever someone raids the sweet jar. Keep other family members out of your treats – or keep yourself on that diet! When the food cupboard door opens, light comes in and the alarm goes off.

- Place the alarm inside your tool shed or workroom, near the door. If someone opens the door during the day, light comes through and the alarm goes off: a great way to ensure that no one touches that complex electronics project you're halfway through.
- Build your own electronic rooster that wakes you up at daybreak. (Who needs an alarm clock?)

Assembling a light alarm parts list

Here's the shopping list for the light alarm project:

- **IC1:** 555 timer IC
- **Q1:** 2N3906 PNP transistor
- **R1:** 100-kΩ potentiometer
- **R2:** 3.9-kΩ resistor
- **R3:** 10-kΩ resistor
- **R4:** 47-kΩ resistor
- **C1, C3:** 0.01-μF ceramic (non-polarised) capacitor
- **C2:** 1.0-μF electrolytic or tantalum (polarised) capacitor
- **Speaker:** 8-Ω, 0.5-watt speaker
- **Photoresistor:** Experiment with different sizes; for example, a larger photoresistor makes the circuit a little more sensitive

'Lil but Loud Amp

Give your electronics projects more volume with this little amplifier designed around parts that are inexpensive and easy to find at most electronics suppliers. The LM386 power amplifier IC boosts the volume from microphones, tone generators and many other signal sources.

Figure 14-11 shows the schematic for this project, which consists of just six parts, including the speaker. You can operate the amplifier at voltages between 5 and about 15 volts. A 9-volt battery does the trick.

To use the amplifier, connect a signal source, such as a microphone or the melody maker circuit in the following section, to Pin 3 of the LM386. Also make sure that you connect the ground of the signal source to the common ground of the amplifier circuit.

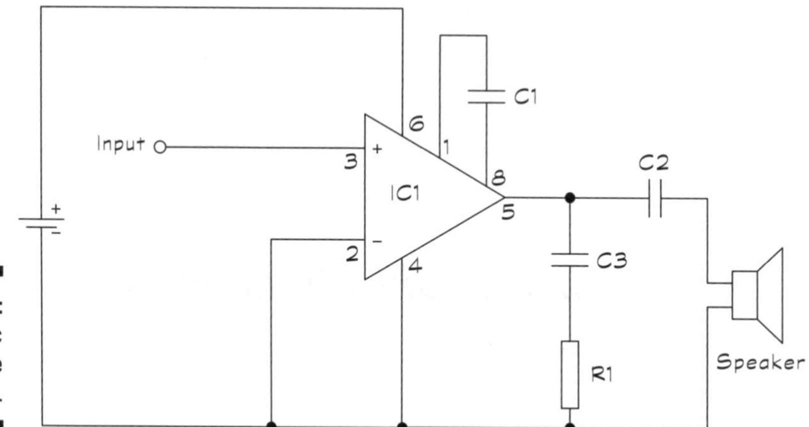

Figure 14-11: Schematic of the little amplifier.

Depending on the source signal, you may find that you get better sound if you place a 0.1- to 10-μF capacitor between the source and Pin 3 of the LM386. For smaller values (less than about 0.47 μF), use a ceramic capacitor; for larger values (1 μF or higher), use a tantalum capacitor. When you use a polarised capacitor, orient the + (positive) side of the component towards the signal source.

This little amp doesn't come with a volume control, and the sound quality may be like a railway station platform tannoy, but this simple circuit packs quite a good little punch in a small and portable package.

Here's a rundown of the parts that you have to gather for this project:

- **IC1:** LM386 amplifier
- **R1:** 10-Ω resistor
- **C1:** 10-μF electrolytic (polarised) capacitor
- **C2:** 220-μF electrolytic (polarised) capacitor
- **C3:** 0.047-μF ceramic (non-polarised) capacitor
- **Speaker:** 8-Ω, 0.5-watt speaker

The better the microphone and speaker, the better the sound!

Making Music with a Melody Maker

Ever wondered how those musical greetings cards work? They use circuits like this one, which is the easiest circuit to build in this chapter and makes a great signal source for the 'Lil but Loud Amp in the preceding section.

Most of the work in this circuit is done by the M66T Simple Melody Generator, a range of CMOS chips costing a little over a pound each. You can buy an M66T that plays songs ranging from 'Rock-a-Bye Baby' to the *Star Trek* theme.

Take one of the M66T chips (the short but sweet *Looney Tunes* theme is a good one), and follow the schematic in Figure 14-12 to connect its three pins to a speaker and a power supply (anything between 1.2 and 3.6-volt; an AA battery works just fine).

To make Looney Tunes louder, simply replace the speaker with the 'Lil but Loud Amp. Connect Pin 1 of the M66T to Pin 3 of the LM386 power amplifier IC and Pin 2 of the M66T to Pin 2 of the LM386.

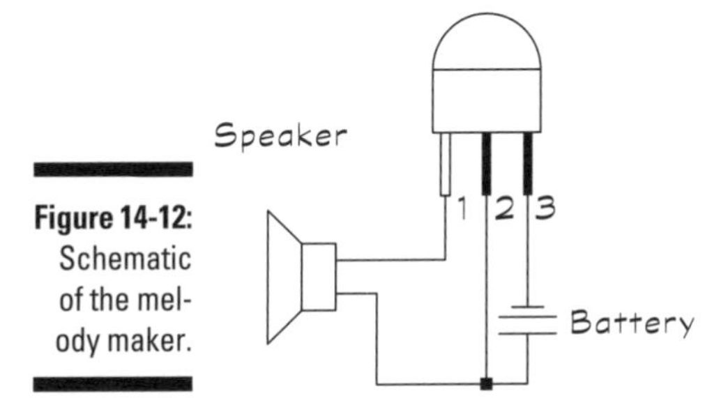

Figure 14-12: Schematic of the melody maker.

The parts for the melody maker are just as simple:

- **IC1:** M66T CMOS device
- **Speaker:** 8-Ω, 0.5-watt speaker

Building the Pocket Water Tester

You may not be able to divine underground water with the water tester circuit in Figure 14-13, but it can help you check for moisture in plants or find water trapped under wall-to-wall carpet.

Understanding how the water tester works

The pocket water tester is deceptively simple and works under the principle of the electrical conductivity of water (a principle that says you don't take a bath with a plugged-in toaster on your lap).

The tester contains two small metal probes. When you place the probes in water, the conductivity of the water completes a circuit. This completed circuit drives current to a transistor. When the transistor turns on, it lights an LED. When the probes aren't in contact with water (or some other conductive body), your tester has a broken circuit, and the LED doesn't light up.

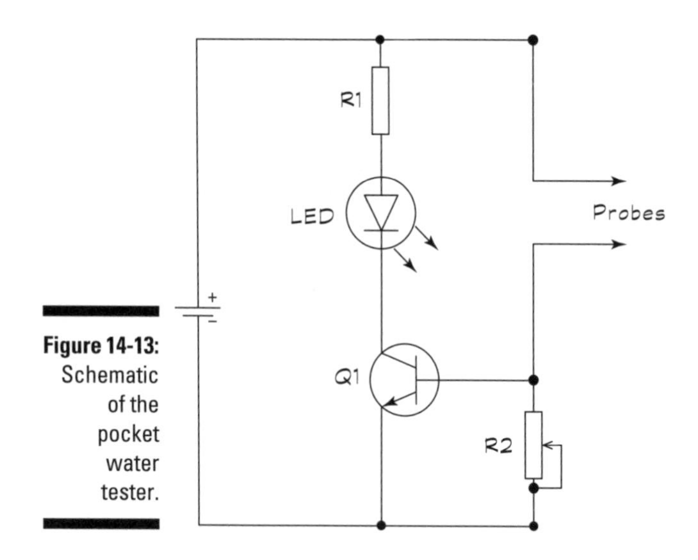

Figure 14-13: Schematic of the pocket water tester.

You make the two probes with small nails. Place the nails about 1 centimetre apart on a piece of plastic (but not wood or metal). The nails should be parallel to one another. File down the tips of the nails to make sharp points. These points help you drive the probes deep into the material that you're testing. For example, you can drive the probes into a carpet and pad to determine if water has seeped under the carpeting after a pipe in the next room burst.

You can adjust the sensitivity of the tester by turning potentiometer R2. Start with the potentiometer in its middle position and turn one way or another, depending on the amount of moisture or water in the object that you're testing.

Power the water tester by using 5 to 12 volts. A 9-volt battery works well.

Gathering water tester parts

Go out and get the following parts to build your water tester project:

- **Q1:** 2N2222 NPN transistor
- **R1:** 470-Ω resistor
- **R2:** 50-kΩ potentiometer
- **LED:** Light-emitting diode (any colour)
- **Probes:** Use two small nails or jump leads with hardened ends

Generating Cool Lighting Effects

If you're a fan of the *Knight Rider* television series from the 1980s, you'll remember the sequential light chaser that the Kitt Car sported. You can build your own (light chaser setup, not car) in the garage.

To build your own mesmerising lighting effects generator (see the schematic in Figure 14-14), you need just two low-priced integrated circuits and a handful of inexpensive parts.

The circuit has two sections:

- **The brains:** A 555 timer IC makes up the first section, on the left of the schematic. You wire this chip to function as an astable multivibrator (in fact, you make the same basic circuit as the LED flasher that we describe in the earlier section 'Building Brilliant, Blinking, Bright Lights'). The 555 produces a series of pulses; turn potentiometer R1 to determine the speed of the pulses.
- **The body:** The second section, on the left of the schematic, contains a 4017 CMOS decade counter chip. The 4017 chip switches each of 10 LEDs on, in succession. The LEDs are switched on when the 4017 receives a pulse from the 555. You wire the 4017 so that it repeats the 1-to-10 sequence over and over again, for as long as the circuit has power.

Arranging the LEDs

You can build the lighting effects generator on a solderless breadboard just to try it out. If you plan to make it into a permanent circuit, give some thought to the arrangement of the ten LEDs. For example, to achieve different lighting effects, you can try the following:

- **Put all the LEDs in a row, in sequence:** The lights chase each other up (or down) over and over again.
- **Put all the LEDs in a row, but alternate the sequence left and right:** Wire the LEDs so that the sequence starts from the outside and works its way inside.
- **Place the LEDs in a circle so that the LEDs sequence clockwise or counter-clockwise:** This light pattern looks like a roulette wheel.
- **Arrange the LEDs in a heart shape:** You can use this arrangement to make a unique Valentine's Day present.

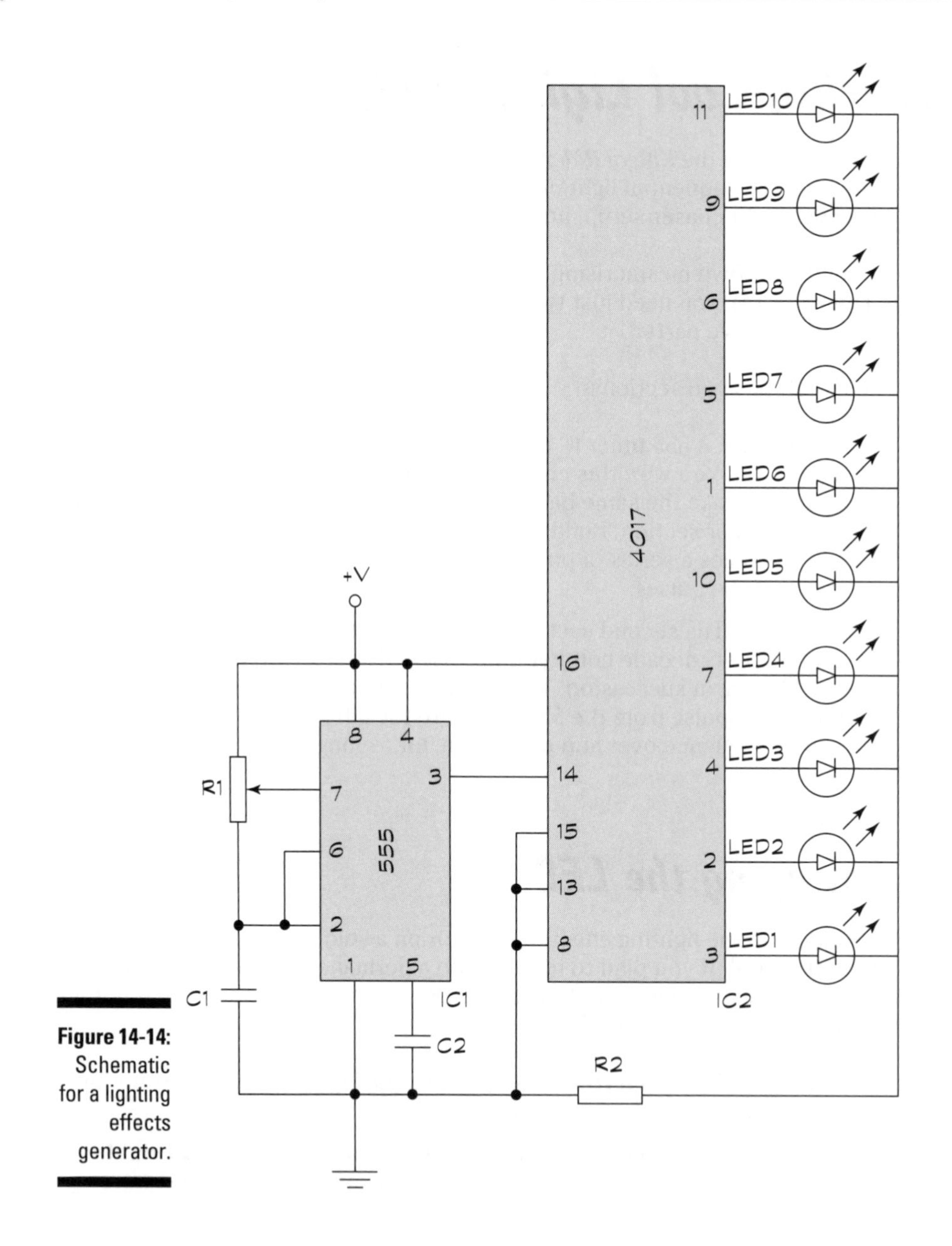

Figure 14-14: Schematic for a lighting effects generator.

Chasing down the parts

To start chasing lights, you need the following parts:

- **IC1:** 555 timer IC
- **IC2:** 4017 CMOS decade counter IC
- **R1:** 1-MΩ potentiometer
- **R2:** 47-kΩ resistor
- **R3:** 330-Ω resistor
- **C1:** 0.47-μF disc (non-polarised) capacitor
- **C2:** 0.1-μF disc (non-polarised) capacitor
- **LED1–10:** Light-emitting diodes (any colour)

Making an Electronic Die

Press the button switch on this circuit to 'throw' a die; the dot patterns look like those on regular dice that you roll physically. The 4017 CMOS decade counter cycles through the numbers one to ten, just as you can see it do with the LEDs in the project in preceding section, 'Generating Cool Lighting Effects'. So how does the die generate the numbers one to six?

This ingenious design uses clever pin combinations until the die resets on the sixth pin. Meanwhile, Pin 12 is on during the count up to five but then switches off. It generates a new pattern in the seven LEDs every time the 555 timer IC pulses.

Here's how to make your electronic die:

1. **Arrange the LEDs:** Follow the pattern shown in Figure 14-15 and connect them together as shown in Figure 14-16. This process may require you to cut the LED pins short. Be careful not to let pins touch and short-circuit the LED pattern.
2. **Build the circuit:** Insert 4017, the diodes and then the rest of the components and jump leads. Double-check the circuit and apply the power supply.

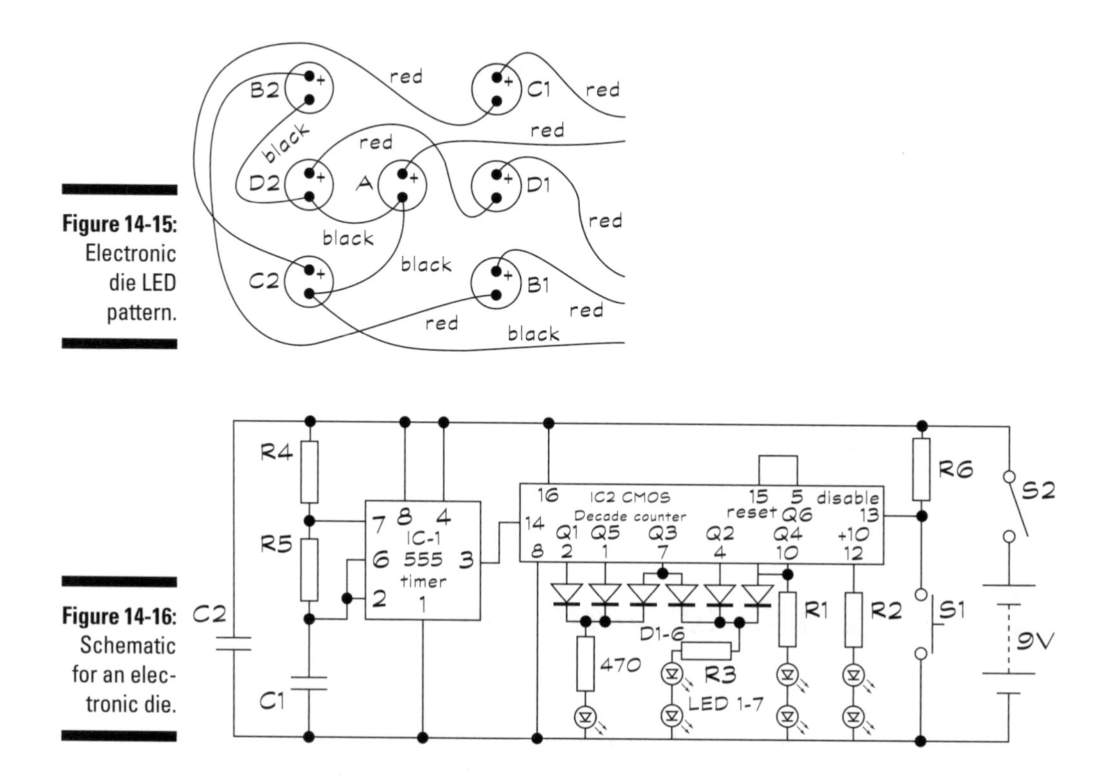

Figure 14-15: Electronic die LED pattern.

Figure 14-16: Schematic for an electronic die.

Here's what you need for the electronic die:

- **IC1:** 555 timer IC
- **IC2:** 4017 CMOS decade counter IC
- **R1-3:** 330-Ω resistors
- **D1-6:** 1N4148 diodes
- **S1:** push switch
- **C1:** 0.01-μF ceramic (non-polarised) capacitor
- **C2:** 0.1-μF ceramic (non-polarised) capacitor
- **LED1–7:** Light-emitting diodes (red, 5-millimetre)

Part IV
The Part of Tens

In this part . . .

This book wouldn't be a bona fide *For Dummies* book without a handful of top-ten lists. These chapters contain golden nuggets that, at the very least, give you something more interesting to browse at the breakfast table than the back of a cereal packet.

We offer some top tips to help you find even more circuits to build, save you time with your circuit design and provide other ideas to feed and foster your habit if (as we expect!) you're now addicted to electronics. And, of course, don't forget our ten terrific sources for electronic parts, tools and other supplies, which appear in Chapter 16.

Chapter 15

Ten Exciting Electronics Extras

In This Chapter

- Jump-starting your electronics experience with a project kit
- Trying out extra equipment
- Finding great deals on testing tools

So you feel ready to get serious about electronics, and you're wondering what else exists, other than basic test equipment, which can help you hone your skills and give you insight into what goes on inside all those wires and components. In addition, you may be anxious to impress your friends and neighbours by showing off cool gadgets you've made *right now* (not next week or next month) and to arm yourself with some impressive tools that come with lots of indicator lights, adjustable knobs and dazzling displays.

If this desire is resonating with you, you're ready to explore some of the toolkits, specialised test equipment and even the useful (gulp) software we describe in this chapter.

You don't absolutely, positively need all these extras just to play around with some LEDs and resistors. A basic multimeter, and maybe a logic probe, are all you need for that. But you may want to consider the ready-made project kits we describe in the next section to jump-start your experience with circuits and look into the additional test gear we list after you've gained some experience and want to graduate to bigger and better projects.

Trying Out Ready-Made Electronics Kits

If you want to make some really cool things happen with electronics, but don't want to start from scratch (at least, not yet), you can purchase one of many different electronics hobby kits. These kits include everything you need to build a functional circuit: all the electronic components, wire, circuit board and detailed instructions for putting the circuit together. Some even include an explanation of how the circuit works.

You can find kits for light-sensitive alarms, simulated traffic signals, electronic combination locks, adjustable timers, decorative light displays and much more. Many of the parts sources mentioned in Chapter 16 provide ready-made kits at reasonable prices. You can practise your circuit-building and analysis skills using these kits, and then move on to designing, building and testing your own circuits from scratch.

Varying Your Voltage

You can use a *variable power supply* instead of batteries to power circuits you build and test at your workbench. A power supply produces a well-regulated (meaning very, very, very steady) voltage, and most models offer voltage outputs ranging from 0 to 20 volts. The model in Figure 15-1 offers a variable output range, controllable by a continuous dial, of about 2 to 20 volts, as well as preset outputs of –5, +5 and +12 volts.

Figure 15-1: A variable power supply.

A power supply is characterised by its voltage range as well as its current capacity. The higher the current rating of the power supply, the heavier the load it can power. Avoid choosing a power supply with only a modest current output – say, less than one amp. You can't adequately drive all circuits with lower currents. Instead, consider a power supply that delivers a minimum of two amps at +5 volts and at least one amp at any other voltage.

Counting Up Those Megahertz

You can use a *frequency counter* (or frequency meter) to help you determine whether or not your AC circuit is operating properly. By touching the leads of this test device to a signal point in a circuit, you can measure the frequency of that signal. For example, imagine you create an infrared transmitter and the light from the transmitter is supposed to pulse at 40,000 cycles per second (also known as 40 kHz). With a frequency counter connected to the output of the circuit, you can verify that the circuit is indeed producing pulses at 40 kHz – not 32 kHz, 110 kHz or some other Hz.

You can use most models, such as the one in Figure 15-2, on digital, analogue and most radio frequency (RF) circuits (such as radio transmitters and receivers). For most hobby work, you need only a basic frequency counter.

Newer multimeters also come with a frequency counter built-in.

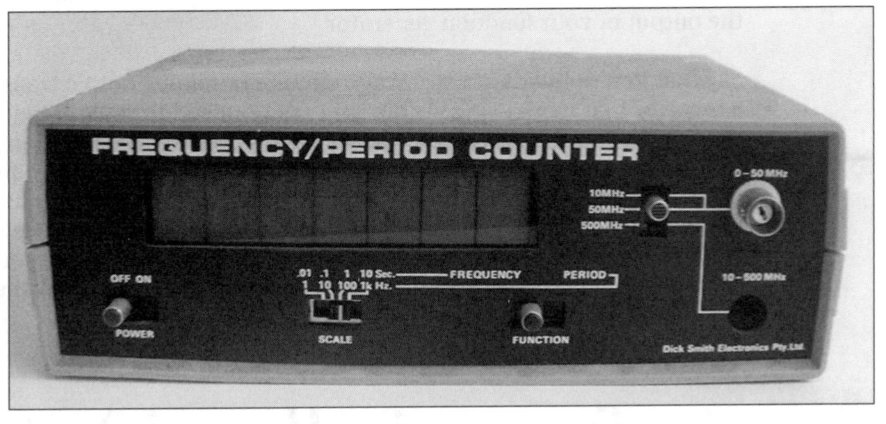

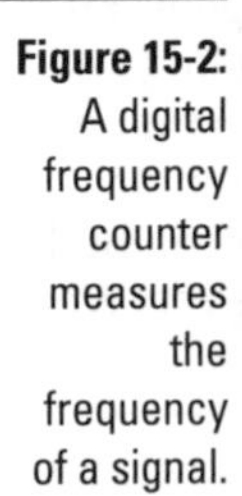

Figure 15-2: A digital frequency counter measures the frequency of a signal.

In digital circuits, signal voltages are limited to a range of zero up to about 12 volts, but in analogue circuits, voltages can vary widely. Most frequency counters are designed to work with analogue voltages ranging from a few hundred millivolts to 12 or more volts. Check the manual that came with your frequency counter for specifics.

Frequency counters display the frequency signal from 0 (zero) Hz to a maximum limit that's based on the design of the counter. This limit usually goes well into the megahertz; some counters have an upper limit of 25 to 50 MHz.

Higher-priced frequency counter models come with or offer a prescaler as an option. A *prescaler* is a device that extends the useful operating frequency of the frequency counter to much higher limits. Go for the prescaler feature if you're working with high-frequency radio gear or computers.

Generating All Kinds of Signals

To test a circuit's operation, a helpful idea is to apply a known signal input to the circuit and observe how the circuit behaves. You can use a *function generator* to create repeating signal waveforms in a variety of shapes and sizes and apply the generated waveform to the input of the circuit under test.

Most function generators develop three kinds of waveforms: sine, triangle and square. You can adjust the frequency of the waveforms from a low of 1 Hz to a high between 20 and 50 kHz. Some function generators come with a built-in frequency counter so that you can accurately time the waveforms you generate. You can also use a stand-alone frequency counter to fine-tune the output of your function generator.

Say that you're building a two-stage circuit containing both a transmitter and a receiver that work using visible light (as opposed to radio frequency – RF – waves). To test the receiver stage accurately, you set the function generator to produce a signal that simulates a transmitted signal with a frequency in the visible light range. Then you attach the two leads from your function generator to the input of your receiver circuit and test that your receiver circuit is indeed receiving the signal. When you get the receiver done, you can build the transmitter, knowing, thanks to the function generator, that the receiver is working properly.

Sweeping Frequencies Up and Down

When you need to test the behaviour of a circuit in response to a whole slew of input frequencies, you may need to use a sweep generator. A *sweep generator* is a type of function generator that produces a signal with a frequency that continuously varies (sweeps) within a range around a specified centre frequency.

Sweep generators typically vary the frequency of the output waveform within pre-selected limits, such as 100 Hz to 1 kHz or 1 kHz to 20 kHz, and allow you to control the sweep rate (how fast the signal frequency changes). Not only does this sweep sound like E.T. calling home (connect a speaker to the output of the sweep generator to hear this effect), but it also helps you identify frequency-related problems in frequency sensitive circuits.

A *frequency sensitive circuit* is designed to operate differently depending on the frequency of the input signal. Filter circuits, resonant circuits and RF transmitter/receiver circuits are examples of frequency sensitive circuits. If you're building a radio receiver circuit, for example, you need to ensure that it operates properly over a range of frequencies. By applying a sweep signal to the input of your circuit, you can observe (in one fell swoop) how your circuit behaves in response to a range of frequencies. A sweep generator is useful in troubleshooting audio and video equipment, where altering the input frequency can reveal faulty components.

Some function generators also have a sweep feature, covering two functions with one tool.

Taking the Pulse

You can use an inexpensive handheld *logic pulser* to help you test and troubleshoot digital circuits. This pen-like device, shown in Figure 15-3, injects a high or low digital pulse into a digital circuit. (A *pulse* is simply a signal that alternates between high and low very rapidly, just like the beating of your heart produces your pulse.) Many logic pulsers allow you to switch between injecting a single pulse and injecting a train of pulses at a desired frequency.

You normally use a logic pulser in conjunction with a logic probe or an oscilloscope to trace the effect of the injected pulse on your circuit. (You can read about both logic probes and oscilloscopes in Chapter 12.) For instance, you may inject a pulse into the input pin of an integrated circuit (IC) while measuring or probing the output of the IC to test whether the chip is operating properly. Logic pulsers come in handy for tracking down circuit problems because you can inject pulses in various portions of a circuit.

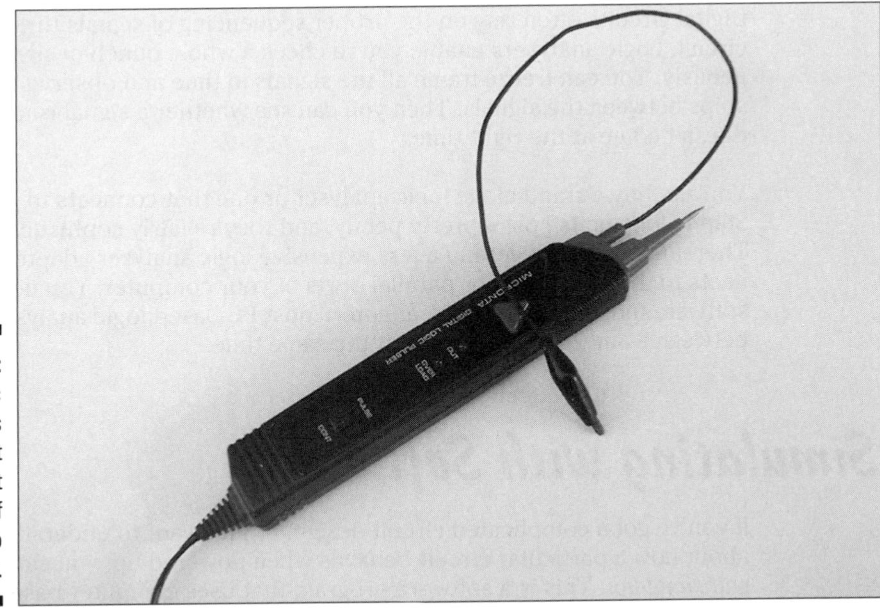

Figure 15-3: A logic pulser feeds a short signal burst or a train of bursts into a circuit.

Most pulsers get their power from the circuit under test, and multiple power supplies are available. You need to be careful which supply you use to power the logic pulser. For instance, if you're testing a chip that's powered by five volts, you don't want to give it a pulse powered by a 12-volt supply or you ruin the chip. Also, some circuits work with split (+, – and ground) power supplies, so make sure that you connect the pulser's supply clips to the correct power points to avoid damage to the components.

Make sure that you don't apply a pulse to an IC pin that's designated as an output but not as an input. Some ICs are sensitive to unloaded pulses at their output stages, and you can destroy the chip by applying the pulse incorrectly. (An *unloaded pulse* produces current that has no way to drain safely to another part of the circuit. If the current is applied to an output of an IC, for example, that output can be damaged, because it's exposed to current that it's not meant to take.)

Analysing Your Logic

To really get the low-down on what's happening in a digital circuit, you need a *logic analyser*, which is like a souped-up oscilloscope (you can read about oscilloscopes in Chapter 12): it shows you the waveforms of several inputs or outputs of a digital circuit at the same time. Many people who are well versed in the black arts of electronics find this analyser much more useful than an oscilloscope for troubleshooting digital circuits.

Digital circuits often rely on the proper sequencing of signals throughout the circuit. Logic analysers enable you to check a whole bunch of signals simultaneously. You can freeze-frame all the signals in time and observe the relationships between the signals. Then you can see whether a signal is missing or doesn't come at the right time.

You can buy a stand-alone logic analyser or one that connects to your PC. Stand-alone units cost a pretty penny, and they're fairly sophisticated. Therefore, consider getting a less expensive logic analyser adapter that connects to the USB, serial or parallel ports of your computer. You need special software that comes with the adapter. Most PC-based logic analysers handle between 8 and 16 digital inputs at the same time.

Simulating with Software

If you've got a complicated circuit design, or just want to understand more about how a particular circuit behaves when powered up, you can use a *circuit simulator*. This is a software program that uses computer-based models of circuit components to predict the behaviour of real circuits. You tell it what components and power supplies you're using and how they should be wired up, and the software tells you whatever you want to know about the operation of the circuit: the current through any component, voltage drops across components, circuit response across various frequencies and so forth.

Many circuit simulators are based on an industry-standard algorithm called SPICE (Simulation Program with Integrated Circuit Emphasis) and can be used for analogue, digital and *mixed-signal* (both analogue and digital) circuits. You can find free circuit simulators online, but you should know that they aren't guaranteed to be accurate, and they don't come with any technical support.

Commercially available simulators can be expensive, but they include lots of value-added features as well as technical support. For instance, Multisim from National Instruments includes the following features (and more):

- An extensive **model library** that contains software models of hundreds of specific manufacturer part numbers, so your simulation shows you exactly how stray capacitance from the Acme Widget #2 affects your sensitive circuit, for example.
- A broad array of **waveform display tools**, including software versions of every test instrument described in this chapter (yes, your computer monitor can be made to look like an oscilloscope screen!).
- A set of in-depth **analysis tools** that help you troubleshoot your circuit and understand exactly how it behaves under various conditions, such as extreme temperature settings, or when every single component in your circuit gangs up on you by varying significantly from its nominal value (this is known as a *worst-case condition*).
- A group of **schematic capture tools** that allow you to control where to place the circuit symbols for your chosen circuit components on a grid displayed on your computer, how to wire-up components and so forth, to build up a circuit diagram.

You can download a free 30-day evaluation of Multisim and try it out yourself by visiting the National Instruments site at `www.ni.com/multisim`. Alternatively, you could download a simulator called Yenka. Produced by a British company called Crocodile Clips, Yenka is available at `www.yenka.com` and is free for you to use at home. Another simulator developed in the UK is Circuit Wizard. You can buy Circuit Wizard for £60 at Maplin stores, and download a trial version from the New Wave Concepts website (at `www.new-wave-concepts.com`) before you buy.

Buying Testing Tool Deals

We aren't going to kid you – electronics test equipment can cost you a lot of money. Much of what you pay for is the accuracy of the device. Manufacturers strive for high accuracy to tout their products to businesses or to meet necessary government regulations. If you're an electronics hobbyist working at home, you don't really need all that extra precision. Usually you can get by with less precise – and less expensive – models. The low-end model of a family of test products is likely to be good enough for most hobby applications, and assuming that you take good care of it, should last many years.

Also, you can buy used or surplus equipment instead of buying everything brand new. In the case of electronics, surplus seldom means used, as it can for other surplus components, such as motors or mechanical devices that have been reconditioned. *Surplus* just means that the original maker or buyer of the goods doesn't need the items any more: surplus is simply excess stock for resale.

Used and surplus items can save you a load of cash, but this stuff doesn't always come with instruction manuals. Sometimes you can buy the manual separately or find it online. Owners of popular test gear often scan the pages of their old equipment manuals and post them online for the benefit of others.

The Appendix lists some useful sources for surplus equipment. Try checking out these sources, too:

- **eBay and other online auction sites:** Before you bid, check the listing carefully for the condition of the equipment and the return policy if the equipment doesn't work as promised. Then check out other auctions, including those that have already ended, to see what the going price is for similar products. And you'd be wise to check the seller's reviews to get an idea of the integrity of the seller you're dealing with.
- **Electronics mail order and local surplus outlets:** These shops are another good source for used test equipment and are handy if you don't want to wait for an auction to finish or you prefer to know the price up front. On the downside, surplus retailers may have limited selections – whatever components the shop was able to purchase. Don't expect to find every value and size of resistor or capacitor, for example.

Whether you use an auction, mail order or a local shop, be sure that the test gear you buy actually works. Have the seller guarantee that the equipment is in working order by giving you a warranty. You may pay a little bit more, but if you don't make sure that it works and you're not so good at fixing broken test gear, you may just be buying an expensive paper weight. If you're brand new to electronics, have a more experienced friend or work associate check out the gear for you.

Pass up sellers, especially on eBay or other auction sites, who aren't willing to guarantee that their products are in working order. Plenty of sellers do take the time to check out their wares and guarantee that the item isn't dead on arrival.

When you're shopping around for parts, you may see the term 'RoHS Compliant' next to some of the items. This term (pronounced *row-haas*) stands for *Restriction of Hazardous Substances* and refers to a recent European Union (EU) directive restricting the placement on the EU market of new electrical and electronic devices that contain more than a certain level of lead and five other hazardous substances. Companies producing consumer and industrial electronics need to worry about RoHS compliance if they want to sell products in EU countries (and China, which has its own RoHS specification). Everything you're likely to buy in the UK should be compliant by now anyway, so you don't need to worry about it.

Boxing Your Tricks – With Knobs On

Having worked out what you want to build, drawn a schematic and built a prototype circuit on a breadboard, you naturally want your creation to look the part. Your gizmo does everything you want it to, but the bits and pieces sticking out make it look vulnerable to bumps, knocks and cups of coffee spilt in the wrong direction. You're ready to make the circuit permanent on a printed circuit board, but if you're going to the trouble of all that soldering you want to put it in a nice, neat-looking box, perhaps with some controls on the front, too. What you need is the right enclosure.

An *enclosure* is just the electronics industry term for a fancy box. Enclosures come in just about any shape or size you can imagine; you just need to find the right one for your project. You can choose from miniature ones, big ones, wide ones, narrow ones, waterproof ones and enclosures designed for keypads and handheld scanners. You can also buy enclosures made of metals, such as aluminium, or plastics, such as the very tough ABS.

To give your gizmo some pizzazz, why not add some knobs and switches on the front? You can choose from big ones, small ones, round ones, waterproof ones, key-operated ones and toggles. So find your enclosure, select the best-looking knobs, switches or lights, and start building your own unique gizmo. OK, so it may not be slick enough to beat the latest iPod in any design awards, but if you're going to all that trouble to build a fully functioning gadget and you need a cover on it for protection anyway, you might as well get the finishing touches right.

You can find a huge variety of enclosures, knobs, switches and the like for sale at the component suppliers we list in Chapter 16.

Chapter 16

Ten Great Sources for Electronics Parts

In This Chapter

- Parts sources from the UK
- Buying from US suppliers

If you're looking for some excellent sources for your electronic parts, this chapter gives you some perennial favourites, both inside and outside the UK. This list is by no means exhaustive; you can find literally thousands of outlets for new and used electronics. But the sources we list here are among the more established in the field, and all have websites for online ordering (some also offer a print catalogue).

Buying British: Suppliers within the UK

Check out these suppliers if you're shopping within the UK. If you live in a different country, you can still consider buying from these outlets because most of them ship worldwide. Remember that shipping costs may be higher, and you may have to pay an import duty, depending on your country's regulations. So it's a good idea to look for local suppliers in your own country first.

Maplin

`www.maplin.co.uk`

These electronics specialist's shops are a familiar sight on the high streets and shopping centres of towns and cities all over Britain. You can buy components at the special components counters that you can usually find towards the back of larger shops. They carry a good stock, but don't be surprised if a store has sold out of your item or has just nine of the LEDs left when you need ten.

Don't be afraid to ask the staff technical questions – someone working in the shop always seems to be willing and able to offer answers when you need them. You can order components online too, and you can even pay by Paypal, the small-payments service that's become popular on eBay. Delivery costs a few pounds for small orders, but you don't have to spend a lot of money on one order to get free delivery.

RS Components

`rswww.com`

RS started life in 1937 as Radiospares Limited. Today the shop is well known in the electronics industry as well as among hobbyists. Posties should be glad for the invention of the CD-ROM and the day when the infamously huge RS catalogue went digital, which gives you some idea of the vast range of components that this distributor can supply. It offers around 160,000 product lines for same-day dispatch and access to another 100,000.

Farnell

`www.farnell.co.uk`

Farnell is based in the UK but supports shoppers from countries worldwide (you can see which countries at a glance by going to the international page at `www.farnell.com`). It stocks around 250,000 products. You can order through Farnell's website but a handling charge applies to small orders.

RSH Electronics

`www.rshelectronics.co.uk/`

Although RSH stocks a smaller range than the giant distributors, it does have one of the lowest delivery charges of just a few pounds with no minimum order and delivers free for orders of over £30 or so. Overseas postage is only £4 too. As well as around 1,500 components, it also supplies the Electronics Club electronic project kits.

Bitsbox

www.bitsbox.co.uk

Bitsbox delivers within the UK for just £1.50 per order and just a pound or two more to other countries. Order on the fun and friendly website, which accepts Paypal as well as card payments. Lots of special offers are always available online, including a pound shop and bargain prices for parts such as the timer chips we use for projects in Chapter 14.

ESR Electronic Components

www.esr.co.uk

This small firm in the North-East of England supplies components to hobbyists as well as schools and colleges. You can order from them by post, phone (at UK local rate), email or online using a credit card. You can then collect your order from their shop in Tyne and Wear or pay for delivery, which costs just a couple of pounds for orders weighing up to 1 kilogram, just over a fiver for orders up to 5 kilograms or free if you spend over £100.

Ordering from across the Pond

For more specialist components, you may have to turn to America, the original heartland of the electronics industry. Here are some of the largest suppliers in the USA that also ship to Europe, but ordering probably costs you more than using a UK supplier.

Digikey

www.digikey.com

If you want a part, Digikey probably has it, carrying thousands upon thousands of items, which it ships all over the world from Minnesota, USA. Go to the main home page and click on Change Country in the top right corner to shop in UK sterling. The firm's online ordering system includes detailed product information, price, available stock levels and even links to product datasheets. The site offers a handy search engine so that you can quickly locate what you're looking for, as well as an online interactive catalogue (with magnification capabilities that you'll certainly need). Digikey will also send a free printed catalogue to the UK, although the text is tiny to fit everything in.

All Electronics

www.allelectronics.com

All Electronics runs a shop in the Los Angeles area and sends mail orders worldwide. Most of its stock is new surplus (see the Appendix for more sources of surplus goods), meaning that the merchandise is brand new but was overstocked. All Electronics has a printed catalogue, which is also available in pdf format on its website. Stock changes frequently, and the latest updates are available only on the website. Be sure to check out the Web Only items.

B.G. Micro

www.bgmicro.com

Selling primarily surplus odds and ends, B.G. Micro has great prices and terrific customer service. Stock tends to come and go quickly, so if you see something you especially like, be sure to order it now!

Mouser Electronics

www.mouser.com

Similar to Digikey, Mouser is a stocking distributor with tens of thousands of parts on hand. If you can't find a part at Mouser, it probably doesn't exist. Mouser carries over 165,000 resistors alone, listed under Passive Components. You can order from its online store or its humungous print catalogue. Mouser now has a UK office serving parts of Europe and South Africa.

Appendix

Internet Resources

In this Appendix, we suggest a selection of interesting and useful Internet resources for all things electronic. Some of them are run by businesses for profit and others by individuals or groups of individuals for nothing but the fame and glory! Loads of good sites are out there, some of which come and go as websites do, but try looking for the latest and most popular ones with search engines such as Google or Yahoo!

Here we introduce you to what we consider to be some of the best websites at the time of writing, to give you a head start.

Finding Guides and Advice

All the websites in this section contain worthwhile information. Browse through them to decide which ones meet your needs. We give the Electronics Club and the North Carolina State University Electronics Tutorial the highest marks, but all these sites have cool and useful information:

- **All About Circuits** (`www.allaboutcircuits.com`): This site contains a series of online books on electronics. They haven't yet posted some sections, but the material they do have is well produced.
- **AOGFB** (`http://library.thinkquest.org/16497/home/index.html`): AOGFB stands for An Online Guide for Beginners, which is just what it provides. The website contains quick guides on basic principles and techniques, such as soldering, and half a dozen beginner circuits to get you going.
- **The Electronics Club** (`www.kpsec.freeuk.com`): This site has a lot of good advice for newcomers to electronics projects, including a tutorial on how to read a circuit diagram, explanations of components and a list of circuit symbols. Also known as the Kelsey Park School Electronics Club, which explains the kpsec part of the web address.
- **Electronics Hobbyist** (`www.amasci.com/amateur/elehob.html`): This American site is for really keen hobbyists, and offers some interesting articles, news snippets, magazine links and links to discussion forums and online user groups. You can even use their links to 'geek dating sites'!

- **Graham Knott's Website** (http://homepage.ntlworld.com/g.knott/): Enjoy exploring this site that Graham Knott, an electronics teacher in Cambridge, organised to make finding information on both beginning and intermediate electronics topics simple.
- **The Institution of Engineering and Technology** (www.theiet.org/education): Browse issues of *E&T Education* magazine to find some fun and amazing ideas for circuits – one even measures the speed of light. Or just print out the poster of UK schematic symbols to put up next to your workbench.
- **The North Carolina State University Electronics Tutorial** (www.courses.ncsu.edu:8020/ece480/common/htdocs): This site contains good explanations of various electronics topics. Many of the illustrations are animated, which makes understanding the concepts easier.
- **Williamson Labs Electronics Tutorial** (www.williamson-labs.com/home.htm): This site provides explanations of basic electronics concepts with good illustrations and fun animations.

Working Things Out with Calculators

You can perform calculations on the sites in this section without having to look up equations or pick up a handheld calculator. Choose a website that covers the particular equation that you want to use:

- **Bowden's Hobby Circuits** (www.bowdenshobbycircuits.info/): The calculators on this site include the standard calculations for Ohm's Law, RC time constants and resistor colour-code conversions. You can also find calculators for functions that you don't find on most other sites, such as a voltage divider calculator.
- **The Electronics Calculator Website** (www.cvs1.uklinux.net/calculators/index.html): Use tools on this site to perform calculations for Ohm's Law, RC time constants and a few other handy equations.
- **Electronics Converters and Calculators** (www.csgnetwork.com/electronicsconverters.html): This site has calculators that perform Ohm's Law calculations, parallel resistance calculations and resistor colour-code conversions, among other operations.

Surfing for Circuits

If you're hungry for even more circuits to build, they're just a mouse click away! Thanks to the magic of the Internet, you can find hundreds – no, make that thousands – of electronic circuits, from basic light and sound demonstrators to advanced projects for your car or boat. Here are a few of the best:

- **Bowden's Hobby Circuits** (www.bowdenshobbycircuits.info/): This personal site from hobbyist Bill Bowden gives you the why, not just the how. Here you find both circuit descriptions and alternative design suggestions.
- **Discover Circuits** (www.discovercircuits.com): This site has thousands of electronic circuits, categorised and cross-referenced, with links back to the original designer's pages, where you can find schematics and so on. The site also has a problem-solving forum.
- **The Electronics Club** (www.kpsec.freeuk.com/proj.htm): You can find some fun projects for beginners on this website. For each project, you see an explanation of how it works and a circuit layout in addition to a schematic.
- **Electronics Project Design** (www.electronics-project-design.com): This well-designed site has concise descriptions and schematics for more cool circuits arranged under subject headings such as amplifiers, timers, motion control and surveillance.

Asking Questions in Discussion Forums

Use the forums on the websites in this section to get answers to your questions about projects or general electronics. Every discussion area has its own style, so spend a little time on each site to decide which forum is right for you. Post your question, and others who've already worked through the same problem may provide the answer that you need.

We found the discussion groups on the following sites especially interesting and helpful:

- **All About Circuits Forum** (http://forum.allaboutcircuits.com/): Here you can find both a general electronics discussion forum and a forum to ask for help from other forum members on any sticky projects.
- **EDA Board** (www.edaboard.com): Explore these active discussions about problems with projects and general electronics, along with several more specialised forums, such as one on PCB design.
- **Electronics Lab** (www.electronics-lab.com/forum/index.php): Another good site with discussions on projects, circuits and general electronics. Check out the Project Q/A section, where readers post questions and get answers on the many projects provided in the projects area of the site.
- **Electronics Zone Discussion** (www.electronic-circuits-diagrams.com/forum/): This site has extremely active discussions on electronic circuits and projects.

Be sure to take the answers that you get on forums with a pinch of salt. Think through the advice that you get before you build a project based solely on some well-meaning stranger's word.

Getting Things Surplus

Looking for some good deals? Try buying surplus or second-hand. Because surplus merchandise comes and goes, you have to be sharp to catch the good stuff – but if you're lucky, you can pick up some bargains.

To check out some suppliers and search their online catalogues, start by typing 'surplus electronic components' into a search engine such as Google or Yahoo! Some suppliers are more geared up to manufacturers that assemble finished circuit boards for industry than hobbyists looking for a few fun parts, but here are a few to start you off:

- **Greenweld** (`www.greenweld.co.uk`): Surplus electronic gadgets and tools as well as some parts.
- **Surplectronics** (`www.surplectronics.co.uk`): A good range of useful and fun components at knockdown prices. Postage is under £4 or free if you spend over a certain amount. Usefully, you can pay by Paypal, the system used by millions of people on eBay.
- **UES&S** (`www.surplus-electronics.co.uk`): Used Equipment Surplus & Storage offers second-hand equipment for your workbench, including multimeters and oscilloscopes. However, you can't pay online; you have to send a personal cheque or have a credit card ready when you call by phone.

In addition to these sources, also make sure that you check out our top-ten list of online electronics outlets in Chapter 16.

Index

• Symbols and Numerics •

• A •

• F •

• G •

• H •

• I •

• J •

• K •

• L •

• M •

• P •

• R •

• S •

• T •

• U •

• W •

• X •

• Z •

FOR
DUMMIES®

A world of resources to help you grow

UK editions

SELF-HELP

978-0-470-01838-5

978-0-7645-7028-5

978-0-470-74193-1

STUDENTS

978-0-470-74047-7

978-0-470-74711-7

978-0-470-74290-7

HISTORY

978-0-470-99468-9

978-0-470-51015-5

978-0-470-98787-2

Motivation For Dummies
978-0-470-76035-2

Overcoming Depression For Dummies
978-0-470-69430-5

Personal Development All-In-One For Dummies
978-0-470-51501-3

Positive Psychology For Dummies
978-0-470-72136-0

PRINCE2 For Dummies
978-0-470-51919-6

Psychometric Tests For Dummies
978-0-470-75366-8

Raising Happy Children For Dummies
978-0-470-05978-4

Sage 50 Accounts For Dummies
978-0-470-71558-1

Succeeding at Assessment Centres For Dummies
978-0-470-72101-8

Sudoku For Dummies
978-0-470-01892-7

Teaching English as a Foreign Language For Dummies
978-0-470-74576-2

Teaching Skills For Dummies
978-0-470-74084-2

Time Management For Dummies
978-0-470-77765-7

Understanding and Paying Less Property Tax For Dummies
978-0-470-75872-4

Work-Life Balance For Dummies
978-0-470-71380-8

Available wherever books are sold. For more information or to order direct go to www.wiley.com or call +44 (0) 1243 843291

FOR DUMMIES®

Helping you expand your horizons and achieve your potential

COMPUTER BASICS

978-0-470-27759-1

978-0-470-13728-4

978-0-470-49743-2

DIGITAL PHOTOGRAPHY

978-0-470-25074-7

978-0-470-46606-3

978-0-470-45772-6

MAC BASICS

978-0-470-27817-8

978-0-470-46661-2

978-0-470-43543-4

Access 2007 For Dummies
978-0-470-04612-8

Adobe Creative Suite 4 Design Premium All-in-One Desk Reference For Dummies
978-0-470-33186-6

AutoCAD 2010 For Dummies
978-0-470-43345-4

C++ For Dummies, 6th Edition
978-0-470-31726-6

Computers For Seniors For Dummies , 2nd Edition
978-0-470-53483-0

Dreamweaver CS4 For Dummies
978-0-470-34502-3

Excel 2007 All-In-One Desk Reference For Dummies
978-0-470-03738-6

Green IT For Dummies
978-0-470-38688-0

Networking All-in-One Desk Reference For Dummies, 3rd Edition
978-0-470-17915-4

Office 2007 All-in-One Desk Reference For Dummies
978-0-471-78279-7

Photoshop CS4 For Dummies
978-0-470-32725-8

Photoshop Elements 7 For Dummies
978-0-470-39700-8

Search Engine Optimization For Dummies, 3rd Edition
978-0-470-26270-2

The Internet For Dummies, 11th Edition
978-0-470-12174-0

Visual Studio 2008 All-In-One Desk Reference For Dummies
978-0-470-19108-8

Web Analytics For Dummies
978-0-470-09824-0

Windows Vista For Dummies
978-0-471-75421-3

Available wherever books are sold. For more information or to order direct go to www.wiley.com or call +44 (0) 1243 843291